T0214619

Lecture Notes
in Business Information Processing 394

More information about this series at http://www.springer.com/series/7911

Witold Abramowicz · Gary Klein (Eds.)

Business Information Systems Workshops

BIS 2020 International Workshops
Colorado Springs, CO, USA, June 8–10, 2020
Revised Selected Papers

 Springer

Editors
Witold Abramowicz 🆔
Poznań University of Economics
and Business
Poznan, Poland

Gary Klein 🆔
University of Colorado
Colorado Springs, CO, USA

ISSN 1865-1348   ISSN 1865-1356   (electronic)
Lecture Notes in Business Information Processing
ISBN 978-3-030-61145-3   ISBN 978-3-030-61146-0   (eBook)
https://doi.org/10.1007/978-3-030-61146-0

This Springer imprint is published by the registered company Springer Nature Switzerland AG
The registered company address is: Gewerbestrasse 11, 6330 Cham, Switzerland

Preface

The Business Information Systems (BIS) workshops give researchers the possibility to share preliminary ideas, first experimental results, and to discuss research hypotheses. It is our great pleasure to contribute to the international discourse on the broader research area of BIS, by enabling the organization of specialized workshops on emerging research themes in parallel to BIS conference sessions. Discussions held during presentations allow for improving the paper and preparing it for publication. From our experience, workshops are also a perfect instrument to create a community around very specific research topics, thus offering the opportunity to promote it. Due to the global travel restrictions, BIS 2020 was held as a virtual conference. In order to allow participants from all around the world to participate in the sessions, we decided to organize recording sessions for workshops prior to the conference. At the time of the actual conference, all presentations were available online. However, live workshop sessions gathered a wide and very active audience. The discussion of presented scholarly work was constructive and provided authors with new perspectives and directions for further research. Based on the feedback received, authors had the opportunity to edit the workshops articles into the current publications. This year, five workshops were organized inconjunction with BIS: BITA (11th edition), BSCT (3rd edition), DigEx (2nd edition), iCRM (5th edition), and QOD (3rd edition). A total of 26 articles were accepted for publication and are included in this volume. We sincerely thank everyone who contributed to the success of the BIS workshops. Most of all, we wish to thank the workshops' chairs, Program Committees, authors, and invited speakers. We acknowledge the contribution of workshops' participants who provided comments and insightful suggestions for the advancement of presented work.

June 2020

Witold Abramowicz
Gary Klein

Contents

DigEx

ICRM

QOD

BITA 2020 Workshop Chairs' Message

Preface

BITA

BITA 2020 Workshop Chairs' Message

Preface

A contemporary challenge for enterprises is to keep up with the pace of changing business demands imposed on them in different ways. Today, there is an obvious demand for continuous improvement and alignment in enterprises, but unfortunately many organizations don't have the proper instruments (methods, tools, patterns, best practices, etc.) to achieve this. Enterprise modeling, enterprise architecture, and business process management are three areas belonging to traditions where the mission is to improve business practice and business and IT alignment (BITA). BITA is manifested through the transition of taking an enterprise from one state (AS-IS) into another improved state (TO-BE), i.e., a transformation of the enterprise and it's supporting IT into something that is regarded as better. Recent development within digitalization, digital transformation, and artificial intelligence (AI) has brought new dimensions to BITA, where BITA becomes an important part of smart business ecosystems. A continuous challenge with BITA is to move beyond a narrow focus on one tradition or technology. There is a need to be able to deal with multi-dimensions of the enterprise in order to create alignment. Examples of such dimensions are organizational structures, strategies, architectures, business models, work practices, processes, and IS/IT structures. IT governance is also a dimension that traditionally has had a strong impact on BITA. There are ordinarily three governance mechanisms that an enterprise needs to have in place: 1) decision-making structures, 2) alignment process, and 3) formal communications.

This workshop aimed to bring together people who have an interest in BITA. We invited researchers and practitioners from both industry and academia to submit original results of their completed or ongoing projects. We encouraged a broad understanding of possible approaches and solutions for BITA. Specific focus was on practices of business and IT alignment, i.e., we encouraged the submission of case studies and experience papers.

The workshop received 10 submissions, and the Program Committee selected 4 submissions for presentation at the workshop. We thank all members of the Program Committee, authors, and local organizers for their efforts and support.

August 2020

Ulf Seigerroth
Kurt Sandkuhl

Organization

Chairs

Ulf Seigerroth (Chair)   Jönköping University, Sweden
Kurt Sandkuhl (Co-chair)   Rostock University, Germany

Program Committee

Dominik Bork   University of Vienna, Austria
Michael Fellmann   University of Rostock, Germany
Hans-Georg Fill   University of Fribourg, Switzerland
Jānis Grabis   Riga Technical University, Latvia
Stijn Hoppenbrouwers   HAN University of Applied Sciences,
                                    The Netherlands
Björn Johansson   Lund University, Sweden
Christina Keller   Jönköping University, Sweden
Marite Kirikova   Riga Technical University, Latvia
Birger Lantow   University of Rostock, Germany
Michael Leyer   University of Rostock, Germany
Geert Poels   Ghent University, Belgium
Nikolay Shilov   SPIIRAS, Russia
Alexander Smirnov   SPIIRAS, Russia
Monique Snoeck   Catholic University Leuven, Belgium
Janis Stirna   Stockholm University, Sweden

Data Quality Assessment – A Use Case from the Maritime Domain

Milena Stróżyna(✉)(iD), Dominik Filipiak(iD), and Krzysztof Węcel(iD)

Poznań University of Economics and Business, al. Niepodległości 10,
61-875 Poznań, Poland
`milena.strozyna@ue.poznan.pl`

Abstract. Maritime transport plays nowadays a key role in the global economy. In this context, assurance of safety and security at sea is of prime importance. To this end, in the maritime domain there exists number of information systems that improve safety, identify hazardous areas and suspicious ships. These systems generate large amounts of data that are characterised with a different, often not sufficient, quality. Assurance of maritime data quality is an important aspect that determines if the data can be used to take informed decision. This paper presents the quantitative assessment of maritime data quality, investigates if the real data meets data quality standards and detects what are the most common quality issues. The presented analysis is conducted on one of the most popular maritime data source – Automatic Identification System (AIS). The paper shows also a potential stemming from utilization of Big Data technologies in a process of data quality assessment.

Keywords: Data quality · Quality assessment · AIS · Maritime awareness

1 Introduction

Nowadays, maritime transport plays a key role in logistics. Around 80% of the global trade volume (70% by value) is carried out by sea [10]. In this context, assurance of safety and security at sea is of prime importance. Therefore, various surveillance systems in the maritime domain were developed and deployed to improve safety, as well as identify hazardous areas and suspicious ships. These systems provide not only the information necessary to take informed decisions by maritime actors, but they also generate large amounts of various data that then needs to be processed, stored, and analysed. In addition to an increased quantity of data, its quality is the most important aspect that determines the potential use. Low-quality data may lead to wrong decision, causing financial losses and posing danger to people and the environment [6].

One of the most commonly used systems in the maritime domain is Automatic Identification System (AIS) – for location and identification of nearby vessels in real time (a tracking system). It has been created as a tool for the

© Springer Nature Switzerland AG 2020
W. Abramowicz and G. Klein (Eds.): BIS 2020 Workshops, LNBIP 394, pp. 5–20, 2020.
https://doi.org/10.1007/978-3-030-61146-0_1

collision avoidance at sea. AIS is based on automatic exchange of data about a ship and its position with other nearby vessels and AIS base stations. Since 2008, AIS signals from ships can also be received by satellites. Thanks to this, the system has a global coverage and creates the possibility to track ships on a worldwide scale. Currently, AIS is mandatory for all vessels above 300 GRT (Gross Register Tonnage).

AIS is very often used in maritime surveillance due to its high update frequency. Depending on the ship speed over ground, the dynamic data should be sent every 2 to 10 s, while the static data every 6 min. Compared to other maritime systems, AIS provides a significant amount of data about the movement of vessels. The exchanged information includes static data (identification numbers MMSI - Maritime Mobile Service Identity and IMO - International Maritime Organization, type, name, callsign, dimensions), dynamic data (location, course, speed, navigational status, heading, draught), and voyage data (destination port, estimated time of arrival). One of the main issue with AIS is the fact that, although it is required to be used by all vessels above 300 GT, the actual use of AIS is at crew's discretion. The ship's captain may decide if an AIS transponder is switched on or off and is responsible for providing and updating some of the actual data being sent by AIS. Therefore, some AIS messages are falsified, sometimes the signal is spoofed, and in some cases errors are committed during the typing [4]. All these issues negatively influence the data quality.

Data quality covers a broad range of concepts and has multiple dimensions [5]. It is often defined as *fitness for use* with respect to a particular application [6]. In case of AIS data, the quality assessment can bring information on the reliability and integrity. Thanks to that, users responsible for surveillance of maritime traffic or crew of other ships may take more informed decisions. The knowledge of the quality of information provided by ships is of prime importance in the situation awareness in the maritime domain [8]. Iphar et al. [4] noticed that although most of the AIS users do not falsify their data, a certain amount of AIS messages is false and vessels emit or receive messages that are not true. This, in turn, may lead to lack of trust in the AIS system or wrong decisions of various maritime actors. Therefore, the maritime society emphasizes the importance of the assessment of the quality of the AIS data.

The aim of this article is to conduct the quantitative assessment of AIS data quality – to investigate which data quality standards are met and detect what are the most common quality issues. This research is also about assessing the scale of the phenomenon that ships do not provide data of a proper quality. The research is conducted on a large amount of data, covering all AIS data generated over a period of one year, and using the state-of-the-art data processing technology – Apache Spark that allows to conduct the quality analysis fast and in real-time. This is a novelty in comparison to other research that so far were conducted on a relatively small samples of data and used traditional technologies, what resulted in a longer processing time.

The rest of the paper is organized as follow. Section 2 presents related research in the area of AIS data quality. Then, in Sect. 3 our approach to data quality assessment is presented, followed by the obtained results (Sect. 4). The paper ends with a discussion of the results and summary of the study.

2 Related Work

Due to its basic characteristics, AIS system is vulnerable to various quality problems. These problems may results firstly from the improper installation of AIS device. The static information, which is entered manually, are not controlled by any authority and thus may be not true. The dynamic information in turn, depends on a proper communication between AIS device and other sensors on board (e.g. GPS antenna). Other problems may arise due to human errors and behaviour of ship's crew. Since AIS is a self-reporting system, some actors may intentionally provide erroneous data in order to hide their activities or may spoof the AIS signal to mislead other actors [4]. Yet another group of quality issues is related to the AIS system itself, namely a limited coverage of AIS system in some areas due to a low reception of satellite AIS in a high vessel density areas, such as the German Bight or English Channel. Moreover, also AIS reception on open sea (outside the terrestrial coverage) is limited due to access gaps, i.e. time periods, when a ship is not in view of any AIS satellite and consequently no vessel position can be acquired. Such gaps can last even a couple of hours.

The existing studies show that the quality issue in AIS data is a common problem in all three mentioned aspects. The analysis conducted by [3] indicated that most often errors concerning the unique identification number (MMSI), defined vessel type (undeclared or default type), ship's name and call sign (no name provided, abbreviations), navigational status (incorrect status), incorrect vessel length, reported draught (non-availability, draught greater then length, inaccurate value), destination and ETA (vague or incorrect entries). Iphar et al [4] paid attention to intentional falsification of AIS signal, identity theft (duplication of MMSI number) and destination masking. They also indicated the problem of switching off the AIS transponder in order to hide some activities.

The quality analysis conducted by [9] focused on completeness and resolution of AIS data. They focused on four aspects: position precision, time interval between two consecutive AIS messages, data completeness and erroneous/corrupted entries. Their results indicated that ships position are rarely invalid but errors in heading and status are quite often. Speed and course over ground data are also sometimes wrong. With regard to completeness of dynamic information, most data contain the necessary kinematic information (i.e. position, time, speed, and course) but the rest of dynamic information is very often missing. They also identified three types of errors: 1) infeasibly large speed values; 2) duplicated AIS messages; 3) missing AIS messages due to errors in data broadcasting. Another problem is AIS spoofing, where three types are usually indicated [1]: falsification of closest point of approach alert, imitation of a fake ship which follows a given path, and simulation of search and rescue alert.

3 Methodology

Our approach to evaluate the AIS data quality is essentially based on statistical analysis of data and assessment of its reliability based on a set of quality attributes. To this end, a sample of real AIS data was retrieved and analysed. The analysed quality attributes included [7]:

- completeness – extent to which data is complete and sufficient for the task at hand,
- free-of-error – extent to which data is correct and reliable,
- ease of manipulation – extent to which data is easy to manipulate,
- timeliness – extent to which data is sufficiently up-to-date for the task at hand,
- reliability – extent to which data is regarded as true and credible.

In order to assess the quality of available AIS data with regard to the above attributes, an analysis of AIS data from two data sources was conducted:

- AIS data received in January-December 2015 from Orbcomm satellites (for the whole globe). The analysed dataset contained 1,390,219,742 messages from 425,166 unique MMSI transponders and the analysis was focused on vessel type, draught, dimensions, and destination.
- AIS data set covering weeks 33–35 of 2018 from the whole globe. The analysed dataset contained 65,896,367 messages. It was used to analyse the following AIS attributes: navigational status, speed over ground, course over ground, true heading, IMO number, call sign, and name.

The aim of these analyses was to assess data quality and thus discover the quality problems that are observed in AIS system. In order to process such a vast dataset, we used Apache Spark, a popular big data processing engine, which can take advantage of in-memory computation. For visualisations, we used R with *ggplot* library.

4 Data Quality Assessment – Use Case

In this section the results of analysis are presented, aiming at checking the quality of data provided in AIS with regard to different static and dynamic AIS parameters.

4.1 Static AIS Parameters

Among the static AIS parameters, vessel identification data (IMO number, call sign, and name), vessel dimensions, and vessel type were analysed.

Vessel Identification. There are several ways of identifying ships: MMSI, IMO, call sign, and name, whereas the last is not unique. The results of the analysis is presented in Table 1. It can be seen that ship identification number (IMO) has a

Table 1. IMOs

IMO	Count
0	36745367
1	50994
303174162	16301
100000000	12717
1048576	12222
999999999	11074
9999999	7183
101	5688
123456789	5574
356515840	5429
111111111	5345
2	5320
1234567	4087
987654321	3951
113	3732
108	3702
30	3467
1000000000	3313
9371531	3188
8888888	3044
103	2876
118	2850
4063232	2831
116	2791
32	2735
139	2735
200000000	2619
11	2602
888888888	2596
11111111	2579

Table 2. Call signs

Call sign	Count
700	36122
300	27028
0	22471
200	22128
600	19345
701	18125
301	13773
-------	12245
NONE	10561
601	9905
0000000	9561
201	8920
305	8491
----	7283
WCZ5857	7223
CH.16	6855
210	6777
617	6689
6666	6350
303	6029
702	5866
703	5768
226	5615
302	5573
603	5546
307	4790
TEST	4779

Table 3. Vessel types

Type	Number of AIS messages [%]	Number of vessels	Number of vessels [%]
Anti-pollution equipment	0.16	615	0.14
Cargo	36.83	141,580	33.30
Diving ops	0.16	417	0.10
Dredging or underwater op	1.66	2,383	0.56
Fishing	6.58	79,783	18.77
High speed craft (HSC)	1.15	2503	0.59
Law enforce-ment	0.76	2,029	0.48
Medical transport	0.02	179	0.04
Military ops	0.37	1,456	0.34
Non-combatant ship	0.02	132	0.03
Not available (default)	7.59	98,101	23.07
Other type	6.43	13,817	3.25
Passenger	7.16	10,743	2.53
Pilot Vessel	1.07	1,709	0.40
Pleasure Craft	2.25	5,564	1.31
Port Tender	0.23	881	0.21
Reserved	0.27	2,948	0.69
Sailing	0.70	3,490	0.82
Search and Rescue Vessel	0.86	2,322	0.55
Spare – Local Vessel	0.06	343	0.08
Tanker	13.99	24,389	5.74
Towing	2.33	5,229	1.23
Tug	7.97	11,663	2.74
Undefined or empty	0.94	10,588	2.49
Wing in ground (WIG)	0.41	2,302	0.54

lot of incorrect values. According to analysed data, there are 47,791 unique IMO numbers, out of which only 45,598 values are 7-digit numbers (as the standard requires). The rest – almost 2200 identifiers – are definitely incorrect. Moreover, the value zero is suspiciously frequent. Regarding the call sign, there were 103,268 unique values. Call signs are missing in 9,463,167 (14.4%) messages. The most popular call signs are presented in Table 2. Normally, call signs for larger vessels should consist of the national prefix plus three letters. Among the most popular ones, there is none that meets this requirement. Instead, there are values obviously wrong which include e.g. dashes, NONE, CH.16. The last row is particularly interesting – "TEST". Name of the vessel, similarly to call sign, is not unique value. We identified 152,473 various names. The name is more often provided than call sign – almost 4 million messages did not contain a vessel name, which represents 6.1% of all messages.

Vessel Type. The next analysed attribute is type of a ship. There are several classes agreed to be used in AIS messages (two-digit value). Nevertheless, in the

Table 4. Default vessel dimensions

Type	Vessels with default A [%]	Vessels with default B [%]	Vessels with default C [%]	Vessels with default D [%]
Anti-pollution equipment	10.79	10.92	11.52	12.73
Cargo	2.32	2.78	2.67	2.30
Diving ops	7.40	3.03	7.71	3.61
Dredging or underwater ops	3.67	4.28	4.65	4.74
Fishing	9.04	9.27	9.80	9.71
High speed craft (HSC)	5.83	5.66	6.19	6.09
Law enforcement	5.45	5.81	5.68	5.89
Medical transport	10.35	10.17	10.30	10.15
Military ops	16.77	14.83	16.87	15.23
Non-combatant ship	21.41	13.71	21.38	22.00
Not available (default)	22.06	24.35	23.73	23.91
Other type	6.77	7.98	7.44	6.96
Passenger	5.46	6.42	6.33	5.19
Pilot Vessel	9.13	9.10	9.43	9.37
Pleasure Craft	7.18	8.44	7.57	8.18
Port Tender	8.60	7.86	7.82	7.80
Reserved	4.71	4.71	5.37	5.54
Sailing	7.44	13.54	9.85	8.12
Search and Rescue Vessel	7.91	7.75	7.73	7.85
Spare – Local Vessel	11.52	11.57	11.54	11.54
Tanker	2.50	3.99	3.00	2.59
Towing	10.88	10.96	11.36	11.07
Tug	6.18	5.60	6.32	5.56
Undefined or empty	18.37	20.31	19.59	20.66
Wing in ground (WIG)	11.10	15.00	14.07	13.85

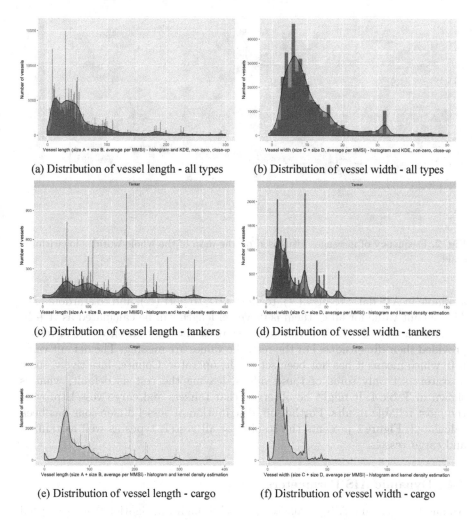

(a) Distribution of vessel length - all types   (b) Distribution of vessel width - all types

(c) Distribution of vessel length - tankers   (d) Distribution of vessel width - tankers

(e) Distribution of vessel length - cargo   (f) Distribution of vessel width - cargo

Fig. 1. Distributions of vessels length and width

sample data we identified 211 different values for type of ship. There were no missing rows – even if there is no IMO or call sign, the type of ship is always filled in. Apart from the types agreed for the AIS standard, there are also unknown three-digit types.

Moreover, almost one fourth of ships (23.07%) provide a default value of ship type, making it is impossible to specify what kind of ship it is. The overall vessel types distribution, calculated for all vessels that sent AIS in 2015, is presented in Table 3. If we analyse the number of vessels, cargo vessels dominate the ranking (33.83% vessels), followed by fishing vessels (18.77%) and tankers (5.74%).

Vessel Dimensions. Vessel dimensions (length and width) are another static parameters that should be provided in AIS. There are four vessel dimensions

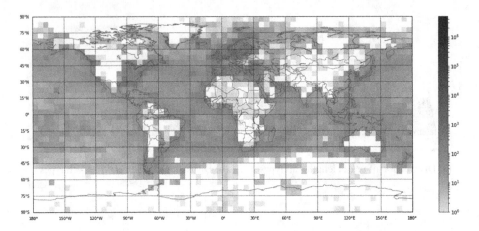

Fig. 2. Frequency of messages visualised on the map of the whole world – logarithmic scale

available in AIS data: to bow (A), to stern (B), to port (C), and to stardom (D). The default value for all of them is 0. Vessel length can be calculated as $A + B$, whereas vessel as $C + D$. The conducted analysis showed that only 70% of vessels provided their dimensions. For the rest, the values are missing. The default value is 0, which means it has not been set by the operator. Counter-intuitively, some operators set only some of these values, leaving the rest as default, what is shown in Table 4. It might be observed that tankers and cargo vessels provide the most reliable results. Further on, distribution of vessel dimensions has been calculated. Figure 1 presents the results for all vessels types as well as tankers and cargo vessels.

4.2 Dynamic AIS Parameters

Among the dynamic AIS parameters, vessel draught, navigational status, speed over ground (SOG), course over ground (COG), destination, and location were analysed.

Location. Ship location is an information that should be reported regularly in AIS. In our study, vessels' locations were analysed based on geographical coordinates. Basically, as expected, the coverage of the analysed data is worldwide (see Fig. 2 that presents vessel traffic in weeks 33–35 of 2018). Data covers almost every point on the globe; however such distribution might be suspicious as not every point on the earth is reachable by vessels, especially terrains close to the North or the South Pole. This might results from phenomenon called AIS spoofing or a wrong configuration of GPS device on board [4].

Vessel Draught. Information about a current draught should be regularly updated by a captain. However, due to the fact that this information is often entered manually (set up statically) it is of low quality. This was confirmed

Table 5. Draught statistics for different vessel types

Type	Draught in metres (mean)	Draught in metres (std. dev.)	Total distinct values of draught (mean)	Days to draught change (mean)
Anti-pollution equipment	3.46	3.99	1.62	45.91
Cargo	7.09	3.98	6.38	13.29
Diving ops	4.28	3.70	2.33	56.53
Dredging or underwater ops	4.21	3.91	3.12	47.28
Fishing	0.90	2.43	1.13	57.88
High speed craft (HSC)	3.49	4.61	1.93	69.56
Law enforcement	2.56	3.62	1.81	56.69
Medical transport	8.05	7.32	1.33	36.56
Military ops	4.09	4.38	1.84	52.11
Non-combatant ship	6.70	7.04	1.77	37.03
Not available (default)	1.25	3.07	1.28	24.58
Other type	4.92	4.08	3.48	34.45
Passenger	4.52	4.13	2.82	33.24
Pilot Vessel	3.34	4.42	2.10	30.34
Pleasure Craft	2.41	3.13	1.83	64.03
Port Tender	4.03	4.60	1.80	29.55
Reserved	1.93	3.03	1.49	47.30
Sailing	2.44	2.98	1.47	58.15
Search and Rescue Vessel	3.06	3.49	1.71	56.46
Spare – Local Vessel	6.12	5.99	1.85	45.08
Tanker	7.84	3.84	13.05	14.06
Towing	3.94	3.85	1.83	53.70
Tug	5.01	3.50	3.40	32.93
Undefined or empty	5.76	6.98	1.79	28.66
Wing in ground (WIG)	4.22	5.08	1.87	32.71

by our analysis. The vast majority of vessels (79.13%) have reported only one draught value in 2015, what can be interpreted that this value has not been updated. The rest of vessels (20.87%) updated this value more or less regularly. The average value of draught for these vessels is 14.68 m (with a standard deviation equals to 15.87 m) – see Table 4 for different vessel types. A quick glance at the results reveals that an average reported draught can vary significantly between different vessel types. Minimum and maximum values were omitted in the table, since virtually for all types they range between 0 and 25.5 m.

Table 5 highlights also some basic statistics about a total number of updates of a draught value, with regard to vessel types. There was a significant difference across vessels' types. Tankers reported around 13 draught values in 2015 on the

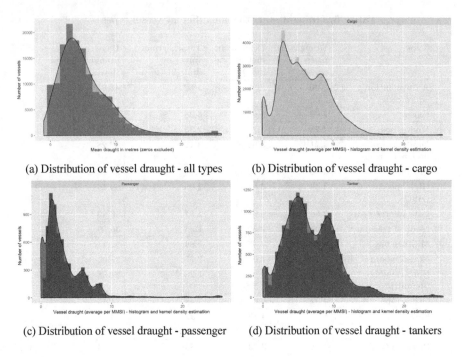

(a) Distribution of vessel draught - all types   (b) Distribution of vessel draught - cargo

(c) Distribution of vessel draught - passenger   (d) Distribution of vessel draught - tankers

Fig. 3. Distributions of vessels draught

average, cargo vessels – 6. It means that for these two types the draught value is updated more often, whereas for fishing vessels update of a draught value is much rarer. This means that a typical cargo vessel or tanker update this value accordingly once in 13 or 14 days, which is the best result in comparison to other types, while High speed craft (HSC) and pleasure crafts update it only once every 69 and 64 days accordingly. However, for the latter vessel types this might result from the fact that their draught actually does not change.

Figure 3 presents distribution of draught values for all vessel types as well as for tankers, cargo, and passenger vessels. These distributions give first information about minimum and maximum draught values reported for a given type. Secondly, they might be used to define a minimum value of draught that a ship should declare in AIS to assess whether the draught is reliable information or not (a limit value).

Navigational Status. The next analysed attribute was navigational status of a ship. As can be seen in Fig. 4, its distribution seems reliable and there is no issue with this attribute in the analysed data. The vast majority of ships travel with status 'under way using engine' (status 0). Other popular statuses are: 'moored' (5), 'at anchor' (1), 'engaged in fishing' (7). However, there are some vessels that sent 'default' status (15).

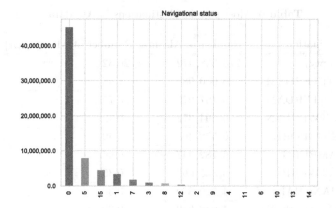

Fig. 4. Navigational status

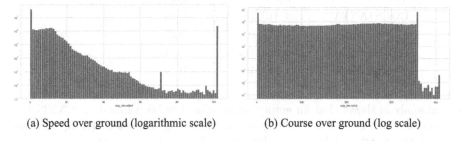

(a) Speed over ground (logarithmic scale)   (b) Course over ground (log scale)

Fig. 5. Distributions of SOG and COG

SOG and COG. The next analysed dynamic attribute was speed over ground (SOG). Its logarithmic distribution is presented in Fig. 5a. We can observe several values that probably are used as a default for missing values (peaks in the chart).

For course over ground (COG), several outliers were identified as can be seen in Fig. 5b, also presented on logarithmic scale. Moreover, 0 and 360 were the most common values, 10x more frequent than any other value (please note the log scale). This may mean that default values have not been replaced. There are also values greater than 360.

Destination. Similarly to draught value, destination should be set up manually and updated regularly by a captain. An initial statistical analysis was conducted to verify whether the provided values of destinations are valid. To explore the declared destinations, firstly the data was cleansed by removing a special character "@" and trimming it (removing leading and trailing spaces). The most popular destinations are presented in Table 6. Notice that a more sophisticated method is needed to obtain more robust results. For instance, "ANTWERP" and "ANTWERPEN" means the same port. The analysis concerned values of a destination and how often this parameter was updated. Unfortunately, the completeness and hence the quality of this variable is not satisfactory. Even such

Table 6. Most popular destinations in AIS data

Destination	Number of AIS msg	Number of AIS msg [%]
(empty)	275, 338, 472	20.32
ROTTERDAM	20, 000, 984	1.48
AMSTERDAM	10, 167, 975	0.75
ANTWERPEN	6, 694, 372	0.49
SINGAPORE	6, 578, 468	0.49
HAMBURG	6, 182, 447	0.46
ANTWERP	5, 335, 668	0.39
SHANGHAI	4, 510, 860	0.33
NOVOROSSIYSK	4, 046, 389	0.30
TIAN JIN	3, 921, 142	0.29
TIANJIN	3, 610, 640	0.27
CONSTANTA	3, 410, 220	0.25
SHANG HAI	3, 374, 164	0.25
BREMERHAVEN	3, 286, 316	0.24
HARLINGEN	3, 222, 434	0.24

initial analysis shows that an empty destination has been found in 275,338,472 messages, which is over 20% of messages, whereas 0 was set in 1,796,596 messages. Interestingly, 'HOME' was declared in 1,815,822 messages.

To explain variability in updating a destination value by a captain, we followed a similar approach as in the case of draught values – we calculated a mean number of unique destinations and total number of destination provided by each ship type as well as average time between destination update in days. The results are presented in Table 7. The obtained mean values show that in case of popular vessel types such as cargo or tankers, these vessels travel on the average to a relatively small number of unique ports (between 3 and 4). In case of passenger vessels and HSC these values are slightly higher. Moreover, destination is updated relatively often (several times a day).

4.3   Comparison of AIS with the Lloyd's Ship Catalog

In the last step of analysis of AIS data quality, ship dimensions provided by a ship were compared with data in an official vessel registry – the Lloyd's database (IHS)[1]. This comparative analysis concerned vessels travelling in three maritime areas: the German Bight, the Baltic Sea, and the North Sea. The aim of this analysis was to assess if values of dimensions provided by vessels in AIS are reliable. In case of the vessels traveling in the German Bight (1,142 vessels in total), the comparison of the overall length and width of vessels provided in AIS with

---

[1] The database provided by "IHS Markit" https://ihsmarkit.com/products/maritime-ships-register.html (accessed in October 2018).

information from *IHS fairplay* database revealed no significant differences with regard to the mean (see Table 8). However, a relatively high standard deviation in case of length may suggest that there are vessels that do not provide correct values in AIS. The difference was calculated as:

$$\text{Difference} = value_{\text{IHS}} - value_{\text{AIS}}.$$

Thus, it might be concluded that data sent by vessels in AIS (where it was provided) does not differ significantly from the data in the official register of ships.

5 Discussion and Summary

The aim of this study was to assess the quality of data provided by vessels in AIS system. The conducted quality assessment concerned both the static and dynamic attributes of AIS messages. The obtained results show that actually for each analysed attribute some problems can be noted, what in turn negatively influences the quality analyses conducted on AIS data. These problems are observed both for the static AIS parameters (vessel identification, vessel type, dimensions) and dynamic attributes (location, draught, destination).

The most common quality issues with AIS data revealed in this study are (Sect. 4):

- Duplicated identification numbers (MMSI). Usually, one can find a number of vessels with the same MMSI number and different types declared. For instance, for MMSI 123456789 there were 22 types assigned, while a vessel with MMSI 2443870000 declared 5 different types during 2015.
- No change of the default values in AIS transponder. This concerns for example value of draught, dimensions, navigational status.
- No update or a relatively rare update of dynamic values during the ship operation (e.g. draught, destination).
- Empty field or providing a wrong value (not meeting the standard requirements) in case of destination or vessel identification attributes.
- Location spoofing or wrong configuration of GPS device, resulting in incorrect positions of ships.

Summarising, there are two main sources of problems with AIS quality: wrong (purposeful or not) configuration of AIS transponder (not changing the default values while installing the transponder on board) or lack of updates of dynamic AIS attributes by a ship captain (intentional or not). Such a situation may result also from a cooperative nature of AIS system – although ships are required to use AIS there is no means to actually control if ships provide correct values. This hampers the situation awareness of various maritime actors and negatively influences the analysis of the situation of decision-makers. The results of this study may thus help maritime actors in anomaly detection and identification of potentially dangerous ships that do not provide correct information while reporting its current status and position.

Table 7. Basic statistics about declared destinations for vessel types

Type	Unique destinations (mean)	Total number of destinations (mean)	Time between destination change in days (mean)
Anti-pollution equip	4.04	60.18	0.55
Cargo	3.62	160.16	0.21
Diving ops	5.37	148.56	0.24
Dredging or underwater	5.55	179.71	0.09
Fishing	3.56	29.92	0.25
High speed craft	6.14	230.97	0.16
Law enforcement	5.03	108.19	0.14
Medical transport	4.37	65.89	0.19
Military ops	5.91	76.65	0.25
Non-combatant ship	8.69	126.62	0.22
Not available (default)	3.84	33.59	0.23
Other type	4.78	129.09	0.23
Passenger	7.28	263.63	0.09
Pilot Vessel	4.97	121.01	0.08
Pleasure Craft	4.27	118.42	0.19
Port Tender	8.54	145.79	0.15
Reserved	3.53	59.24	0.63
Sailing	3.10	59.40	0.35
SAR Vessel	3.84	76.74	0.29
Spare – Local Vessel	6.71	83.99	0.29
Tanker	3.31	234.40	0.25
Towing	4.18	85.69	0.23
Tug	5.12	183.64	0.13
Undefined or empty	5.33	57.49	0.57
Wing in ground	4.15	56.12	0.36

Table 8. Observed differences in vessels' length and width for the German Bight

Parameter	Mean	Median	Std. Dev.
Difference in length [m]	−0.175	0	31.868
Difference in width [m]	0.257	0	2.708

Table 9. The differences between the ship's length in the AIS and in the Lloyd's database for the whole analysed area

Percentage deviation	Length		Width	
	vessels	[%]	vessels	[%]
Up to 3%	38917	86.70%	18595	73.80%
From 3% to 5%	1324	3.00%	2602	10.30%
From 5% to 10%	1981	4.40%	2182	8.70%
From 10% to 20%	1270	2.80%	1003	4.00%
From 20% to 30%	365	0.80%	273	1.10%
From 30% to 40%	206	0.50%	139	0.60%
From 40% to 50%	152	0.30%	83	0.30%
From 50% to 60%	133	0.30%	46	0.20%
From 60% to 70%	98	0.20%	28	0.10%
From 70% to 80%	79	0.20%	25	0.10%
From 80% to 90%	46	0.10%	22	0.10%
From 90% do 100%	36	0.10%	23	0.10%
More than 100%	258	0.60%	178	0.70%
Total	44 865		25 199	

Data quality issues on the low level can be linked to real-world application cases of AIS. Businesses that base on reliability of AIS messages need to be able to understand basic caveats of this IT system. Particularly sensitive to errors are applications that aggregate data to identify meaningful patterns. One of the examples is reconstruction of maritime traffic networks [2]. For this task only correct data should be used. Once traffic networks are identified we can more precisely identify risks and more reliably predict time of arrival at destination port (ETA).

References

1. Balduzzi, M., Wilhoit, K., Pasta, A.: A security evaluation of AIS. Trend Micro, 1–9 (2014)
2. Filipiak, D., Węcel, K., Stróżyna, M., Michalak, M., Abramowicz, W.: Extracting maritime traffic networks from AIS data using evolutionary algorithm. Bus Inf. Syst. Eng. **62**(4), 435–450 (2020)
3. Harati-Mokhtari, A., Wall, A., Brookes, P., Wang, J.: Automatic identification system (AIS): a human factors approach. J. Navig. **60**(3), 373–389 (2007)
4. Iphar, C., Napoli, A., Ray, C.: Data quality assessment for maritime situation awareness. ISPRS Ann. Photogramm. Remote Sens. Spatial Inf. Sci. **II–3/W5**, 91–296 (2015). https://doi.org/10.5194/isprsannals-II-3-W5-291-2015. https://www.isprs-ann-photogramm-remote-sens-spatial-inf-sci.net/II-3-W5/291/2015/

5. Lewoniewski, W., Węcel, K., Abramowicz, W.: Multilingual ranking of wikipedia articles with quality and popularity assessment in different topics. Computers **8**(3), 60 (2019). https://doi.org/10.3390/computers8030060. https://www.mdpi.com/2073-431X/8/3/60
6. Nahari, M.K., Ghadiri, N., Jafarifard, Z., Dastjerdi, A.B., Sack, J.R.: A framework for linked data fusion and quality assessment. In: 2017 3th International Conference on Web Research (ICWR), pp. 67–72. IEEE (2017)
7. Pipino, L.L., Lee, Y.W., Wang, R.Y.: Data quality assessment. Commun. ACM **45**(4), 211–218 (2002). https://doi.org/10.1145/505248.506010
8. Stróżyna, M., Eiden, G., Abramowicz, W., Filipiak, D., Małyszko, J., Węcel, K.: A framework for the quality-based selection and retrieval of open data - a use case from the maritime domain. Electron. Markets **28**(2), 219–233 (2017). https://doi.org/10.1007/s12525-017-0277-y
9. Tu, E., Zhang, G., Rachmawati, L., Rajabally, E., Huang, G.B.: Exploiting AIS data for intelligent maritime navigation: a comprehensive survey from data to methodology. IEEE Trans. Intell. Transp. Syst. **19**(5), 1559–1582 (2017)
10. UNCTAD: Review of Maritime Transport (2017). http://unctad.org/en/PublicationChapters/rmt2017%7B%5C_%7Den.pdf

Digitalization of Small and Medium-Sized Enterprises: An Analysis of the State of Research

Katharina Klohs(✉) ⓘ and Kurt Sandkuhl ⓘ

University of Rostock, Albert-einstein-str. 22, 18059 Rostock, Germany
{katharina.klohs,kurt.sandkuhl}@uni-rostock.de

Abstract. This paper gives an overview of the existing research topics in the field of SME digitalization. Digitalization commonly causes severe changes in both, organizational structures, business models and the IT landscape of an enterprise, i.e., there is a need for business and IT-alignment. By means of a literature analysis and a subsequent systematic mapping study the paper examines on which areas current research work in the field is focused and where potential research gaps exist. The literature analysis has its focus on already existing literature analyses in order to get a comprehensive overview. All identified papers are subsequently classified and visually represented in a diagram. The classification is done in two dimensions. Dimension 1 shows to which step of the digital transformation process the papers refer, dimension 2 refers to the success factors of the digital transformation. The result shows that the focus of recent research mainly was on the analysis of the current situation in companies and that the other steps of the digital transformation are largely ignored. The paper also concludes that there is no step-by-step guide for SMEs that shows how to go through the digital transformation.

Keywords: Digitalization · Digital transformation · SME · Literature analysis · Systematic mapping

1 Introduction

More than 99% of all companies in Europe [11] are small and medium-sized enterprises (SMEs). Research in digital transformation indicates that SMEs consider digitalization and digital transformation as major challenges, in particular due to the effects on both, business and IT. In order to remain competitive and open up new fields of business, the use of digital technologies is often unavoidable. Digital transformation often causes severe changes in business models, forms of organization and work, products and processes [9]. This transformation requires both the use of innovative technologies, such as those emerging in the field of artificial intelligence (AI), and the management and social shaping of organizational change processes [8]. In this context, the present article aims to determine the current state of research on this topic. For this purpose three research questions were formulated.

© Springer Nature Switzerland AG 2020
W. Abramowicz and G. Klein (Eds.): BIS 2020 Workshops, LNBIP 394, pp. 21–33, 2020.
https://doi.org/10.1007/978-3-030-61146-0_2

RQ1: *Which areas of digitalization in connection with SMEs did previous research address in literature analyses?* The areas of digital transformation can be very diverse. The aim is to find out which ones the researchers focus on. Examples include new technologies, such as the Internet of Things or virtual reality, or the changes in the company brought about by digitalization, such as the new corporate culture or digital leadership.

RQ2: *What steps in the process of digital transformation are researchers focusing on?* There are various approaches that structure the process of digital transformation step by step. In this paper the process of Klasen [Kl19] is used. The paper aims to determine which of the steps in the process the researchers are concentrating on.

RQ3: *Which weaknesses or gaps in research on the digitalization of SMEs can be identified?* The weaknesses or gaps can be, among other things, areas of digital transformation that are not or only marginally addressed by researchers or process steps of digital transformation that are not sufficiently addressed.

The paper starts with a short background on digitalization and business and IT alignment (Sect. 2). It follows a systematic literature analysis (Sect. 3) with primary focus on already existing literature analyses. In Sect. 4, a systematic mapping study is conducted to categorize the literature found and to reveal gaps in research. On the one hand, the steps of the transformation process (Sect. 4.2) are categorized and on the other hand success factors of digital transformation (Sect. 4.3) are considered. The result (Sect. 4.3) shows that most scientific contributions focus on the first transformation process step, analysis. Section 5 discusses the limitations of the work. In the last section, the main findings of the paper are summarized and an outlook on future research is given (Sect. 6).

2 Background: Digitalization and Business/IT-Alignment

Digitalization and digital transformation are topics receiving substantial attention in research and industry. The general expectation is that they will have substantial effects on markets, companies and their operations. Digitalization can be seen, in simplified terms, as a generic term for efforts to convert information, documents, processes, products or services into a form that can be processed or supported by IT. In research, digitalization historically is subdivided into phases which, depending on scientific background, are more technologically characterized, consider the socio-economic change or investigate specific industries [35]. The current phase, denoted "digital transformation", is often referred to as the "third" or "fourth industrial revolution" [35]. The focus of DT is on the disruptive social and economic consequences, which, due to new technological application potentials, lead to changes in economic structures, qualification requirements for employees and working life in general [34].

Business and IT-alignment (BITA) in general is a continuous process aiming at aligning strategic and operational objectives and ways to implement them between the business divisions of an organization and the organization's information technology division [7]. Many challenges are linked to BITA since the business environment continuously changes and so does the IT in an enterprise, but the pace of change and the time frame needed to implement changes are different. Digital transformation is considered as major driver of changes in both, business environment and IT. As a consequence; BITA is of high relevance in digital transformation processes. In this context, the success factors for digital transformation and the different DT phases investigated in the literature study can serve as contributions to structuring the BITA.

3 Literature Review

A systematic literature analysis is carried out to review the existing work. The following section describes the exact procedure of the literature analysis and presents the results.

3.1 Method and Search Term

For the literature review the approach of Kitchenham [13] is used. Kitchenham says that before a systematic review is undertaken it must be ensured whether it is necessary at all. In particular, already existing reviews should be identified and evaluated. This reflects an aim of this paper. For the literature analysis the following steps are taken from Kitchenham's approach: (1) Formulation of research questions, (2) Selection of resources to be searched, (3) Definition of search terms, (4) Definition of selection criteria, (5) Checking the relevance of the results, (6) Analysis of the results found to be relevant. The research questions are already defined in the previous chapter (1). In the next step the resources to be searched are selected (2). Besides Scopus (www.scopus.com), Web of Science (www.webofknowledge.com) and EBSCOhost (*search.ebscohost.com*) the database AIS eLibrary (https://aisel.aisnet.org) is searched. The following search strings are used (3), which refer to the title, abstract or keywords:

- digi* AND "literature review" AND (sme OR "Small and Medium Enterprise")
- Digitalisierung AND Literaturanalyse AND (KMU OR "kleine und mittlere Unternehmen")

As described by Kitchenham [13] different terms and synonyms are used for the search. Synonyms used for "literature review" are literature analysis, systematic review and structured review. Selection criteria for the results are defined (4). Since the current state of research is of particular interest, all publications before 2016 will be excluded. Furthermore, the search will be limited to German and English language contributions. Based on these criteria a first number of contributions can be identified. For a preliminary evaluation the title and the abstract are read to decide on relevance.

3.2 Evaluation of the Literature Review

Table 1 gives an overview of the results found with the help of the search terms as well as those subsequently found to be relevant. Especially with EBSCOhost a large number of contributions must be declared as irrelevant. The reason for this is that the database also lists so-called "news" that have no scientific background. In addition, EBSCOhost and Web of Science also provide articles that are already included in the Scopus results set. The abstract is read of those articles that cannot be sorted out at first sight. With the help of the abstract it can be recognized whether the articles explicitly refer to SMEs and whether a literature analysis is carried out. If both points are not evident, the contribution is found to be irrelevant and thus sorted out. Thus, 18 out of initially 70 found contributions can be considered relevant.

Table 1. Results of the literature analysis

Database	Findings	Results of relevance
Scopus	15	10
Web of Science	7	2
EBSCOhost	45	4
AISeL	3	2
Summe	**70**	**18**

The 18 literature analyses focus on different topics that are repeated in different contributions. Based on these recurring topics, the areas shown in Table 2 were identified. The main focus of research in connection with digitalization of SMEs is mainly on Industry 4.0 and readiness/maturity models.

Table 2. Topics of the articles

Topics	
Industry 4.0	[6, 10] * [18, 19, 23, 27]
Readiness/Maturity	[18, 19, 27, 29, 33]
Servitization[a]	[17, 25]
Big Data	[22] * [23]
Marketing/Social Media	[5] * [12] *
IT-Governance	[15]
Other Topics	[21, 28, 30, 32]

[a]"Servitization is the innovation of an organization's capabilities and processes to shift from selling products to selling integrated products and services that deliver value in use." [2].

Due to access restrictions[1], 4 of the 18 papers found to be relevant cannot be read and analyzed in full. As a result, only the results of the available 14 papers are comprehensively examined and included in the Systematic Mapping.

---

[1] Articles that were not available in full text are marked with an *.

4 Systematic Mapping

In the following section the theoretical basics of Systematic Mapping are presented. This includes the description of the procedure of Systematic Mapping as well as the explanation of the dimensions "steps in the transformation process" and "success factors for digital transformation".

4.1 Systematic Mapping Process Based on Petersen et al.

Systematic mapping studies are used to categorize research contributions and results and to present them visually [26]. Petersen et al. [26] propose the following process for this study. (1) Definition of the research question; (2) Search for primary literature; (3) Screening of the papers found; (4) Keywording of the abstracts and preparation of the classification scheme und (5) data extraction and mapping

As a literature analysis already was carried out, we continue with step 4. Since the papers identified in the literature analysis are themselves literature analyses, all papers found in the analyses are classified. The classification is based on the abstracts. Deviating from Petersen et al.'s [26] suggestion, classes are defined in advance, i.e. the papers found are only sorted in. The classes are based on business transformation process according to Klasen [14] on the one hand and on the success factors for digitalization based on the dimensions of the Digital Maturity Check of St. Gallen [3] on the other.

4.2 Process of Transformation

According to Klasen [14], business transformation is the strategic reorientation and organizational transformation of a company or one of its parts in order to secure its long-term performance in the market [14]. This reorientation of the entire company can also be caused by the introduction of digital technologies, for this reason the process steps according to Klasen [14] can also be used for digital transformation. Klasen's approach was chosen because the author not only has a scientific background, but also has practical experience and was able to shape important transformation processes in an entrepreneurial manner. This practical relevance was important to us.

In the first step of the process, the problem that will arise or has already arisen due to certain factors must be recognized and analyzed. Examples of these factors can be: new customer needs, substitute products that make the own product redundant and the related new development of products and services. The next step is to develop a strategy to counteract these problems. According to Klasen [14] the company asks itself the following questions, among others: How will we differentiate ourselves and our products in the market in the future? Which skills must be developed and used in the future?

Before the developed strategy can finally be implemented, a project for its implementation is started. The most important activities for starting a project include setting up the project team, project planning. After project planning, the actual implementation of the goal set by the strategy takes place. The implementation is divided into three phases. In concept planning, a bundle of different solution variants for the implementation of the strategy (or parts of it) is developed, evaluated and the best idea for the company is proposed for further detailed planning [14]. In the detailed planning phase, the agreed solution variants are worked out and concrete plans for implementation are drawn up so that execution can take place in the third phase. Here it is important that the progress of the measures is regularly measured, evaluated and communicated. Especially if the new strategy to be implemented has an impact on the company's employees, it is inevitable to include them. Since the acceptance of the new strategy by the employees is of crucial importance, early involvement is essential. This can be done during the project start in order to be able to react to resistance and wishes.

In addition to employees, there are other stakeholders who should be informed about the transformation. This includes, for example, the customers who, for marketing purposes, should also be made aware that something is about to change. At the end of the transformation process is the anchoring. Anchoring includes the documentation of the results and experiences of the transformation process. The results are recorded by the project documentation. A knowledge management system ensures that the acquired knowledge is passed on to colleagues within the project.

The following chart shows the transformation process according to Klasen [14].

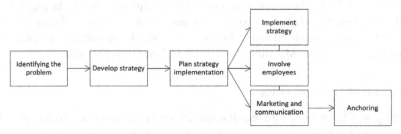

Fig. 1. Transformation process based on Klasen [14]

4.3 Success Factors of Digital Transformation

To be successful as a company in the digital transformation, several factors play a role that must be taken into account. The so-called success factors. The success factors used in this article are taken from the Digital Maturity Model of the University of St. Gallen [3]. The categories addressed in the Digital Maturity Check and listed in Table 3 support the company in reflecting on the effects of digital transformation and identifying possible fields of action and are therefore also suitable as success factors.

Other authors have also dealt with success factors of digital transformation and identified similar factors. However, leadership, governance and network partners (e.g. suppliers) were also mentioned [1, 16] (Table 4).

4.4 Evaluation of the Systematic Mapping

Based on the transformation process and the success factors, a graphic is now created and the identified papers are sorted. The result is shown in Fig. 2.

Table 3. Success factors of digital transformation based on St. Gallen [3]

Success factor	Description
Customer Experience	The needs and requirements of the customers are known and it is possible to react to changes
Product innovation	New products and services are developed through the use of digital technologies
Strategy	Digital technologies are firmly anchored in the corporate strategy
Organization	The organisation is adapted to the new challenge and has the necessary resources
Process digitalization	Processes are adapted to the digital structures and, if possible, are automated
Collaboration	Digital technologies are used for more efficient collaboration
Information technology	IT infrastructure and information systems enable the use of new digital technologies
Culture & Expertise	The employees are open to new technologies and missing digital expertise is being built up
Transformation management	Digital transformation is a supported, planned and controlled process

Table 4. Additional success factors of the digital transformation

Success factor	Description
Leadership	All managers are involved in the implementation of the digital strategy
Governance	The digital activities can be viewed and controlled by the company
Network partners (suppliers)	New digital technologies are examined to determine whether a connection to the supplier is possible

The mapping shown does not contain all identified papers. For some papers it was not possible to sort them into the previously defined classes because they deal with other topics of digitalisation, such as ecological factors or the development of different software. Furthermore, a classification of the contributions of Michalik et al. [17] and Peillon and Dubruc [25] on the topic of servitization into the scheme is not possible. In the remaining 12 contributions examined, a total of 208 papers were identified and sorted into the mapping. It is possible that a paper was sorted into several categories.

The graph shows that most researchers are concerned with the analysis of SMEs. This includes in particular the often mentioned readiness or maturity models. The most

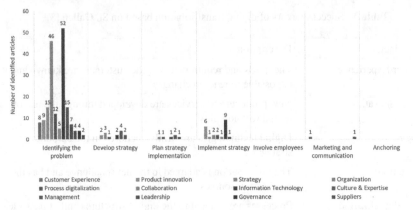

Fig. 2. Systematic Mapping

frequently mentioned dimensions/factors in these models are strategy, organization, processes, technologies and culture and expertise in the company.

Pirola et al. [27] conclude that the existing Maturity models have limitations with regard to their use in SMEs because the structure of the models does not always fit the organizational forms of SMEs. Mittal et al. [18] also state in their research that the models identified in the literature are mainly oriented towards the needs of larger companies and also identify a research gap.

But before measuring the maturity of a company, the question arises: Which factors and resources are relevant for the willingness of a company to respond to the digital challenge [29]? In this context, Sanchez and Zuntini [29] have derived a framework for digital literacy in SMEs. The framework aims to assess organizational readiness in terms of its ability to create value in the new digital environment. For this purpose, the internal and external components are analyzed. Internally, the success of the company in creating value clearly depends on the definition of a digital strategy and the ability to implement it [29]. But although the definition of a strategy is so important, there is no support for SMEs for the next step after the maturity level has been assessed [18]. This is also clearly shown in the graph, the steps after the analysis of the company are not very much discussed in research. Only seven publications[2] deal with the development of a strategy for digital transformation.

One of the factors addressed in the process step of strategy development is the factor "culture and expertise". This tendency also coincides with the strategic questions for strategic reorientation mentioned by Klasen [14]. For example, it says "What skills must the company develop and use in the future? Constant technological development means that the capabilities of organizations and people must be developed further in order to be able to respond to current and future challenges. In this context, Sousa and Wilks [30] cite disruptive technological capabilities such as artificial intelligence, robotics, the Internet of Things and digitalization. However, it is important to develop not only technological skills, but also skills that contribute to the development of society as a whole. Sousa and Wilks [30] give the following examples: critical thinking and

---

[2] Multiple answers are possible in the matrix.

problem-solving skills; network collaboration; adaptability; effective oral and written communication; information evaluation and analysis; curiosity and imagination.

The next process step is "project planning". According to Klasen [14], this step includes an assessment of the risks of the project. According to Moeuf et al. [20], the biggest risks in the introduction of Industry 4.0 in SMEs include a lack of expertise and a short-term strategic mindset. The contribution of Moeuf et al. [20] also shows that training is the most important success factor, that managers have a prominent role in the success and/or failure of a project and that SMEs should be supported by external experts. Birkel et al. [4] also discuss the risks of introducing Industry 4.0 in their contribution, which the authors divide into 5 categories. Economic risks include the risks associated with high or wrong investments. From an ecological point of view the increased waste and energy consumption as well as possible ecological risks are described in connection with the concept of "batch size one". From a social point of view, the loss of jobs, the risks of organizational restructuring and the re-qualification of employees, as well as internal resistance are considered, among other things. In addition, risks can be associated with technical risks, e.g. technical integration, information technology (IT) risks such as data security, and legal and political risks, e.g. unclarified legal clarity with regard to data ownership [4].

The second most common process step, yet one that is treated very little, is implementation. In the implementation of digitalization projects, researchers focus primarily on the technology factor. Moeuf et al. [19] identify reasons for the introduction of new technologies. Flexibility, cost reduction and improvement of productivity play an important role here. Nevertheless, Moeuf et al. [19] note that SMEs use newer technologies only to a limited extent. First and foremost, the low-cost but least revolutionary technologies such as cloud computing and simulation are being introduced, while those that enable profound business transformations, such as collaborative robots or machine-to-machine communication, are still neglected. The reason for the hesitant introduction may be that SMEs have a lack of resources compared to larger companies. Among other things, there are no resources to finance research and development activities or to manage computer solutions [19].

The step of involving employees, which Klasen [14] cited as a very important point, as well as all subsequent steps could hardly be identified in the analysis. Only Veile et al. [31] mention in their contribution the communication with suppliers and customers. It is shown that the digital networking of suppliers and customers contributes to the optimization of the global value chain.

5   Limitations

Limitations in the present work should not be left unmentioned. The main limitations lie mainly in the literature analysis itself. On the one hand, only three databases were searched and thus perhaps important contributions were not found. On the other hand, many papers were already sorted out on the basis of their abstract. There is also the possibility that important publications were sorted out by mistake. The limitation to articles after 2015 may also lead to the fact that important contributions were not found. Another limitation of the work may be that some papers may have addressed each other.

For example, Mittal et al. [18] may have referred to the work of Moeuf et al. [19] in the literature analysis for Industry 4.0 and therefore achieve similar results. These references were not reviewed. In addition, papers on the topic of marketing/social media were not available as full-text, which meant that an essential topic was lost. In the case of systematic mapping, the sorting of the papers in the classes was very subjective and was done only on the basis of the abstracts. As a result, contributions may have been sorted incorrectly or not at all.

6   Conclusion and Outlook

This paper has provided an overview of current research on digitalization in SMEs. It is noted that in the last 4 years 18 literature analyses have been carried out on this topic by different researchers from different countries. The focus of the research is mainly on the development and evaluation of readiness and maturity models and Industry 4.0 (RQ 1). This trend is also evident in Systematic Mapping. Research is mainly concentrated on the transformation process step analysis (RQ 2). However, not all success factors of digital transformation are addressed in the same way here, but primarily the organization of a company and the technologies used or to be used are considered, which confirms the trend towards Industry 4.0. Especially neglected at the stage of analysis are the management, IT governance, the cooperation of employees among each other and the connection to or cooperation with suppliers. Although the main driver in change processes is primarily the management of the company [24], this plays hardly any role in research on digital transformation in SMEs. Except in the analysis phase, the factor leadership is no longer addressed. Beyond the process step of the analysis, only isolated publications can be found. Strategy development and the implementation of these are only dealt with in a small number of contributions. However, not all success factors are covered here either. For example, the factors leadership, as described above, IT governance and suppliers are not addressed at all in these stages of the transformation process. In the implementation phase, technologies play a particularly important role. In addition, current research only addresses project planning with regard to the risks of implementing digital transformation, in particular Industry 4.0. In addition to the lack of expertise, management and technologies are also addressed here. The involvement of employees and the anchoring of the implemented strategy are not addressed in any way. Only the step marketing and communication, here especially the communication to suppliers and customers, is mentioned in an article. The present paper thus shows that the main focus of research is on how to measure the digital maturity of a company. After the measurement, however, companies do not receive any support or advice on how to improve the level. It could be recognized that there is a large gap in research on the development of digitalization strategies and their successful implementation (RQ 3). This finding is supported by the fact that very few contributions have been identified in this area.

Future research should focus primarily on how SMEs can go through the process of digital transformation step by step. Here, the first step should be to examine whether and how many general models for digital transformation exist and whether these have already been successfully used before they are adapted to the structures of SMEs.

References

1. Azhari, P., Faraby, N., Rossmann, A., Steimel, B., Wichmann, K.: Digital Transformation Report 2014 (2014). https://www.wiwo.de/downloads/10773004/1/dta_report_neu.pdf. Accessed 11 Mar 2020
2. Baines, T.S., Lightfoot, H.W., Benedettini, O., Kay, J.M.: The servitization of manufacturing. JMTM **20**(5), 547–567 (2009). https://doi.org/10.1108/17410380910960984
3. Berghaus, S., Back, A., Kaltenrieder, B.: Digital maturity & transformation report (2017). https://crosswalk.ch/digital-maturity-and-transformation-report. Accessed 16 Jan 2020
4. Birkel, H., Veile, J., Müller, J., Hartmann, E., Voigt, K.-I.: Development of a risk framework for industry 4.0 in the context of sustainability for established manufacturers. Sustainability **11**(2), 384 (2019). https://doi.org/10.3390/su11020384
5. Centobelli, P., Cerchione, R., Esposito, E., Raffa, M.: Digital marketing in small and medium enterprises: the impact of web-based technologies. Adv. Sci. Lett. **22**(5), 1473–1476 (2016). https://doi.org/10.1166/asl.2016.6648
6. Contreras Pérez, J.D., Cano Buitrón, R.E., García Melo, J.I.: Methodology for the retrofitting of manufacturing resources for migration of SME towards industry 4.0. In: Florez, H., Diaz, C., Chavarriaga, J. (eds.) ICAI 2018. CCIS, vol. 942, pp. 337–351. Springer, Cham (2018). https://doi.org/10.1007/978-3-030-01535-0_25
7. Seigerroth, U.: Enterprise modeling and enterprise architecture: the constituents of transformation and alignment of business and IT. IJITBAG **2**(1), 16–34 (2011)
8. ESF: ESF-Bundesprogramm Zukunftszentren (2019). https://www.esf.de/portal/DE/Foerderperiode-2014-2020/ESF-Programme/bmas/2019_03_13_zukunftszentren.html. Accessed 21 Feb 2020
9. Franzetti, C.: Digitalisierung, digitale transformation. In: Franzett, C. (ed.) Essenz der Informatik, pp. 223–240. Springer, Heidelberg (2019). https://doi.org/10.1007/978-3-662-58534-4_15
10. Gamache, S., Abdulnor, G., Baril, C.: Toward industry 4.0: Studies and practices in Quebec SMES. In: How digital platforms and industrial engineering are transforming industry and services. In: CIE 2017. Proceedings of International Conference on Computers and Industrial Engineering, 11.10 bis 13.10 2017 (2017)
11. iwd: Die EU vernachlässigt den Mittelstand (2017). https://www.iwd.de/artikel/die-eu-vernachlaessigt-den-mittelstand-342416/. Accessed 21 Feb 2020
12. Jokonya, O., Mugisha, C.: Factors influencing retail SMEs adoption of social media for digital marketing. In: ECSM 2019 - Proceedings of The 6th European Conference On Social Media, Brighton, 13 June–14 June 2019, pp. 145–153. ACPIL, [S.l.] (2019)
13. Kitchenham, B.: Procedures for Performing Systematic Reviews (2004). https://pdfs.semanticscholar.org/2989/0a936639862f45cb9a987dd599dce9759bf5.pdf. Accessed 10 Feb 2020
14. Klasen, J.: Business Transformation. Springer, Wiesbaden (2019). https://doi.org/10.1007/978-3-658-25879-5
15. Levstek, A., Hovelja, T., Pucihar, A.: IT governance mechanisms and contingency factors: towards an adaptive IT governance model. Organizacija **51**(4), 286–310 (2018). https://doi.org/10.2478/orga-2018-0024
16. Leyh, C., Meischner, N.: Erfolgsfaktoren von Digitalisierungsprojekten - Einflussfaktoren auf Projekte zur Digitalen Transformation von Unternehmen. ERP **2018**(2), 35–38 (2018). https://doi.org/10.30844/ERP18-2_35-38
17. Michalik, A., Besenfelder, C., Henke, M.: Servitization of small- and medium-sized manufacturing enterprises: facing barriers through the Dortmund management model. IFAC-PapersOnLine **52**(13), 2326–2331 (2019). https://doi.org/10.1016/j.ifacol.2019.11.553

18. Mittal, S., Khan, M.A., Romero, D., Wuest, T.: A critical review of smart manufacturing & Industry 4.0 maturity models: implications for small and medium-sized enterprises (SMEs). J. Manuf. Syst. **49**, 194–214 (2018). https://doi.org/10.1016/j.jmsy.2018.10.005

19. Moeuf, A., Pellerin, R., Lamouri, S., Tamayo-Giraldo, S., Barbaray, R.: The industrial management of SMEs in the era of Industry 4.0. Int. J. Prod. Res. **56**(3), 1118–1136 (2018). https://doi.org/10.1080/00207543.2017.1372647

20. Moeuf, A., Lamouri, S., Pellerin, R., Tamayo-Giraldo, S., Tobon-Valencia, E., Eburdy, R.: Identification of critical success factors, risks and opportunities of Industry 4.0 in SMEs. Int. J. Prod. Res. **58**(5), 1384–1400 (2019). https://doi.org/10.1080/00207543.2019.1636323

21. Molinillo, S., Japutra, A.: Organizational adoption of digital information and technology: a theoretical review. Bottom Line **30**(1), 33–46 (2017). https://doi.org/10.1108/BL-01-2017-0002

22. Moonen, N., Baijens, J., Ebrahim, M., Helms, R.: Small business, big data: an assessment tool for (big) data analytics capabilities in SMEs. Proceedings **2019**(1), 16354 (2019). https://doi.org/10.5465/ambpp.2019.16354abstract

23. Müller, J.: How SMEs can participate in the potentials of Big Data within Industry 4.0. In: International Conference on Information Systems 2019 Special Interest Group on Big Data Proceedings. (2019)

24. Nyhan, B.: Competence development as a key organisational strategy - experiences of European companies. Ind Commer. Train. **30**(7), 267–273 (1998). https://doi.org/10.1108/00197859810242897

25. Peillon, S., Dubruc, N.: Barriers to digital servitization in French manufacturing SMEs. Procedia CIRP **83**, 146–150 (2019). https://doi.org/10.1016/j.procir.2019.04.008

26. Petersen, K., Feldt, R., Mujtaba, S., Mattsson, M.: Systematic mapping studies in software engineering. In: 12th International Conference on Evaluation and Assessment in Software Engineering (EASE), 26–27 June 2008. BCS Learning & Development (2008). https://doi.org/10.14236/ewic/ease2008.8

27. Pirola, F., Cimini, C., Pinto, R.: Digital readiness assessment of Italian SMEs: a case-study research. JMTM **ahead-of-print**(ahead-of-print) **61** (2019). https://doi.org/10.1108/jmtm-09-2018-0305

28. Riera, C., Iijima, J.: The role of it and organizational capabilities on digital business value. PAJAIS, 67–95 (2019). https://doi.org/10.17705/1pais.11204

29. Sanchez, M.A., Zuntini, J.I.: Organizational readiness for the digital transformation: a case study research. G&T **18**(2), 70–99 (2018). https://doi.org/10.20397/2177-6652/2018.v18i2.1316

30. Sousa, M.J., Wilks, D.: Sustainable skills for the world of work in the digital age. Syst. Res. **35**(4), 399–405 (2018). https://doi.org/10.1002/sres.2540

31. Veile, J.W., Kiel, D., Müller, J.M., Voigt, K.-I.: Lessons learned from Industry 4.0 implementation in the German manufacturing industry. JMTM **ahead-of-print**(ahead-of-print) (2019). https://doi.org/10.1108/jmtm-08-2018-0270

32. Waradaya, A., Sasmoko, So, I.G.: Mediating effects of digital technology on entrepreneurial orientation and firm performance: Evidence from small and medium-sized enterprises (SMEs) in Indonesia **8**(5C). https://doi.org/10.35940/ijeat.e1098.0585c19 (2019). https://doi.org/10.35940/ijeat.e1098.0585c19

33. Williams, C., Schallmo, D., Lang, K., Boardman, L.: Digital Maturity Modelsfor small and medium-sized enterprises: a systematic literature review. In: The ISPIM Innovation Conference. Celebrating Innovation: 500 Years Since daVinci, 16–19. Juni 2019 (2019)

34. Rifkin, J.: The Third Industrial Revolution: How Lateral Power Is Transforming Energy, the Economy, and the World. St. Martin's Griffin (2013). https://www.amazon.com/Third-Industrial-Revolution-Lateral-Transforming/dp/0230341977

35. Hirsch-Kreinsen, H., ten Hompel, M.: Digitalisierung industrieller Arbeit: Entwicklungsper-
spektiven und Gestaltungsansätze. In: Vogel-Heuser, B., Bauernhansl, T., ten Hompel, M.
(eds.) Handbuch Industrie 4.0 Bd.3. SRT, pp. 357–376. Springer, Heidelberg (2017). https://
doi.org/10.1007/978-3-662-53251-5_21

Incremental Modeling Method of Supply Chain for Decision-Making Support

Szczepan Górtowski$^{(\boxtimes)}$ and Elżbieta Lewańska$^{(\boxtimes)}$

Institute of Informatics and Quantitative Economics,
Poznań University of Economics, al. Niepodleglośi 10, 61-875 Poznań, Poland
{szczepan.gortowski,elzbieta.lewanska}@ue.poznan.pl

Abstract. The paper presents the concept of Incremental Modeling Method of the supply chain. Research is conducted in accordance with the Design Science Research paradigm. The work presents an innovative concept of supply chain modeling using Data Science (DS) methods. Its tasks are to mitigate several problems of supply chain modeling such as time required to change the model, model accuracy, wide range of model outputs. As supply chains consist of numerous, heterogeneous data sources that fulfil BD definition, it is postulated to model the supply chain as coherent with BD technology. Requirements and procedure for the construction of such a model are presented in a framework form. The issue of including the analysis results in the enterprise management system was also taken into consideration.

Keywords: Supply chain · Modeling · Big Data · Data Science · Decision-making process · Decision support system · Analytics · Data-driven supply chain

1 Introduction

The concept of supply chain management (SCM) analytics initially referred to the process of obtaining and presenting information about the supply chain in order to provide measurement, monitoring, forecasting and management of the chain. O'Dwyer and Renner [13] developed this idea into Advanced SCM analytics. They noticed that simple methods are insufficient for complex supply chain analysis. Among the challenges faced by SC analyzes, they distinguish, among others speed-to-analysis, the ability to drill into data, using broader sets of internal and external information, including both structured and unstructured data. Advanced SCM analytics can help supply chain professionals analyze growing data sets using proven analytical and mathematical techniques [13]. One of the most famous call for research [18] regarding SCM and DS proposes to implement Data Science techniques to solve supply chain problems and predict outcomes, taking into account data quality and availability issues. Research goals are:

- concept of Incremental Modeling Method of the supply chain, which aims to gain advantage of rapid adaptation of the model to changes in the environment,

W. Abramowicz and G. Klein (Eds.): BIS 2020 Workshops, LNBIP 394, pp. 34–44, 2020.
https://doi.org/10.1007/978-3-030-61146-0_3

– analysis of ways to present the supply chain model,
– taking modelling into account in decision-making processes.

This article presents theoretical assumptions and phases in the Incremental Modeling Method. In the next step, the authors will present a case study using this method.

Paper is structured as follows. Research methodology is presented in Sect. 2. Literature review is conducted in Sect. 3. Proposed Incremental Modeling Method of supply chain is presented in Sect. 4. It has been divided into sub-sections, which address the issues raised during modeling. The work ends with conclusions and a discussion of future goals.

2   Research Methodology

An extensive selection of databases has been chosen as a way to access various range of publications regarding the subject (e.g. journal articles, conference materials, dissertations, theses, books, magazine articles). Databases such as Google Scholar, SpringerLink, Emerald and Scopus were searched to ensure that the most relevant issues will have been identified. The journals reviewed are for instance, International Journal of Logistics Management, Logistics Research, Journal of Industrial Engineering International, Journal of Business Logistics, Supply Chain Management Review etc. Literature analysis was conducted in the area of using Big Data technology in SCM and methods of supply chain modeling. Interviews with business experts supplemented the knowledge gained from the literature. The literature review focused on several aspects of SC modeling, such as: supply chain presentation methods, difficulties in model creation, weaknesses of current methods, and the use of BD in modeling. The area of interest of this publication is supply chain modeling. This research follows the principles of Design Science Research (DSR) guidelines by Hevner et al. [7].

Creating an artifact requires the use of available resources to achieve the desired goals. The goals meet the requirements of the environment. The result of research according to the Design Science Research (DSR) paradigm is the development of technology-based solutions which are important and significant for business. The environment presents the space of the research problem, determines the resources of the organization, including people along with their skills, capabilities, organizational systems and available and possible technologies. Within the organizational system group, organizational strategy, organizational culture, structure and implemented processes stand out. The analysis of available resources is crucial to determine the possibilities of research. On the other hand, the environment also includes needs that will be met by the results of research. The analysis of business needs allows determining the potential research areas. Provides opportunities to determine the expected artifacts. Observation of the environment provides requirements for artifacts. Needs in conjunction with an overview of the current state of knowledge and available technologies allows to plan research. The Fig. 1 shows the relationship between the environment, the knowledge base and artifact. Proposed artifact creates an

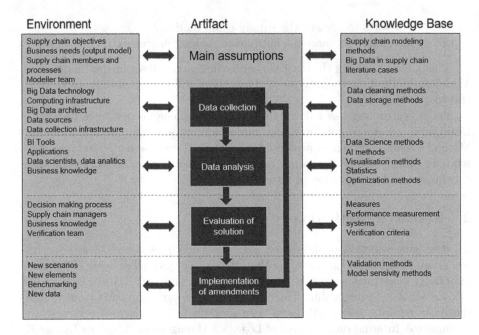

Fig. 1. Dependencies between the environment and knowledge base in the Incremental Modeling Method

opportunity for enterprises to improve processes and reduce the risk of making decisions.

The proposed method applies DSR methodology. The method consists of a preparatory phase and four successively following repeated phases. The preparation phase contains the basic assumptions of the created model. In this phase, the resources and the purpose or objectives of the model are determined. In the first phase, data is collected and cleaned. In the second phase, data is used for analysis. The results of the analyzes are properly presented to decision makers. In this way, it is possible to verify and validate the model, which occurs in the third phase. Finally, in the last phase, the cycle is summarized, deviations determined, potential model improvement opportunities and, consequently, the implementation of corrections.

The problem explication phase has been completed and this paper presents requirements gathering for artifact solving the problem. This artifact is the Incremental Modeling Method consisting of four phases occurring in the cycle. The artifact is depicted in the center of Fig. 1. One of the phases of which the artifact consists of, regarding data analysis, is discussed in detail in the following chapters and shown in Fig. 3.

3   Related Works

Supply chain model is one of the basic tools for optimizing and researching the supply chain. The model is a simplified representation of a complex real object. Relationships between SC members must be presented in an understandable form for the recipients. Graphical presentation of the supply chain in the form of a set of nodes, processes, KPIs etc. is more intuitive for the decision maker. The process of building a supply chain model is called modeling. Supply chain modelling allows the faster and more accurate answering to manager questions related to the supply chain which would not be possible otherwise [9].

Modeling is a complex process involving many stages. In the literature, SC modeling is divided according to the approach used and consideration of randomness. The main approaches are simulation and analytical. Simulation systems are very often part of the decision support system, especially at the strategic level [1]. Unlike simulations, an analytical approach creates a goal function that is attempted to be optimized using mathematical methods. The most common analytical approaches found in the literature are:

- Deterministic - The result is fully determined by the initial conditions and parameters. In this model, the data analyst has full knowledge of variable values and time series data. Once configured, the parameters will not be changed during analysis [12,15].
- Stochastic - This allows you to add uncertainty to the model. The initial condition can be changed during the run of the model depending on the partial result. Randomness is often used to simulate unexpected changes. This model is often used when some variables cannot be fully modeled or the data analyst does not know exactly what value the [8] parameters can have.
- Hybrid - This model uses some deterministic techniques in which some variables will be modeled with the allowable failure of iteration of the [2] solution.

An important problem raised by [14] is the choice of the goal function to be optimized. Supply chain models are generally multiple criteria models. Hence, various methods of selecting metrics are used [10]. The decisions of the management team are usually focused on minimizing costs, while the proper objective function should be focused on maximizing profit or other performance measures. Review of Big Data analytics (BDA) in logistic and supply chain management (LSCM) done by Gang et al. specifies the applications focused on strategy and operations [20]. This review shows how widely logistics managers are interested in data analysis technologies. However, in order to gain the full advantage of BDA, it must be involved into the company's decision-making system. The Sanders book provides insight into the management implications of implementing BDA in organizations [16].

Duan et al. wrote that rapid advancement and availability of Big Data technologies also allowed the development of Artificial Intelligence (AI) [3]. BDA plays an increasingly important role in business operations [19]. It provides information that could not be obtained using existing methods and the analyzes themselves are created faster. BDA enables the organization to handle supply chain disruption, better synchronize supply and demand, reducing the cost of inventory and transportation.

One of the leading areas of using AI in business is decision support. AI techniques mentioned by Duan et al. are rule-based inference, semantic linguistic analysis, Bayesian networks, similarity measures, artificial neural networks especially convolutional ANN, genetic algorithms [3]. In addition to the technical requirements of such a system, the authors point out the problems associated with inclusion of its results in decision-making processes. As Mishra and Sharma show, there is a strong relationship between supply chain strategy and selected performance measures [11].

According to authors' literature review and knowledge there is a lack of work addressing the issue of using Big Data technologies for modeling supply chains. This research gap was found in other papers [17]. Another problem in management is the inclusion of the results created using Big Data technology, Data Science, and AI methods in the decision support system.

4   Incremental Modeling Method

SC modeling is a complex and multi-step task. Understanding the business needs and purpose of creating a model is essential. Then describe them in a formalized manner using appropriate notation. Crucial model building decisions are determination of the model complexity and description of dependencies. Many models could be built as stochastic without the random element. Nevertheless, including elements of randomness and multiple calculation of values allows to run calculations in conditions as close to the real environment as possible. Results of the model analysis should be legible to end users so they can use them in decision-making processes. Therefore, the model must include a specification of the SC's objectives and criteria for measuring them, list of elements (i.e. SC members) with their type and relations between all these elements (i.e. configuration of the SC network). In addition, data sources must be identified at the initial design stage and assigned to specific processes. Big Data technology is becoming a game changer in several aspects of model building. This technology removes the limitation associated with the computing power necessary to create mathematical models. Thanks to it, one can use more advanced Data Science methods for analysis. It also allows to start collecting additional data that was not gathered before because of the high cost, e.g. from sensors, the Web, etc. Similarly, this technology can overcome the problem of high model maintenance costs that is discussed in the literature [4]. Complex Data Science methods allow to search for dependencies in data, detect trends, etc. Excessive simplification of the model leads to limited possibilities of using the model.

4.1   Techniques for Presenting the Model

Choosing SC participants and the SC facilities is the basic decision in the modelling process. SC participants are suppliers, producers, wholesalers etc. One participant could consist of many facilities, i.e. factories, warehouses, truck terminals, point of sales etc. The nodes are connected with relations which could be processes. A model that contains too many relations and nodes can be difficult to interpret. On the other hand, generalization leads to the loss of relevant information. An example of such a model for a chain of stores has been presented in an earlier work [5]. The supply chain can be presented as:

- Graph – The nodes of the graph are other organizations forming the supply chain. In a more extensive version, they can be specific facilities. The edges are types of business processes and determine the direction of the flow of goods, however the data flow is symmetrical (i.e. bi-directional).
- Relationship matrix – The relationships in the supply chain are determined and presented as matrix. A separate matrix is created for each process.
- List of processes and their participants – This type of presentation is useful for representing edges in computer systems. The set of nodes is obtained as the second list of participants involved in the process.
- Flow diagram – Further business processes are presented along with participants and designated information creation sites.

Objects related to the same process are grouped into one class. Entities that perform a similar role in the supply chain but in different ways, should be treated separately. For example, in a sales network a producer and a retail supplier should be treated as two separate nodes.

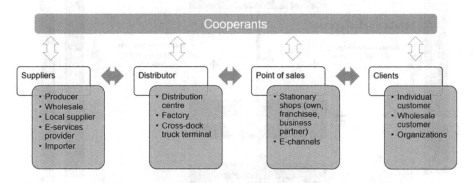

Fig. 2. Supply chain participants

Figure 2 illustrates an exemplary model for an extended supply chain, presented in graphical form. Each node and process generates a large data volume. Determining data sources is an important element of supply chain modeling. When participants and processes are defined, performance measures and model restrictions should be specified. Supply chain participants have limited logistic, financial, sales and technological resources. Restrictions are considered both in the context of the entire supply chain and individual participants at the general level and at each node.

4.2 Data Analysis

In the process of modeling the supply chain, analysis that transforms data into information plays a central role. As shown in Fig. 1, it is the stage between the phase of data collection and the use of obtained information. At this stage, technical and business knowledge is also translated. The authors research and experience shows that the ability to translate business needs into technical requirements and vice versa is a significant communication barrier between teams with different competences. Hence the role of the analyst.

The environment, in accordance with the scientific research design paradigm, provides business knowledge that gives interpretation and creates a space for research and analysis value. Knowledge stored in business departments is translated into the level of data. Thus, data context limits the decision space. It also allows to reject solutions that cannot be applied at an early stage. The knowledge base provides data analysis methods along with techniques for finding solutions. The resulting information is visualized, described and systematized. The relationship between the phases of data collection and analysis as well as between data analysis and the use of the solution is presented in Fig. 3.

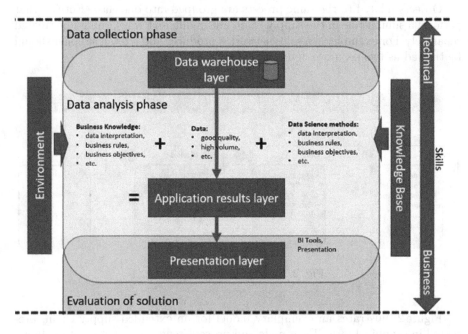

Fig. 3. Data analysis phase

The analysis phase follows the data collection phase. The element connecting both phases is the data warehouse. Analysis phase requires the availability of high-quality structured and semi-structured data. Effective sharing of a data warehouse resource is the responsibility of the warehouse architect. A properly built data warehouse makes remodeling easier.

Appropriate data sets should be addressed to calculate the metrics. The construction of metrics provided from a knowledge base with a specific business complement allows for a proper assessment of a given phenomenon or process. Similarly, the environment sets limits for modeled phenomena. Benchmarking values or environmentally desirable solutions can be used to evaluate phenomena or processes.

The environment also provides strategic goals for the supply chain that should be expressed in a form of performance measures. The selection of relevant performance measures for a given supply chain is the result of translating the strategy into measurable and quantifiable values that constitute a compass for the company's activities. The choice of performance measures is a non-trivial task and is specific to the organization. Due to the superior nature of performance measures in supply chain performance measurement systems, their determination is a strategic decision. These measures should be referred to and be consistent with metrics of subsequent levels. The potential goal of creating a supply chain model may be the selection of performance measures and metrics to describe them. Research on performance measures and their selection based on the AHP method for a chain of stores was presented in an earlier work [6].

The performance measurement system is a framework to measure the efficiency of the supply chain. The key decision when choosing a measurement system is to define the company's strategy and supply chain goals. Another partial goal is to set performance measures that can be used for the organization. Then the designated measures are selected and prioritized. In the next stage, weights for performance measures are determined. The final task is to include a performance measurement system in the company management system. This requires appropriate presentation of analysis results for managers at all levels. Integration of the supply chain requires communication between its participants regarding selected performance measures and measurement methods.

4.3   Incorporation of the Model in the Enterprise Decision Support System

Literature review showed that the basis for building a decision support system in organization can be a supply chain model [1]. On the other hand analysis of both scientific and business publications indicates that the problem in enterprises is to use the potential of advanced analytics. For example, extensive measurement systems based on a complex model are difficult to interpret. Therefore, it is important to skillfully incorporate analytics into the enterprise management system, successively creating a data-driven supply chain.

The use of some DS methods, for example artificial neural networks, hides layers of inference from the decision maker. Decision-making processes based on the use of these models require decision-makers to understand the principles of their functioning and how they arise. Lack of information on the selection method, data taken into account for the model and the possibility of partial verification of the correctness of the model may lead to a lack of confidence in the resulting tools.

After completing the modeling cycle, the possibilities of extending or making changes to the model in accordance with the guidelines from the use phase of the results are analysed. It is necessary to distinguish several possible directions of changes, which include extending the model with new phenomena, improving the model, and extending the data set. Possible directions of changes might be identification of deviations, change of data analysis techniques, introduction of modifications based on business knowledge, modeling of new elements, testing of alternative scenarios.

The proposed method supports communication between people from the technical and business zone. First of all, delimiting the roles of people in the modeling process. They are divided into strictly technical persons, analysts and business experts. Introducing the position of analysts who, on the one hand, are able to use data and use technical formalisms. On the other hand, they have business knowledge and soft skills that allow presentation of results and drawing business conclusions. The stage in which this takes place is the evaluation of the solution phase. The conclusions of the evaluation stage can be divided into business, which support business decisions and feedback to the model itself. Hence, the final stage of the Incremental Supply Chain Modeling Method cycle is the implementation of patches. It is a summary of the previous phase in which the list of expectations for the model was developed. At this point the possibilities of extending or making changes to the model in accordance with the guidelines from the previous phase should be considered. It is necessary to distinguish several possible directions of changes, which include extending the model with new phenomena, improving the model, and increasing the data set. Examples of directions of changes are as followed:

- Elimination of deviations - if the model did not pass the verification tests of the correctness of the model's operation.
- Change of analysis techniques - if current techniques are ineffective, or improvement of results is sought, analysis checking the previous method.
- Introducing modifications based on business knowledge - if a deviation was found that could be justified and described in business terms.
- Modeling of new elements - if they were not included in previous considerations due to lower priority or considered irrelevant.
- Alternative scenarios - if alternative business methods used in the supply chain were identified. Increasing the set of potential participants in the supply chain.

In the absence of the postulated changes, the model should be implemented in its current version.

5   Conclusions

Based on the analysis of the revised literature, a research gap in modeling supply chain using Big Data technology has been found. Moreover, research contributions to SC modeling are sought by organisations.

This work presents the concept of Incremental Modeling Method. Future work will focus on describing the example implementation of such a model for an enterprise with an extensive supply chain. After completing the construction of such a model, an experiment will be carried out consisting of checking the correctness and sensitivity of such model.

References

1. Campuzano, F., Mula, J.: Supply Chain Simulation: A System Dynamics Approach for Improving Performance, p. 104. Springer, London (2011). https://doi.org/10.1007/978-0-85729-719-8

2. Chiadamrong, N., Piyathanavong, V.: Design of supply chain network under uncertainty environment using hybrid analytical and simulation modeling approach. J. Ind. Eng. Int. **13**(4), 465–478 (2017). https://doi.org/10.1007/s40092-017-0201-2

3. Duan, Y., Edwards, J.S., Dwivedi, Y.K.: Artificial intell. Decis. Making Era Big Data Evol. Challenges Res. Agenda. Int. J. Inf. Manage. **48**, 63–71 (2019). https://doi.org/10.1016/j.ijinfomgt.2019.01.021

4. Goetschalckx, M.: Supply Chain Engineering. International Series in Operations Research & Management Science. Springer, Boston, MA (2011). https://doi.org/10.1007/978-1-4419-6512-7

5. Górtowski, S.: Supply chain modelling using data science. In: Abramowicz, W., Paschke, A. (eds.) Business Information Systems Workshops. BIS 2018. Lecture Notes in Business Information Processing, vol. 339, pp. 635–645. Springer, Cham (2019). https://doi.org/10.1007/978-3-030-04849-5_54

6. Górtowski, S., Lewańska, E.: Incremental modeling of supply chain to improve performance measures. In: Abramowicz, W., Corchuelo, R. (eds.) Business Information Systems Workshops. BIS 2019. Lecture Notes in Business Information Processing, vol. 373, pp. 637–648. Springer, Cham (2019). https://doi.org/10.1007/978-3-030-36691-9_53

7. Hevner, A.R., et al.: Design science in information systems research. MIS Q. Manage. Inf. Syst. **28**(1), 75–105 (2004). https://doi.org/10.2307/25148625. https://www.jstor.org/stable/10.2307/25148625

8. Malikia, F., et al.: The use of metaheuristics as the resolution for stochastic supply chain design problem: a comparison study. Int. J. Supply Oper. Manage. **4**(3), 193–201 (2017). http://search.proquest.com/docview/2063813572/

9. Márquez, A.C.: Dynamic Modelling for Supply Chain Management: Dealing with Front-End, Back-End and Integration Issues, pp. 1–297. Springer, London (2010). https://doi.org/10.1007/978-1-84882-681-6

10. Mishra, D., et al.: Supply chain performance measures and metrics: a bibliometric study. Benchmarking **25**(3), 932–967 (2018). https://doi.org/10.1108/BIJ-08-2017-0224

11. Mishra, P., Sharma, R.K.: Benchmarking SCM performance and empirical analysis: a case from paint industry. Logist. Res. **7**(1), 1–16 (2014). https://doi.org/10.1007/s12159-014-0113-0

12. Bidhandi, H.M., et al.: Development of a new approach for deterministic supply chain network design. Eur. J. Oper. Res. **198**(1), 121–128 (2009). https://doi.org/10.1016/j.ejor.2008.07.034. http://search.proquest.com/docview/204146678/

13. O'Dwyer, J., et al.: The promise of advanced supply chain analytics. Bus. Econ. Manage. **15**(1), 6 (2011)

14. Ravindran, A.R., Warsing, D.P.: Supply Chain Engineering: Models and Applications. CRC Press, Boca Raton, London, New York (2016)
15. Sadghiani, N.S., Torabi, S.A., Sahebjamnia, N.: Retail supply chain network design under operational and disruption risks. Transp. Res. Part E Logist. Transp. Rev. **75**, 95–114 (2015). https://doi.org/10.1016/j.tre.2014.12.015. http://search.proquest.com/docview/1662053939/
16. Sanders, N.R.: How to use big data to drive your supply chain. Calif. Manage. Rev. **58**(3), 26–48 (2016). https://doi.org/10.1525/cmr.2016.58.3.26
17. Vieira, A.A.C., et al.: Supply chain hybrid simulation: from big data to distributions and approaches comparison. Simul. Model. Pract. Theor. **97**, 101956 (2019). https://doi.org/10.1016/j.simpat.2019.101956
18. Waller, M.A., Fawcett, S.E.: Data science, predictive analytics, and big data: a revolution that will transform supply chain design and management. J. Bus. Logist. **34**(2), 77–84 (2013). https://doi.org/10.1111/jbl.12010
19. Wamba, S.F., et al.: The performance effects of big data analytics and supply chain ambidexterity: the moderating effect of environmental dynamism. Int. J. Prod. Econ. **222**, 107498 (2019). https://doi.org/10.1016/j.ijpe.2019.09.019
20. Wang, G., et al.: Big data analytics in logistics and supply chain management: certain investigations for research and applications. Int. J. Prod. Econ. (2016). https://doi.org/10.1016/j.ijpe.2016.03.014

Impact of New Mobility Services on Enterprise Architectures: Case Study and Problem Definition

Mark-Oliver Würtz[1,2] and Kurt Sandkuhl[1(✉)]

[1] Institute of Computer Science, University of Rostock, Rostock, Germany
{mark-oliver.wuertz,kurt.sandkuhl}@uni-rostock.de
[2] Jellyco, Stuhr, Germany
mow@jellyco.de

Abstract. New mobility solutions (NMS), such as car sharing, city bikes or e-scooters, have become increasingly popular in large cities. NMS have started to affect the way how people move in urban areas, but they can also be expected to affect public transportation companies and in particular their IT-landscape. The purpose of this paper is to investigate this topic from an enterprise architecture (EA) perspective. The purpose of our research is to contribute to this field by investigating the relevance of the topic from a business perspective. The main contribution of our work is (a) a literature analysis of EA use in public transportation and for NMS integration, (b) results from expert interviews investigating the business perspective from inside a large regional transportation organization and (c) the analysis of problems of NMS integration into EA.

Keywords: Enterprise architecture · New mobility services · Public transportation

1 Introduction

Car sharing, city bikes, e-scooters, bike sharing – and many more mobility solutions offered in larger cities in pay-as-you-go or subscription models – have become increasingly popular and have also started to affect the way how people move within short and medium distances in urban areas [4]. These solutions often are categorized as new or innovative mobility solutions (NMS) and have been subject to research, for example with respect to new architectures [1], business models [2], platforms [3], acceptance by end users [4] or required standards. However, the way new mobility solutions affect public transportation companies and in particular the IT-landscape in these companies has not attracted much research so far.

Enterprise architecture management (EAM) is an established discipline in many large companies and aims at a coordinated and long-term development of the business and IT-aspects of an enterprise [5]. Although EAM has been acknowledged as a relevant approach for the sector of public transportation as well, there is a lack of research on how new mobility solutions affect the enterprise architecture of public mobility providers.

© Springer Nature Switzerland AG 2020
W. Abramowicz and G. Klein (Eds.): BIS 2020 Workshops, LNBIP 394, pp. 45–56, 2020.
https://doi.org/10.1007/978-3-030-61146-0_4

So far, guidance or proposals on how to best integrate services and information systems facilitating or connecting NMS into an existing EA are rare. The purpose of our research is to contribute to this field by – as a first step – investigating the state of research and the relevance of the topic from a business perspective. The guiding question is "How do new mobility solutions affect the EA of regional public transportation organizations?". By "regional organizations", we refer to public transportation providers in large cities or districts. The main contribution of our work is (a) a literature analysis of EA use in public transportation and for NMS integration, (b) results from expert interviews investigating the business perspective from inside a large regional transportation organization and (c) the analysis of problems of NMS integration into EA.

The rest of the paper is structured as follows. Section 2 introduces the research methods used in the paper. Section 3 summarizes the background of our work about NMS and EAM. Section 4 also discusses related works on EAM in public transportation. Section 5 is focused on the expert interview we performed as part of our study with a leading representative of a large mobility provider and presents the findings drawn from the interview analysis. Section 7 summarizes our findings and discusses future work.

2   Research Approach

The work presented in this paper is part of a research aimed at method and tool support for integrating NMS into EA of public transportation organizations. It follows the paradigm of design science research [6]. This study concerns a step towards the explication of problems and elicitation of requirements for the envisioned design artifact: EA method support for integrating NMS including tool support. The work presented in this paper started from the following research question: *RQ: In the context of NMS introduction, what business challenges are visible in industrial practice and what problems have to be addressed in method support?*

The research method used for working on this research question is a combination of literature study, expert interview and argumentative-deductive work. Based on the research question, we started identifying research areas with relevant work for this question and analyzed the literature in these areas. The purpose of the analysis was to find existing approaches or theories supporting NMS integration and real-world case studies allowing us to study the problem relevance in detail. Since the literature study showed a lack of established approaches (see Sect. 4), we decided to focus on an expert interview in a real-world case for investigating the problem (see Sect. 4). The case is used to explore the existence and shape of the challenge of NMS integration.

The interview was based on guidelines and questions developed based on the above RQ. The participant was the head of EAM of a public transportation organization in a large German city. The results obtained and written down were subjected to a qualitative content analysis according to Mayring [9]. This method serves the purpose of empirical data evaluation and includes recommendations for a systematic and verifiable text analysis. Mayring's approach includes 6 steps: Step 1 is to decide what material to analyze, which in our case consists of the notes taken during the interview. The interviewed person has many years of experience in the field of EA and public transportation, which makes him a suitable expert for the RQ. Step 2 is to make explicit how the data collection (i.e. in our case the interview) was arranged and prepared. The purpose of this step is to make all factors transparent which could be relevant for interpreting the data. The interviewee was selected based on his position and from existing contacts of the researchers involved. As preparation for the interview, hypotheses and interview guidelines were outlined (see Sect. 5). The interview took 115 min.

Step 3 is to make explicit how the transcription of the material was done. The material was analyzed step by step following rules of procedure by dividing the material into content analytical units. Step 4 concerns the subject-reference of the analysis, i.e. ascertaining the connection to the concrete subject of the analysis. The subject-reference was implemented by (a) defining the research question and corresponding sub-questions in the interview guidelines and (b) using the subjects of these sub-questions as categories during the analysis. Step 5 recommends a theory-guided analysis of the data, which is supposed to balance the vagueness of qualitative analysis with theoretical stringency. For theory-guidance, we took the state-of-the-art into account during the formulation of the sub-questions as well as the analysis of the material. Step 6 defines the analysis technique, which in our case was content summary. This attempts to reduce the material in such a way as to preserve the essential content.

Based on the findings from the interview, we argue that method support is needed for a successful integration of NMS into EA of public transportation companies. This argumentation needs more case material to support our conclusions in future work.

3 Background and Related Work

3.1 New Mobility Services

In addition to classic local public transport such as buses, trams, underground trains, ferries and taxis, so-called New Mobility Services (NMS) have established themselves since the beginning of digitalization, partly within the framework of so-called "Mobility as a Service" (MaaS) economies. These expanded the classic local passenger transport by transport concepts like car sharing, ridesharing, bike and scooter sharing. It can be expected that the future will add other technologies to this, such as autonomous driving and flight services. The above concepts will be discussed in more detail:

Car-sharing programs occur in two main forms: A-to-B free floating or A-to-A station-based car-sharing providers. In the first case, cars can use the city's parking infrastructure. Renting and returning of the cars takes place within the boundaries of a city or at designated traffic junctions such as airports, train stations or other important urban locations (examples: Car2go; DriveNow, ZipCar). In the second case, cars can be picked up at specially constructed parking stations and returned there after use (examples: Cambio, Ubeeqo). Here, data is the essential component of the business model, because this is the only way that customers are able to use the service and operators are able to bill. Ridesharing refers to carpooling services like the US company Uber that provide a vehicle and driver for short distances. The service is booked by the user via a smartphone app, which is used for requests, notifications and invoicing. Ridesharing is available in a wide variety of forms (ViaVan, FREE NOW, Uber, etc.) and represents a hybrid approach of taxi and public transport, a demand-controlled area operation. Various studies have shown that these services complement public transport [10] and help to reduce private car ownership [11] in urban regions.

Bike and scooter sharing operators offer bicycles and scooters in conventional and electric form. Usually intended for short trips from A to B, this type of locomotion - apart from walking - is the most economical way to move around in an urbanized environment. These services have experienced an enormous boom in Europe (and worldwide) in the last 2 years. The reason for this rapid expansion were the innovations that the product has undergone. Intelligent locks and developments in geo-localization technologies have made it possible that bicycles and pedal-scooters can not only be offered stationary at certain points, but that free-floating fleets which can be operated fully functionally without a large additional infrastructure have emerged in cities as well. This saves public money and offers users a versatile service. Bicycles and scooters are the ideal means of transport for the first/last mile as an emotion-free means of transport.

The market is just beginning to roughly divide NMS transport concepts into two categories, station-based and non-station-based ("dockless") offers. This can be applied to all concepts presented above. The new services mentioned above are based on existing means of transport, which are either digitized or are being replaced by new ones that are used by many different people, decoupled in time, or that individualized offers are made transparent in the short term to as large a target group as possible and lead to several people sharing a means of transport (car-sharing services). The challenge here is twofold: a. making these services as simple as possible (e.g. APP design/operating sequence) and b. integrating them into an existing Mobility Information Infrastructure to reach and persuade as many potential customers in a given region as possible. In the second case, public transport companies in particular will be called upon to implement this requirement in the future. Currently, the NMS offer is not yet fully integrated (see also MaaS below) into the customer information, local transport planning, ticketing and billing systems of the public providers. Only with extensive integration will local public transport organizations be able to make multimodal mobility simple and convenient for all providers and customers.

NMS in a broader sense also includes areas that are only of secondary importance to local public transport, namely intelligent parking solutions, intelligent speed adjustment and intelligent traffic management. These tasks, which are more closely related to traffic monitoring, could also become interesting for public transport organizations in the future. The optimization of park & ride systems and traffic light control in the context of speed adjustment and priority-controlled traffic management could become a sub-task of public transport or strongly influence it in the future. This is already being implemented to some extent. In Berlin, for example, the clearing of the bus lane is carried out by the regional service provider, Berliner Verkehrsbetriebe.

Mobility-as-a-Service (MaaS) is an integrated mobility service that integrates different forms of transport services. MaaS provides added value to the user by providing a single application (app) for the use of different mobility services, providing the option of a single ticketing and payment system instead of several different ticketing and payment systems (the NMS providers). In doing so, the MaaS operator integrates various transport options under its platform (e.g. as a mobility platform) for greater customer satisfaction. One of the success factors of such a platform is the integration of as many transport concepts as possible, such as public transport, car-sharing, car or bicycle sharing, taxi, car rental or leasing, which can be considered either independently or in combination. A further success factor is the creation of new business models and opportunities for companies to optimally design the operation and organization of transport options. In this context, access to current or better information (public transport data, traffic data, weather data, etc.) for the customer is the focus of attention.

The goal of MaaS is to provide the highest possible value for its users, so that it offers a real alternative to using a car, an alternative which can be more efficient, more sustainable and in many cases also cheaper. During the ongoing transition to mobility as a service (MaaS), the integration of as many transport providers as possible in an urban area is therefore essential to enable the users to plan and execute multimodal city trips. A central challenge that MaaS providers or companies of joint mobility have to face is a fragmented regulatory landscape in the municipal sector. This does not make it easy for such companies to develop new business models and operate them successfully. Promoting such MaaS approaches at the local level as best practices would therefore be beneficial. It should be taken into account that especially in German-speaking countries, public transport companies are mostly public law institutions - with mostly good relations to regional politics - and have an optimal overview of local public transport in their respective region. These arguments suggest that the likeliest operators of MaaS approaches in a region are the public transport companies themselves.

3.2 Enterprise Architecture Management

The field of Enterprise Architecture (EA) [5, 12] has been developed for more than a decade as a discipline with a scientific background and useful decision supporting functions and models for forward-thinking enterprises and organizations [13]. Enterprise Architecture aims to model, align and understand important interactions between business and IT to set a prerequisite for a well-adjusted and strategically oriented decision-making framework for both digital business and digital technologies [14].

Enterprise Architecture Management, as defined by several standards such as Archi-Mate [15] and TOGAF [16] today, uses a relatively large set of different views and perspectives for managing and documenting the business-IT-alignment (BITA) [13]. EAM represents a management approach that establishes, maintains and uses a coherent set of guidelines, architecture principles and governance regimes that offer direction and support in the design and development of an architecture to realize the enterprise's transformation objectives. An effective architecture management approach for digital enterprises should additionally support the digitalization of products and services [17] and be both holistic and easily adaptable.

4 Literature Analysis on EAM in Public Transportation

The search for relevant literature on the topic was mainly done with Google Scholar. The topic includes books and articles from the field of regional public transport (RPT) and Enterprise Architecture Management, which were published internationally mainly in English. In addition, there are recent publications from the area of "New Mobility

Services" (NMS), "Mobility Platform" and "Mobile as a Services" (MaaS), which partly overlap with each other.

At the beginning of the research, the authors relied on the publication of Scholz "IT-Systems in Public Transport" [18], which contains the industry model (domain model) ITTC Core Model (ITVU) for public transport. This model describes business processes and fundamental data structures (classes), which are represented in packages (building blocks) and explained in their mode of operation.

In order to show since when the discussion about "New Mobility Services" (NMS) has been going on, the authors have checked the sources for older publications on this topic and found the following: The debate on the substitution of inefficient passenger transport systems, such as the car by other forms of transport, is as old as the car itself. It should be emphasized that the authors Heinze and Kill in their recommendation for action "Sustainable strategies for public transport in Berlin-Brandenburg" [19] in 1994 and the authors Beutler and Brackmann in their working paper "New mobility concepts in Germany: ecological, social and economic perspectives" [20] in 1999 already discussed all the current topics that are still being discussed today - more than 20 years later - in the same form, with the exception of a few technological components such as autonomous driving and air travel taxes. It should further be emphasized that in the publication by Beutler and Brackmann, the two areas of mobility management software (information) and mobility management hardware (transport systems) are already divided. A comparable discussion still exists today, whereby this discussion is divided into new transport systems and platform services (customer information).

In addition to the discussion of NMS, products based on new technologies have emerged that follow the "Mobility as a Service" (MaaS) approach. The focus for the customer is no longer on the means of transport and its provider, but on the service itself, in order to get from A to B quickly and cost-effectively. In some cities, this has now gone so far that customers can plan, book and pay for a trip from their front door to the front door of their destination. In 2016, this mobility integration was critically assessed in a review paper by the authors Kamargianni, Li, Matyas and Schäfer entitled "A critical review of new mobility services for urban transport" [21]. This paper concluded that MaaS is a "...promising mobility solution and is expected to make a significant contribution to future urban reform". [22]

Finally, the sustainability of new mobility services should be mentioned in the technical analysis of the topic. Here, the authors Hildebrandt et al. 2015 have outlined a business model that is as sustainable as possible using a car sharing service [22]. The aim of the paper was to reconcile economic and ecological aspects in an alternative business model. Another publication that points in the same direction is the 2017 paper "Smart mobility and smart environment in the Spanish cities" [23] by the authors Aletà, Alonso and Ruiz.

As a second step, after the classification of NMS, MaaS and Mobility Platform, the IT-relevant topics related to Enterprise Architecture Management were examined in the technical discussion, with the following publications being particularly notable in the context of the work. Pflüger et al. outlines in his 2016 concept paper "A concept for the architecture of an open platform for modular mobility services in the smart city" [24] the architecture of an open and modular service platform for mobile services in a future smart city. The paper concludes that the provision of data and services for the operation of mobility services will trigger even more innovation among mobility service providers. Another article from 2013 deals with a part of NMS, namely with the parking of cars in urban environments [25] and the possible control options in a modern mobility concept. This text is limited exclusively to parking in urban areas.

Yet another publication from 2015 deals with an intelligent mobility platform [26], which uses a map-based platform as its core component. In one project, an architecture was developed which is able to collect, update and process heterogeneous information and real-time data from different sources and actors in order to optimize the offers for the users of public transport. The publication that comes closest to the topic is from 2017 by the authors Levina, Dubgorn and Iliashenko with the title: "Internet of Things within the Service Architecture of Intelligent Transport Systems" [27]. The text discusses an enterprise architecture for the St. Petersburg situation center based on IoT technology. This paper is primarily concerned with monitoring traffic and road users and mostly refers to IT components rather than business processes that would have to be established. In addition, there is work on NMS in China, for example on Architecture and Implementary Scheme of Intelligent Public Transportation [28], Intelligent Dispatching Management System [29] and Intelligent Public Transportation Control [30].

All articles and papers presented herein contain important partial solutions for managing an integrative mobility concept for public transport companies. A comprehensive platform that covers all important functions from the customer's point of view, such as offer presentation, search for offers, offer selection/optimization, operational planning/implementation, fare calculation, conclusion of the contract of carriage, connection information and safety assurance across all modes of transport was not found in the course of the research - even the abstracts of the Chinese texts did not indicate this.

As presented above, the literature we researched on the structure and architecture of NMS or mobility platforms was mainly about the platforms itself. However, we found no information about by whom and how these central services are developed and operated and who is responsible for them or how a public transport company could integrate them into its existing business architecture.

5   Expert Interview in Public Transportation

The literature study carried out in the previous section allows for the conclusion from the publications found that the general situation of Enterprise Architecture Management based on e.g. classical frameworks such as, e.g. TOGAF, in the context of New Mobility Services as a holistic approach in a public transport company is missing. Therefore, the authors decided to conduct an interview with the head of IT-Division (CIO) of the Berlin public transport company (BVG). In this section, we will present the hypotheses on which the study is based, the interview study itself, the methods used and the results.

At this point, the authors would like to point out that the interview with just one respondent did not have the goal of representativeness but rather to find out and deal with a specific challenge in the context of public transport and EAM.

The following 3 hypotheses are defined on the basis of the literature work (see previous chapter) and examined in the interview study:

– Hypothesis 1: Digitalization projects such as New Mobility Services affect the enterprise architecture of public transport companies.
– Hypothesis 2: NMS form a disruptive approach to classical public transport.
– Hypothesis 3: EAM helps with digitalization projects or the introduction of new mobility services in public transport companies.
– Hypothesis 4: TOGAF is a suitable architecture framework for the BVG.

However, it is not certain whether classical public transport companies will be able to maintain their unrestricted leadership role in local public transport in the future. This thesis deals with the challenges of digitization in German public transport companies and whether these challenges can be overcome, at least partially, with the help of Enterprise Architecture Management.

5.1 Approach

This study is based on the implementation of a guideline-based expert interview, which is based on the premise that public transport organizations have no or insufficient EAM expertise to date and why this expertise is important for the industry in the course of digitalization. The interview partner in this case was the CIO of Berlin public transportation (Berliner Verkehrsbetriebe - BVG) who on the one hand answered questions on the topic of digitization and on the other hand on the use of EAM in the (BVG) and the public transport sector. At the time of the interview, the CIO was mainly responsible for the digital restructuring of the IT. The guideline-based expert interview was structured as shown in Fig. 1.

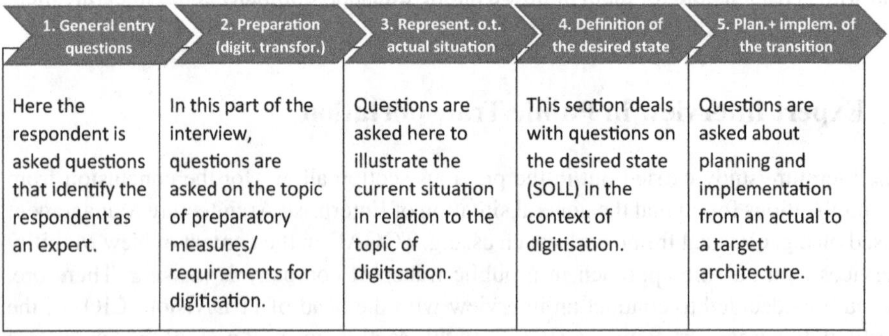

1. General entry questions	2. Preparation (digit. transfor.)	3. Represent. o.t. actual situation	4. Definition of the desired state	5. Plan.+ implem. of the transition
Here the respondent is asked questions that identify the respondent as an expert.	In this part of the interview, questions are asked on the topic of preparatory measures/ requirements for digitisation.	Questions are asked here to illustrate the current situation in relation to the topic of digitisation.	This section deals with questions on the desired state (SOLL) in the context of digitisation.	Questions are asked about planning and implementation from an actual to a target architecture.

Fig. 1. Structure of the Interview Process

Table 1 shows sample questions for each bullet point, some of which are used in the context of the discussion of results under 5.3:

Table 1. Selected questions from interview guideline

1.	1 How long have you been CIO at Berliner Verkehrsbetriebe (BVG)? 2 How much expertise do you have in the field of digitization? 3 Which of your knowledge of methods helped with the BVG's digitization plans?
2.	1. Do you see disruptive approaches in the course of digitization that [lastingly] endanger the core business model of the BVG? 2. To what extent has digitalization changed the requirements for the architecture? 3. Which state [of the architecture documentation] did you first consider? a. Illustration of the ACTUAL state, or b. Definition of the target state (TARGET)
3.	1 Which organisational areas of the SNB were particularly affected by the digital transformation (especially which part of IT)? 2. Which architectural principles were important during transformation in the SNB? 3. How important is the use of an appropriate framework for management?
4.	1 Which information is most important for the creation of the desired situation? 2. What are the biggest mistakes in the development/creation of a target architecture? 3. Have changes to the business architecture been identified in addition to the IT architecture?
5.	1 How important do you consider an appropriate migration strategy? 2. Do you think a consolidated architecture roadmap is useful? 3. Could the desired solution (migration of ACTUAL/SOLL architecture) be completed in a comprehensive project or did a program have to be set up?

During the guideline-based expert interview, the questions did not have to be answered chronologically or completely. It turned out in the course of the interview that the expert himself took up the desired content of the work and incorporated it into his answers in advance.

5.2 Results

After evaluating the interview protocols, identical answers or identical tendencies were assigned to a group of statements and thus taken into account. Similarities, contradictions and differences were identified. Strongly divergent statements were not taken into account for reasons of robustness or they were suitable for elimination, especially tendentious statements. The official or personal circumstances of the respondent were taken into account.

The result of the expert interview shows that hypothesis 3 and 4 were confirmed. Hypothesis 1 and 2 can only be confirmed in parts in a dedicated way. The core results can be summarized as follows in the context of the hypotheses:

Hypothesis 1: The introduction of New Mobility Services in a public transport area, whether initiated by the local public transport company itself or by an external provider, always leads to an impact on the public transport company itself - even if it only leads to new or replanned services offered by the public transport company. The question is whether this already has a significant impact on the business or IT processes and thus affects the enterprise architecture. The expert pointed out that "...at the beginning of digitization, the areas of sales (incl. offer of traffic services), IT and administration were affected within BVG ...". In addition, "in the future, core business and personnel will increasingly be included in digitization projects..." [30].

Hypothesis 2: The expert did not want to confirm whether BVG's core business was threatened by a disruptive collapse with the introduction of further New Mobility Services, such as UBER, (future) flight tax or also white sharing providers, such as Fleetbird[1] with emmy[2], but the question arose for him: "Is it a threat to my core business if someone offers a white sharing service? Or is it more likely to threaten taxis or how is this related to classic public transport? There are different opinions." [30] The interviewee did not take a clear position on this.

The interviewee sees this differently in the case of B2B business models, such as mobility platforms with their own pricing models. Here, the division manager sees the public transport sector in danger, because if you look at the providers "...for example, the NOW Group together with Transdev, who offer completely new price models in the B2B business, then that is, in his "...assessment, an immediate threat to the original BVG business model. [30].

Hypothesis 3: Enterprise Architecture Management and in particular TOGAF as EAM framework are, according to the interviewee, "...a good tool to bring together strategic applications..." or IT landscapes" [30] to manage them.

It brings IT and the business departments together and is a proven means of achieving a common view (IT and business department). In addition, it creates transparency between the current situation (ACTUAL) and the future IT requirements of the business (TARGET) - so that the IT alignment in the company can be controlled and guaranteed according to the specifications of the business strategy.

It does not matter whether the applications/IT landscape is operated in-house or off-site. Even applications hosted off-site leave "footprints" in the EAM through support /processes or at process level at the least.

Hypothesis 4: The BVG has chosen TOGAF as EAM framework because it provides a structure that can be used as a basis for a uniform approach. This was very important, because the system landscape of the BVG was not or not strategically thought "forward". In addition, the overview and interrelationships (dependencies/interfaces etc.) of various large IT projects were difficult to recognize and unmanageable.

Finally, the expert revealed that "...he personally would just always take TOGAF. Why? Because in my perception most people have had contact with it. It's not even that I'm saying it's the best, but it's the most pragmatic, because many have had contact with it before." [28].

---

[1] Fleetbird is the operator of emmy (Berlin); https://fleetbird.com/;

[2] emmy is an electric scooter sharing service operated by Fleetbird; https://emmy-sharing.de/;

6 Conclusions and Future Work

The results of the expert interview described in 5.3, including the interview itself, lead to the conclusion that the successful introduction of new mobility services is currently taking place less in public transport companies, but is being established by newcomers (such as emmy, etc.). The issues driving the established public transport companies are more the orientation towards the classic public transport business. In order for public transport companies to survive in the long term, they themselves must be able to integrate and manage mobility platforms into their companies. It makes no difference whether they operate the platform themselves or have it operated by service providers. The important thing here is that the public transport company's role of being the leading mobility provider must be secured.

To achieve this goal, the classic public transport companies should know what a possible target architecture for mobility platforms ideally looks like, which processes are associated with it and how you can design the interfaces, products and services so openly that other providers of mobility systems can join in at any time. We will deal with the topic in future publications in order to collect further results, e.g. central requirements for such a mobility platform, illumination of organizational aspects, proposal of processes and recommendation of architecture models and technical solutions.

References

1. Pflügler, C., Schreieck, M., Hernandez, G., Wiesche, M., Krcmar, H.: A concept for the architecture of an open platform for modular mobility services in the smart city. Transp. Res. Procedia **19**, 199–206 (2016)
2. Hildebrandt, B., Hanelt, A., Piccinini, E., Kolbe, L., Nierobisch, T.: The value of IS in business model innovation for sustainable mobility services-the case of Carsharing. In: Wirtschaftsinformatik, pp. 1008–1022 (2015)
3. Marchetta, P., Natale, E., Pescapé, A., Salvi, A., Santini, S.: A map-based platform for smart mobility services. In: 2015 IEEE Symposium on Computers and Communication (ISCC), pp. 19–24. IEEE, July 2015
4. Kamargianni, M., Li, W., Matyas, M., Schäfer, A.: A critical review of new mobility services for urban transport. Transp. Res. Procedia **14**, 3294–3303 (2016)
5. Lankhorst, M.: Enterprise Architecture at Work. Modelling, Communication and Analysis. The Enterprise Engineering Series. Springer, Heidelberg (2017). 10.1007/978-3-662-53933-0
6. Johannesson, P., Perjons, E.: An Introduction to Design Science. Springer (2014). 10.1007/978-3-319-10632-8
7. Mayring, P.: Qualitative Inhaltsanalyse. In: Handbuch qualitative Forschung in der Psychologie, pp. 601–613. VS Verlag für Sozialwissenschaften (2010)
8. Research Service, European Parliament: Uber under the hood, 2016; UITP, 2016 (2016)
9. APTA: CityAM, 2016; Orb International, 2017; ADEME, 2016 (2016)
10. Nurmi, J., Pulkkinen, M., Seppänen, V., Penttinen, K.: Systems Approaches in the enterprise architecture field of research: a systematic literature review. In: Aveiro, D., Guizzardi, G., Guerreiro, S., Guédria, W. (eds.) EEWC 2018. LNBIP, vol. 334, pp. 18–38. Springer, Cham (2019). https://doi.org/10.1007/978-3-030-06097-8_2
11. Simon, D., Fischbach, K., Schoder, D.: Enterprise architecture management and its role in corporate strategic management. IseB **12**(1), 5–42 (2013). https://doi.org/10.1007/s10257-013-0213-4

12. Niemi, E., Pekkola, S.: Using enterprise architecture artefacts in an organisation. Enterp. Inf. Syst. **11**(3), 313–338 (2017). https://doi.org/10.1080/17517575.2015.1048831
13. The Open: ArchiMate® 3.0.1 Specification, 1st ed. Van Haren Publishing (2017)
14. The Open Group: The TOGAF Standard, version 9.2. TOGAF series. Van Haren Publishing, Zaltbommel (2018)
15. Urbach, N., Ahlemann, F.: Transformable IT landscapes: IT architectures are standardized, modular, flexible, ubiquitous, elastic, cost-effective, and secure. In: Urbach, N., Ahlemann, F. (eds.) IT Management in the Digital Age, Management for Professionals, vol. 6, pp. 93–99. Springer, Cham (2019). 10.1007/978-3-319-96187-3_10
16. Scholz, G.: IT systems in public transport: Information technology for transport operators and authorities. Heidelberg: dpunkt.Verlag (2016)
17. Heinze, G.W., Kill, H.H.: Zukunftsfähige Strategien für den ÖPNV in Berlin-Brandenburg, pp. 0722–8287 (1994)
18. Beutler, F., Brackmann, J.: Neue Mobilitätskonzepte in Deutschland: Ökologische, soziale und wirtschaftliche Perspektiven, WZB Berlin (1999)
19. Kamargianni, M., Li, W., Matyas, M., Schäfer, A.: A critical review of new mobility services for urban transport. Transp. Res. Procedia, Jg. **14**, 3294–3303 (2016)
20. Hildebrandt, B., Hanelt, A., Piccinini, E., Kolbe, L.: The value of IS in business model innovation for sustainable mobility services-the case of carsharing (2015)
21. Aletà, N.B., Alonso, C.M., Ruiz, R.M.A.: Smart mobility and smart environment in the Spanish cities. Transp. Res. Procedia **24**, 163–170 (2017)
22. Pflügler, C., Schreieck, M., Hernandez, G., Wiesche, M., Krcmar, H.: A concept for the architecture of an open platform for modular mobility services in the smart city. Transp. Res. Procedia **19**, 199–206 (2016)
23. Barone, R.E., Giuffrè, T., Siniscalchi, S.M., Morgano, M.A., Tesoriere, G.: Architecture for parking management in smart cities. IET Intell. Transp. Syst. **8**(5), 445–452 (2013)
24. Marchetta, P., Natale, E., Pescapé, A., Salvi, A., Santini, S.: A map-based platform for smart mobility services. In: 2015 IEEE Symposium on Computers and Communication (ISCC), pp. 19–24 (2015). https://doi.org/10.1109/iscc.2015.7405448
25. Levina, A.I., Dubgorn, A.S., Iliashenko, O.Y.: Internet of Things within the service architecture of intelligent transport systems. In: 2017 European Conference on Electrical Engineering and Computer Science (EECS), 2017, pp. 351–355
26. Zhaosheng, Y., Jianming, H.: Research on the architecture and implementary scheme of intelligent public transportation system in China. Commun. Transp. Syst. Eng. Inf. **1** (2001)
27. Guowu, Z.: Build up and development of intelligent dispatching management system of Beijing Public Transport. J. Northern Jiaotong Univ. **5** (1999)
28. Geng, J.-H., Hu, N.-P., Tong, G.: Systematic design of Qingdao intelligent public transportation control system. J. Qingdao University of Science and Technology **6**, 2003
29. Menge, F.-W.: Digitale Transformation in einem ÖPNV-Unternehmen (BVG). Berliner Verkehrsbetriebe (BVG) AÖR, Usedomer Straße 24, D-13355 Berlin-Wedding. Zugriff am: 6. Februar 2020

BSCT

BSCT 2020 Workshop Chairs' Message

Preface

The Third Workshop on Blockchain and Smart Contract Technologies (BSCT 2020) was unprecedented. It was planned to be held during June 8–10, 2020, in Colorado Springs, USA. The event was co-organized with the 23rd International Conference on Business Information Systems (BIS 2020). Due to the COVID-19 pandemic, the workshop and the main conference were organized online. Therefore, it can be said that it was not only international but also decentralized. The participants joined the event from many parts of the world. The presented volume contains the selected and revised papers prepared for the workshop.

The workshop covered a wide area of topics. One of the more financial papers explores the volatility of cryptocurrencies compared to their fiat counterparts. Another paper brings an original contribution to the problem of the current winner-takes-all approach in the mining rewards process. There is a text that provides a good approach for using probabilistic logic programming for smart contract modeling. One of the papers gives an interesting overview of existing considerations concerning the introduction and how to organize the usage of Central Banks Digital Currency, which is a hot topic at the moment. The authors of another submission propose a solution to verify the trustworthiness of blockchain users and blacklist those who commit financial frauds. The aspects of security are also raised in another text that evaluates what would have been saved from identity theft and fraud by using blockchain-based ID management. There is also a paper which aims to analysis the adoption of anonymous transactions in cryptocurrencies. Finally, two more papers present a new voting system for use in fundraising and address the risks of smart contracts in energy markets. Evidently, the range of topics is diverse. The organizers are strongly convinced that the workshop as a platform for discussion and sharing thoughts is therefore vital and will be even more so in the years to come.

As mentioned, the third edition of the workshop was prepared and organized in very specific circumstances. In fact, for a long time, it was not certain whether the event would take place at all, as the situation worldwide was very serious. In the end, we received 17 submissions. The papers were carefully reviewed by the dedicated members of the Program Committee. In total 52 reviews were prepared. It was the basis for deciding on the selection of the most promising set of presented research. Together, they totaled 9 papers accepted for publication after additional editorial review. Consequently, the acceptance rate in this year's edition was 52%.

The chairs perceive the organization of the second workshop as a great success. On this occasion, we would like to express our heartfelt gratitude to all the participants, and especially authors, whose attendance and contribution allowed us to organize this very successful event. We believe that the workshop is all about the community. Therefore, the voice of the participants is of utmost importance to us. Program Committee members are also extremely important to us. That is why we would like to state our deepest gratitude towards them. It is only thanks to their knowledge, sound judgment, and kind support that the workshop took place in its final shape. Without the hard work of the Program Committee the organizers would not be able to make the proper selection of the works.

We would also like to direct our sincere appreciation to the organizers of the hosting conference (BIS 2020). Without a doubt, their organizational know-how and assistance were one of the key factors of success. We would like to invite all this year's participants, authors, and newcomers to the planned 4th workshop on blockchain and smart contract technologies in 2021.

August 2020

Saulius Masteika
Erich Schweighofer
Piotr Stolarsk

Organization

Chairs

Saulius Masteika (Chair)   Vilnius University, Latvia
Erich Schweighofer (Co-chair)   University of Vienna, Austria
Piotr Stolarski (Co-chair)   Poznań University of Economics and Business, Poland

Program Committee

Jan Heinrich Beinke	University of Osnabrück, Germany
François Charoy	Université de Lorraine, LORIA, Inria, France
Nicolas T. Courtois	University College London, UK
Stefan Eder	Benn-Ibler Rechtsanwälte GmbH, Austria
Adrian Florea	Lucian Blaga University of Sibiu, Romania
Jaap Gordijn	Vrije Universiteit Amsterdam, The Netherlands
Aquinas Hobor	National University of Singapore, Singapore
Constantin Houy	DFKI - Institute for Information Systems, Germany
Monika Kaczmarek	University of Duisburg-Essen, Germany
Kalinka Kaloyanova	University of Sofia, Bulgaria
Gary Klein	University of Colorado Boulder, USA
Saulius Masteika	Vilnius University, Lithuania
Raimundas Matulevicius	University of Tartu, Estonia
Kouichi Sakurai	Kyushu University, Japan
Erich Schweighofer	University of Vienna, Austria
Piotr Stolarski	Poznań University of Economics and Business, Poland
Davor Svetinovic	Masdar Institute of Science and Technology, UAE
Herve Verjus	Université de Savoie, LISTIC, Polytech'Savoie, France
Hans Weigand	Tilburg University, The Netherlands, and Leipzig University, Germany
Jakob Zanol	Working Group Legal Informatics, Austria

Crowdfunding with Periodic Milestone Payments Using a Smart Contract to Implement Fair E-Voting

Anwar Alruwaili[✉] and Dov Kruger[✉]

Department of Electrical and Computer Engineering, Stevens Institute of Technology, Hoboken, NJ, USA

{aalruwai,dkruger}@stevens.edu

Abstract. This paper develops a new voting system for use in fundraising that increases fairness for all investors in approving multi-stage funding for a project. A voting system is a method for people to choose one from two or more possibilities. For electronic voting, it is vital to secure the vote against fraud, so selecting a robust security protocol is critical. For online projects, there are issues of requiring performance for continued payment, and fairness to all investors regardless of the money they invest. To avoid fiascos where projects request vast sums of money and then do not deliver, we define smart contracts supporting projects with milestones and a sequence of opportunities for the investors to vote on whether to continue. For a project with multiple milestones, a new algorithm Anwar gives each investor an equal say and equal risk until their money runs out. This protocol works in a smart contract which requires no central authority, parties, or private channel.

Keywords: Fundraising · Crowdfunding · Voting system · Smart contract · Blockchain · Ethereum · Solidity · Decentralized applications

1 Introduction

Fundraising is the activity of collecting money to support a project. On platforms such as Kickstarter, projects are publicized often involving substantial sums of money, and a promise that the creators will work on the project until it is successful. While success is not guaranteed, none of the participants are guaranteed refunds, and there have been notable cases where outright fraud has occurred [1, 2].

Raising funds use different types of models which do not particularly represent investor interest well. Kickstarter crowdfunding platform uses an all-or-nothing funding model [3]. The Indiegogo platform uses the receiving funding as it comes [4, 5].

Blockchain financial systems use those traditional models on smart contracts. Blockchain technology uses different tokenized crowdfunding models. Initial Token Offering (ITOs) or often called Initial Coin Offerings (ICOs) model, use smart contracts to collect funding and to support a new project via digital assets or cryptocurrency [6]. Smart contracts use the traditional all-or-nothing model. Security Token Offering (STOs)

© Springer Nature Switzerland AG 2020
W. Abramowicz and G. Klein (Eds.): BIS 2020 Workshops, LNBIP 394, pp. 61–72, 2020.
https://doi.org/10.1007/978-3-030-61146-0_5

are similar to (ICOs), and (STOs) manage to increase protection for the investors to get security tokens and appoints an investment contract into an underlying investment asset [7]. Initial Exchange Offerings (IEOs) models are also similar to the (ICOs), and (IEOs) are managed and run by online trading platforms (often called crypto exchanges) on behalf of the startup. Thus, the creator has to sign up on the exchange's platform where the (IEO) is conducted [8]; however, the IEO does not run under smart contracts, and the creator does not own the (IEO) token [9, 10]. The exchange's platform does not protect the interests of investors [11]. Thus, there is no protection against unintentional threats and guarding against intentional threats in the market.

These models help to quickly raise funding money, and some projects have success stories; for example, Basic Attention Token (BAT) raised $35 million in less than 30 s [12]. However, there are a lot of failed and scammed projects that have caused investors to lose their money. These projects are categorized as deceased, hack, scam, and parody [13].

Most crowdfunding services are developed in a centralized way requiring trust of the service [3, 4]. Kickstarter requires creators to show a prototype of what they are making before submitting a fundraising proposal, and the goal fund can be raised only after the project has been delivered [3]. Also, white papers show a prototype that informs investors about solving a complex issue. Showing a prototype is both difficult and does not guarantee that a final product can be made. Unfortunately, Kickstarter and Indiegogo have no guarantee to return money if the project fails, scams, or misses the delivery date [3, 4]. For example, in Kickstarter, the Zano project raised over $3 million, and, in Indiegogo, the PopSLATE 2 project raised around $1 million. Zano promised to deliver a device for autonomous, intelligent, and swarming [sic] nano drones to capture high-definition photos and videos; unfortunately, the project failed without refunds and due to poor quality [14, 15]. PopSLATE 2 promised to deliver a smart case that adds a smart second screen, e-reader, and extra battery for iPhone. The project's last update was that there is no money available for refunds [16].

For companies looking to use blockchain to raise capital, a smart contract can be defined with all the terms of the deal, but investors have been easily scammed. PlusToken scammers were able to steal funds of 20,000 BTC, worth more than $1 billion at the time [17]. The amount of money scammed on Pincoin and iFan was 660 million [18]. In 2017 more than 80% of ICOs were identified as scams, and the scammed ICOs raised to $11.9 billion, and in 2018 between January and May new scammed ICOs reached $13.7 billion [19]. In January 2020, a fake Libra token was created in the Ethereum platform; fortunately, there were not many investors investing in this project [20]. There is very little transparency when the models of most fundraisers use all-or-nothing funding, and that helps to scam investors easily.

Since the ICOs do not provide any principles and agreements with promises, it mostly subjects or causes to launch fraudulent crowdfunding projects. The ICO needs to be upgraded and designed to provide fair services to the investors and to avoid high-risk situations of losing anything.

According to the federal trade commission's advice, in order to avoid crowdfunding scams an investor has to research the creator's background before investing [21], and that is true; however, this advice does not offer logical solutions. Because of the lack of trust, the current funding models are inefficient for new startups because investors are understandably skittish about participating. The result is that worthwhile engineering efforts are inhibited because of a lack of funding.

In devising a fair voting system for investors, two issues are key to fairness. First, voters should not have a project canceled due to a small number of people, and second, investors who put up a large amount of money should not be forced into paying it by small investors.

This paper describes a decentralized voting protocol that can raise funds with a sequence of milestones and presentations, after each of which the investors vote whether to release funds (a tranche) for the next stage of development.

This paper is structured as follows. Section 2 gives related work. Section 3 describes in detail on how proposed architecture. Section 4 shows the platform architecture. Section 5 proffers the results and discussions, and finally, Sect. 6 provides the conclusion.

2 Related Works

The blockchain platforms and smart contracts provide a strategy to host voting protocols without the need for a centralized trusted party to control the voting process. A survey and a literature review of a crowdsourcing approach to software engineering [22].

Ante, L. and Fiedler, I provide an overview of the (STO) model and compare the advantages of (STOs) over (ICOs). They support the idea of (STO) model, which has a promise to be a bitter fit for companies' and investors' needs [23]. However, the (STO) are like the (IPO) and still underdeveloped [24, 25]. Furthermore, the (IEO) model has short-term viability than the (ICO) and (STO) models [26], since the online platform or centralized exchange does not guarantee the success of the project [9].

H, Zhu and ZZ, Zhou propose to use crowdfunding that could be on blockchain platforms [27]. Cynthia Weiyi Cai proposes a review on crowdfunding and blockchain. The (ICO) replaces the Initial Public Offering (IPO) from centralized services to decentralized services by offering digital value [28]. M. Zichichi et al. [29] implement a crowdfunding mechanism that is called LikeStarter to combine social interactions, which let users raise funds while becoming popular on the social network. F. Fusco et al. [30] propose an e-voting system based on blockchain technology. The system is called a crypto-voting system that uses permissioned sidechains—the sidechain links to and from two blockchain networks. The crypto-voting aims to improve the efficiency of the voter verification and voting operation records by linking two blockchains. Voters and voting procedures are on the first record, and counting votes and showing the voting results are on the second record. A. Alimoğlu and C.Özturan [31] provide a smart contract based on voting to operate virtual organizations for software development communities and users. S. Pandey et al. [32] implemented a smart contract for crowdfunding campaigns. The contributors send the funds to a smart contract. Then the project manager generates a payment request to a vendor. The contributors have to vote if they agree on that vendor. I. Vakilinia et al. [33] present a smart contract for insuring a cyber-product to insurers participating in a sealed-bid auction. When a vendor requests insurance for a cyber-product, the insurers participate in a sealed-bid auction. The auction winners are selected as the insurers. Other implementations are [34–36]; however, none of them have considered blockchain to support funding a project with a series of tranches of payments, each predicated on passing milestones and voting.

3 Proposed Architecture

The challenge of defining a multiparty project funding algorithm is defining who is allowed to vote and how to determine the payment of the tranches. A multiparty invest-ment contract must have a party who is promising to implement a project (the creator). It may have several required limits such as minimum total investment (like Kickstarter), a minimum number of investors, a stated goal of the project, and a sequence of due dates/milestones to be met, and corresponding payments to be made. The flow chart for a multiparty investment smart contract is shown in Fig. 1.

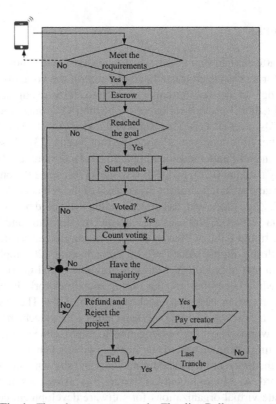

Fig. 1. Flowchart to represent the Timeline Ballots system.

Let a finite set $E_n = \{v_1, v_2, \ldots, v_n\}$ of eligible electorate participants to vote. The notation n is for an individual voter $v_n = \pm 1$ for the participant's vote. The majority of the votes in each tranche can be done by applying the percentage formula by x is (1).

$$(1 + (x/100)E) < \sum_{v=1} v \qquad (1)$$

The smart contract holds the investors' funds, and when a contract is agreed by all parties, the investment begins, as shown in Fig. 2. The total amount of funds invested is stored in the smart contract address; this represents the fund from which all future payments are made. The first vote by investors is taken, and if the threshold is achieved, the vote passes, and the first tranche of money is given to the creator. The creator then works until the next milestone due date. When the date arrives, the creator presents its results.

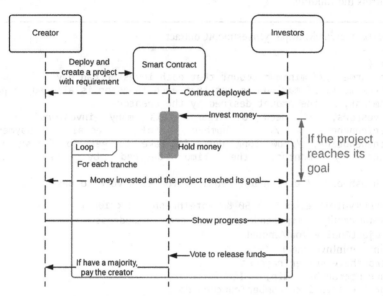

Fig. 2. Sequence Diagram of the Proposed System.

At that point, investors vote on whether the current milestone is met or not, and the process continues until either the project is complete, or a vote of investors fails in which case all remaining funds are returned to the investors. This means that if investors lose confidence in the creator, they can vote to terminate. For simplicity, it must be considered that the funds are divided into n equal tranches, but in practice, this need not be the case.

This model is called show-me-the-progress. The creator satisfies the investors. For voting on each tranche to be distributed to the creator, the investors who wish to approve vote yes; an abstention counts as a no.

All investors with funds remaining are eligible to vote for a tranche. The smart contract contains the threshold required for a passing vote. After each round of tranche and based on voting results, the smart contract takes its actions to either release the escrow to the creator or refund back all money to the participants.

4   Platform Architecture

The contract has one structure that represents the tranche, and a mapping represents the investment amount to the participant account. The structure of each tranche collects the number of participants, tranche's deadline, tranche's total amount, number of approval, and a boolean to check if the tranche is completed. Furthermore, the tranche struct has two mappings to check if the participant is an investor and if the participant has voted.

To generate each tranche is in a timeline sequence, each tranche is obtained from the previous tranche by only adding a fixed number in the timeline of a project that needs to be delivered.

Algorithm createMultipartyInvestmentContract defines the terms of a smart contract that controls the funding.

Algorithm 1 createMultipartyInvestmentContract

```
Input (
minInvestment,// minimum amount that each investor can pay
maxInvestment, // maximum amount that each investor is allowed to pay
goalAmount, // the amount desired by the creator
minInvestors,  //  need  at  least  this  many  investors  to  go
numberTranches,      //      number      of      equal      payments
trancheDeadLine, //How long does it take to get to finish each
tranche   by   months,   the   time   desired   by   the   creator

voteThreshhold // percentage required for each vote to pass)

1. require(voteThreshhold > 50 && voteThreshhold < 100 )
2. creatorIdentity ← msg.sender, // the creator address
3. projectGoal ← goalAmount
4. min ← minInvestment
5. itranche ← numberTranches
6. currenttranche ← min;
7. for i ← 1 to i <= numberTranches do
8.     tranches[i].numberTranche ← i
9.     tranches[i].completedtranche ← false
10.    tranches[i].trancheDeadLine ← (i * trancheDeadLine * 1 month) +
       now
15. raisedAmount ← 0
```

The fundProject algorithm defines the participant input's data. It does not allow more funds to be raised once the goal amount reaches, and the contract begins (after the first tranche's deadline). The smart contract address holds the funds under each participant's identity. This function uses a payable keyword, which is a modifier for receiving funds in the Solidity and Vyper languages.

```
Algorithm 2 fundProject

Input amount
1. require(numberTranche > 0)
2. require(projectGoal >= raisedAmount)
3. require(now < tranches.[1].termDeadLine)
4. require(amount >= minInvestment || participant[msg.sender] >=
   minInvestment)
5. require(participant[msg.sender] + amount <= maxInvestment)
6. require(amount < maxInvestment)
7. if (participant[msg.sender] >= minInvestment) {
8.    for i ← 1 to i <= itranche do {
9.       res = amount/numberTranche
10.      tranches[i].trancheTotalAmount += res
11.      }
12.      raisedAmount += amount
13.      participant[msg.sender] += amount
12.   }
13.   else{
14.   for i ← 1 to i <= itranche do {
15.      res = amount/numberTranche
16.      tranches[i].numOfParticipant += 1
17.      tranches[i].trancheParticipant[msg.sender] = res

18.      tranches[i].trancheTotalAmount += res}
19.      tranches[i].isParticipant[msg.sender]← true
20.      counter= i}, // count how many tranches
21. participant[msg.sender] += amount

22. raisedAmount += amount
```

The submitVote algorithm allows only the participants to submit a vote. The function requires participants to vote before the tranche's deadline; rejects counting votes if the participants have costed their vote by using a mapping from the participant's address to a boolean data type. To avoid processing, a tranche more than once, a boolean is implemented to check if a tranche is completed or not.

```
Algorithm submitVote

Input tranche
1. require(tranches[tranche].trancheDeadLine > now)
2. require(tranches[tranche].completetranche == false)
3. require(tranches[tranche].isVoter[msg.sender] == false)
4. require(tranches[tranche].isParticipant[msg.sender] == true)
5. tranches[tranche].isVoter[msg.sender] ← true
6. tranches[tranche].numOfApprovals++
```

The paymentToCreatorEscrow algorithm shows that only the creator address (creatorIdentity) has permission to receive the money if the project reaches its goal; the request must occur after each deadline in the tranche.

Algorithm paymentToCreatorEscrow

```
Input tranche
1. require(msg.sender == creatorIdentity)
2. require(tranches[tranche].trancheDeadLine < now)
3. require(counter > 0)
4. counter -= 1
5. require(projectGoal <= raisedAmount)
6. tranche[tranche].trancheTotalAmount ← 0
7. require(tranche[tranche].completedtranche == false)
8. require(1+(voteThreshhold/100)*tranches[tranche].numOfParticipan) <
   tranches[tranche].numOfApprovals)
9. address payable receiver ← address(uint160(cryptoContract))
10. receiver.teransfer(tranches[tranche].trancheTotalAmount)
11. tranches[tranche].completedtranche ← true
12. tranches[tranche].trancheTotalAmount ← 0
```

The refundBack algorithm defines the terms of a smart contract to refund the participants if they decide to stop funding the project. It is required that the project's goal amount and deadline still have not been reached.

Algorithm refundBack

```
1. require(now < tranches[1].trancheDeadLine)
```
```
2. require(raisedAmount < projectGoal)
3. require(participant[msg.sender] > 0)
4. require(tranches[1].isParticipant[msg.sender] == true)
5. amount = participant[msg.sender]
6. if (tranches[1].isVoter[msg.sender] == true) {
7.      tranches[1].isVoter[msg.sender] ← false
8.      tranches[1].numOfapprovals -=1 }
9. for i ← 1 to i <= itranche do {
10. tranches[i].numOfParticipant -= 1
11. tranches[i].trancheTotalAmount -= amount/numberTranche}
12. address payable receiver ← address(uint160(msg.sender))
13. raisedAmount -= amount
14. receiver.transfer(amount)
15. participant[msg.sender] ← 0
16. tranches[1].isParticipant[msg.sender]← false
```

The getTheRefundBack algorithm defines the terms to refund the participant if the project failed on one of the tranches due to no more voters.

Algorithm 3 getTheRefundBack

```
input tranche
1.  require(now < tranches[tranche].trancheDeadLine)
2.  require(participant[msg.sender] > 0)
3.  require(tranches[tranche].isParticipant[msg.sender] == true)
4.  amount ← tranches[tranche].trancheParticipant[msg.sender] * counter
5.  address paypal recipient ← msg.sender
6.  recipient.transfer(amount)
7.  participant[msg.sender] ← 0
8.  tranches[tranche].isParticipant[msg.sender] ← false
```

---

5 Result and Discussion

To avoid high-risk situations of losing money, participants have to vote multiple times to complete fundraising a project on its timeline. The smart contract helps creators continue funding their projects by showing their progress to the participants. It is also to predict if a creator is a scammer. When a creator misbehaves and finds a way to scam customers by paying a small amount from several accounts to fund the project, it is very important to raise the percentage of the majority more than 90%. Otherwise, this platform can also be run by a trusted party, so if there is no progress, the trusted party has to take an action to terminate the entire project. Other impacts may happen, the misbehaves of creators, when they can create more fake IDs to announce the project for implying to serve the participants' interests in the future or to trick the participants into voting even if the plans show misleading results about their ability to continue their job successfully.

The properties that the paper considers include correctness property, safety properties, and fairness properties. The correctness property supports two bases: partial and total correctness. The partial is to compute the result of voting correctly without being changed. Furthermore, the voting result of each tranche influences the next tranches. The total correctness is that each tranche refers to the previous tranche and together make the project completed. The model allows only the participants who have invested in the project to vote. If the majority decided not to continue funding the project, then the smart contract allows the participants to request their investments back by using their account address. The safety property is to stop losing money if the creators have shown no results. Double voting is not allowed when each ballot is stored by checking whether participants are approved in each tranche. To achieve fairness, the project must have progressed toward a specific goal. If the project's process never gets implemented, the creator cannot raise funds. The fairness constraint is a condition. The creators must demonstrate the participants with the project's achievements. Thus, the fairness constraint is imposed on the scheduler of the system timeline to continue funding their project.

The Solidity language is still young and cannot implement a sophisticated algorithm. So, running a math formula such as applying the percentage formula (1), it causes an error due to a fixed point arithmetic type in the language [37]. Instead of that formula, the number of all voters must be greater than the result of dividing all participants by 2 or 1.25.

6 Conclusion and Future Work

This method reduces the likelihood that scammers can steal large amounts from investors and creates a framework for startups to build trust by delivering partial results over time. The participants have more say in a voting system if the results of a project are unacceptable. It protects all participants from losing their money at once, forcing project creators to demonstrate progress to the participants. Defining a fair voting process allows participants to collectively decide on whether to terminate the project. For future work, in this smart contract, the period is fixed and is not variable in this smart contract. The creators may request for extending the deadline of each tranche due to some complexities occurring such as mission-critical.

References

1. Strickler, Y., Chen, P., Adle, C..: Kickstarter, Accountability on Kickstarter (2012). https://www.kickstarter.com/blog/accountability-on-kickstarter. Accessed 25 Fe2020/2/25, (2012)
2. Mukherjee, A., Yang, C., Xiao, P., Chattopadhyay, A.: Version.: Does the crowd support innovation? innovation claims and success on kickstarter, 05 September 2017 (2017)
3. Kickstarter Handbook, Kickstarter Funding Rules (2020). https://www.kickstarter.com/help/handbook/funding. Accessed 25 Feb 2020
4. Indiegogo Funding Rules. https://www.support.indiegogo.com/hc/en-us. Accessed 25 Feb 2020
5. Carmen, C.: Digital Curation Projects Made Easy: A Step-by-Step Guide for Libraries, Archives, and Museums. Rowman & Littlefield, NY (2018)
6. U.S. Securities and Exchange Commission. Investor Bulletin: Initial Coin Offerings (2017). https://www.investor.gov/introduction-investing/general-resources/news-alerts/alerts-bulletins/investor-bulletins-16. Accessed 25 Feb 2020
7. Pauw, C.: What Is An STO, Explained. Cointelegraph (2019). https://cointelegraph.com/explained/what-is-an-sto-explained. Accessed 26 Feb 2020
8. U.S. Securities and Exchange Commission. Investor Bulletin: Initial Exchange Offerings (2020). https://www.sec.gov/oiea/investor-alerts-and-bulletins/ia_initialexchangeofferings. Accessed 10 Mar 2020
9. Spiring, M.: Initial Exchange Offerings: The Benefits And Limitations| Coinspeaker (2020). https://www.coinspeaker.com/initial-exchange-offerings-benefits-limitations. Accessed 24 Apr 2020
10. Initialcoinlist.com.: Pros And Cons Of IEO (2019). https://www.initialcoinlist.com/pros-cons-ieo. Accessed 24 Feb 2020
11. Tkachenko, O.: ICO, IEO, STO - Description, Advantages And Disadvantages Of Investing (2020). https://www.liteforex.com/blog/for-investors/ico-ieo-sto—description-advantages-and-disadvantages-of-investing. Accessed 7 Mar 2020
12. Rob, P.: Why Floyd Mayweather, Startups, and Practically Everyone Else Are Betting on Digital Currencies (2020). https://www.inc.com/business-insider/cryptocurrency-mainstream-bitcoin-ethereum-ico-floyd-mayweather-times-square.html. Accessed 27 Feb 2020
13. Deadcoins curated list of projects and ICOs (2020). https://deadcoins.com. Accessed 10 Mar 2020
14. Torquing Group Ltd: Project overview of the torquing group of the creator, ZANO - Autonomous. Intelligent. Swarming. Nano Drone (2016). https://www.kickstarter.com/projects/torquing/zano-autonomous-intelligent-swarming-nano-drone/posts. Accessed 24 Feb 2020

15. Harris, M.: How Zano Raised Millions On Kickstarter And Left Most Backers With Nothing (2016). https://medium.com/kickstarter/how-zano-raised-millions-on-kickstarter-and-left-backers-with-nearly-nothing-85c0abe4a6cb. Accessed 28 Feb 2020

16. popSLATE Support Section, popSLATE 2 - Smart Second Screen for iPhone (2016). https://www.indiegogo.com/projects/popslate-2-smart-second-screen-for-iphone#/updates/all. Accessed 24 Feb 2020

17. Hill, E.: PlusToken scammers blamed for dramatic crypto sell-off (2019). https://coinrivet.com/plustoken-scammers-blamed-for-dramatic-crypto-sell-off. Accessed 2 Feb 2020

18. Jack, F.: The 10 biggest ICO scams swindled $687.4 million (2018). https://www.finance-monthly.com/2018/10/the-10-biggest-ico-scams-swindled-687-4-million. Accessed 27 Feb 2020

19. Dowlat, S.: Crypto Asset Market Coverage Initiation: Network Creation (2018). https://research.bloomberg.com/pub/res/d28giW28tf6G7T_Wr77aU0gDgFQ. Accessed 2 Mar 2020

20. Yahoo! Finance, Fake Libra ICO scam appears on Twitter (2020). https://finance.yahoo.com/news/fake-libra-ico-scam-appears-150013977.html. Accessed 29 Feb 2020

21. Consumer Information Federal Trade Commission.: Avoid crowdfunding scams (2019). https://www.consumer.ftc.gov/blog/2019/05/avoid-crowdfunding-scams. Accessed 22 Feb 2020

22. Mao, K., Capra, L., Harman, M., Jia, Y.: A survey of the use of crowdsourcing in software engineering. J. Syst. Softw, **126**, 57–84 (2017)

23. Ante, L., Fiedler, I.: Cheap Signals in Security Token Offerings (2019). https://doi.org/10.2139/ssrn.3356303. Accessed 4 Mar 2020

24. Hofer, L.: The status Quo of tokenomics: ICOs are dead and STOs not yet alive (2019). https://www.ico.li/the-status-quo-of-tokenomics. Accessed 6 Mar 2020

25. Hackernoon.com: Making Sense Of The Difference Between IPO, ICO, IEO, And STO? (2019). https://hackernoon.com/what-is-the-difference-between-ipo-ico-ieo-and-sto-0s19c38rg. Accessed 4 Mar 2020

26. Fries, T.: IEOs are stealing the show from security tokens, but not for long (2019). https://thetokenist.io/ieos-are-stealing-the-show-from-security-tokens-but-not-for-long. Accessed 6 Mar 2020

27. Cai, C.W.: Disruption of financial intermediation by FinTech: a review on crowdfunding and blockchain. Account. Finance **58**(4), 965–992 (2018)

28. Zhu, H., Zhou, Z.Z.: Analysis and outlook of applications of blockchain technology to equity crowdfunding in China. Finance Innov. **2**, 29 (2016)

29. Zichichi, M., Contu, M., Ferretti, S., D'Angelo, G.: LikeStarter: a Smart-contract based Social DAO for Crowdfunding," In: IEEE INFOCOM 2019 - IEEE Conference on Computer Communications Workshops (INFOCOM WKSHPS), pp. 313–318, Paris, France, (2019)

30. Fusco, F., Lunesu, M.L., Eros Pani, F., Pinna, A.: Crypto-voting a Blockchain based e-Voting System. In: 10th International Conference on Knowledge Management and Information Sharing (2018)

31. A. Alimoğlu and C. Özturan, "Design of a Smart Contract Based Autonomous Organization for Sustainable Software," 2017 IEEE 13th International Conference on e-Science (e-Science), Auckland, 2017, pp. 471–476

32. S. Pandey, S. Goel, S. Bansla and D. Pandey, "Crowdfunding Fraud Prevention using Blockchain," 6th International Conference on Computing for Sustainable Global Development (INDIACom), New Delhi, India, 2019, pp. 1028–1034

33. I. Vakilinia, S. Badsha and S. Sengupta, "Crowdfunding the Insurance of a Cyber-Product Using Blockchain," 2018 9th IEEE Annual Ubiquitous Computing, Electronics & Mobile Communication Conference (UEMCON), New York City, NY, USA, 2018, pp. 964–970

34. J. Lyu, Z. L. Jiang, X. Wang, Z. Nong, M. H. Au and J. Fang, "A Secure Decentralized Trustless E-Voting System Based on Smart Contract," 2019 18th IEEE International Conference On Trust, Security And Privacy In Computing And Communications/13th IEEE International Conference On Big Data Science And Engineering (TrustCom/BigDataSE), Rotorua, New Zealand, 2019, pp. 570–577

35. L. V. Thuy, K. Cao-Minh, C. Dang-Le-Bao and T. A. Nguyen, "Votereum: An Ethereum-Based E-Voting System," 2019 IEEE-RIVF International Conference on Computing and Communication Technologies (RIVF), Danang, Vietnam, 2019, pp. 1–6

36. D. Khoury, E. F. Kfoury, A. Kassem and H. Harb, "Decentralized Voting Platform Based on Ethereum Blockchain," 2018 IEEE International Multidisciplinary Conference on Engineering Technology (IMCET), Beirut, 2018, pp. 1–6

37. Solidity.readthedocs.io. Solidity 0.6.5 Documentation. [online] Available: https://solidity.rea dthedocs.io/en/develop/types.html#fixed-point-numbers, Last accessed 2020/3/13

SharedWealth: Disincentivizing Mining Pools Through Burning and Minting

Thomas H. Austin[1]([📧]) [ID], Paul Merrill[1] [ID], and Justin Rietz[2] [ID]

[1] Computer Science Department, San José State University, San Jose, USA
thomas.austin@sjsu.edu, pdm@pdm.me
[2] Department of Economics, San José State University, San Jose, USA
justin.rietz@sjsu.edu

Abstract. Bitcoin has provided a framework for a decentralized currency system. However, the irregular payouts of its winner-takes-all reward strategy have proven to be unacceptable for the miners who are responsible for verifying transactions. To address this problem, miners have formed mining pools to ensure more regular rewards, but these mining pools undermine the decentralization that is one of Bitcoin's key benefits. Previous work has sought to make it impossible for mining pools to form, but it is not clear that the situation would improve without addressing the economic incentives that lead to the formation of mining pools in the first place.

This work introduces SharedWealth, a protocol designed to reward miners who find "near misses" to the required proof-of-work target to make a Bitcoin block. We show how this approach provides miners with more regular pay, decreasing their incentives to join a mining pool. We use a *burn-and-mint strategy*, where new rewards are paid out based on the expected number of near misses rather than dividing up the rewards according to the actual number of near misses, thus eliminating any incentive for a block producer to discard the near misses of other miners.

1 Introduction

Since its introduction, Bitcoin has revolutionized the world of digital currency. At the core of its design is the idea of rewarding miners for their work in verifying and producing new blocks. By making it more profitable to support the network rather than cheat the system, or so the original paper claims [8], the Bitcoin protocol provides a resilient decentralized payment system.

A key piece of the architecture is a proof-of-work that connects blocks of transactions to form a blockchain. In exchange for finding a valid proof, a miner may propose a block, thereby gaining the rewards in the form of newly minted bitcoins and whatever transaction fee rewards are offered by clients proposing transactions.

Unfortunately, as Bitcoin's popularity has risen, the reward paid to any single miner has become highly irregular. Especially since bitcoin mining currently

W. Abramowicz and G. Klein (Eds.): BIS 2020 Workshops, LNBIP 394, pp. 73–85, 2020.
https://doi.org/10.1007/978-3-030-61146-0_6

requires specialized ASIC mining rigs, miners find this infrequent payout unacceptable.

Mining pools have helped to address this issue by allowing miners to join together. A mining pool operator coordinates the group's efforts in exchange for a share of the rewards. Whenever a proof is found, the reward is divided up amongst the group according to their contribution in terms of hashing power.

The success of mining pools has lead to more even payouts for miners, but has also lead to increasing centralization as operators of large mining pools control large percentages of the total hashing power of the network.

Previous work has shown that mining pools may be broken up through nonoutsourceable puzzles [7]. However, it is not clear that this approach would lead to decreased centralization without addressing the economic challenges that lead to the formation of mining pools in the first place.

In this work, we seek to redesign the Bitcoin protocol to address the problems with Bitcoin's economic incentives. Our model, called SharedWealth, rewards miners for finding lesser proofs[1], which we refer to as *near misses*. A miner who succeeds in finding a near miss can write a *near-miss transaction* to the network to gain a share of the rewards once a valid block proof is finally discovered.

The idea of these near misses is already used today to coordinate efforts of mining pools, but typically requires coordination of a centralized pool operator. Integrating this reward system into the protocol itself in a non-trivial task. A naive approach would divide the block reward among all miners who find either a proof or a near miss. We refer to this approach as the *pie-splitting model*. Unfortunately, a miner in this model has an incentive to exclude the near-miss transactions of others, since every near miss reported decreases their own reward.

We address this problem by using a *burn-and-mint* model. All transaction fees offered in the block are burned, and new coins are issued for each near miss found. If the expected amount of near misses are reported, the total amount of coins created will equal the total of the transaction fees plus whatever additional coins should be created according to the protocol. However, if additional near misses are discovered, additional coins are minted so that no other miner's payout is affected. Likewise, if fewer near misses are discovered than expected, the coins that should have been generated are simply lost.

By our design, miners have no dis-incentive to report others' near-miss transactions. Therefore, they will include these transactions according to the same rules as other transactions – that is, they will be included if the transaction fee is greater than the fee for other transactions offered.

In this paper, our discussion focuses on Bitcoin, since it is by far the largest cryptocurrency by market capitalization. However, we note that our model is

---

[1] In this paper, we refer to a nonce that will satisfy the requirement for producing a valid new block as a *block proof*. A nonce that does not satisfy this requirement, but which produces a hash that meets a lower threshold requirement is referred to as a *near-miss proof*, or simply as a *near miss*. We refer to both types generically as "proofs".

relevant for other proof-of-work blockchains, and perhaps for other styles of blockchains that have issues with groups analogous to mining pools.

2   Background and Terminology

The Bitcoin protocol introduced the concept of the *blockchain*. We define a *block* as a triple of a data payload, a cryptographic hash of its parent block (if there is one), and a *proof* of some form (discussed later in this section). These blocks form a tree, which is the blockchain. The root of this tree is the *genesis block*.

Unlike most trees, a blockchain has rules to eliminate branching as much as possible. When a branch extends for more than a few blocks, it tends to have a name (e.g., Bitcoin Cash, Ethereum Classic, etc.).

Blockchain protocols can be distinguished by their proof mechanism, such as *proof-of-work* (PoW), *proof-of-stake* (PoS), etc. Bitcoin proofs rely on the hashcash algorithm [1], where the cryptographic hash value of the block with its "proof" field must produce a value below a given threshold.

Miners in proof-of-work systems collect transactions into blocks and search for a proof value that will produce a valid hash value for the block. Once a miner finds a proof, it announces its block to the mining network. The other miners verify the transactions and the block proof; if all is valid, they accept the block and work on producing a new child block of this block. This system provides a form of probabilistic consensus known as *Nakamoto consensus*.

In exchange for their work, miners are rewarded with both transaction fees paid by the clients, and by newly generated coins in a special *coinbase transaction*. The amount of coins produced per block is specified by the Bitcoin protocol, and is set to reduce over time until eventually being eliminated altogether. We collectively refer to the transaction fees and the rewards from the coinbase transaction for a given block as the *block reward*.

Due to the difficulty of finding Bitcoin proofs, miners have formed *mining pools*, where all miners work together and divide the rewards. In order to validate the work done for the pool, each miner must submit a *share* whenever they come close to finding a valid proof. In the context of a mining pool, a share is a proof value that (when hashed with the block) produces a hash value that exceeds the value for a valid block proof, but is lower than a second threshold.

Most mining pools are organized and run by an *operator* who collects transactions and specifies the block for the pool miners to work on. The operator is responsible for tabulating the shares and rewarding miners for their work appropriately. In exchange, the operator takes a cut of the reward.

Several different forms of mining pools exist. In *proportional reward* (PROP) mining pools, miners do not receive any reward for their shares immediately, but instead receive a portion of the block reward according to the number of shares they have found out of the total number of shares discovered.

A downside of PROP mining pools is that the rewards paid per share diminish over the time it takes the mining pool to find a proof, since each additional share discovered reduces the value of each share. As the value of shares drop, the miners have an incentive to stop mining for the pool and start instead mining for a different pool where the shares are more valuable; miners switching between pools based on the changing value of shares within the pool is referred to as a *pool hopping attack*.

In pay-per-share (PPS) mining pools, the operator immediately pays a reward to any miner that finds a valid share. When a proof is found, the operator claims the entirety of the reward. Since the miners in a PPS pool always receive the same value for a share, pool hopping attacks are not an issue. However, there is more risk for the operator.

The most common style of mining pools in recent years is pay-per-last-N-shares (PPLNS). In this approach, miners are paid for their shares when a valid block proof is discovered, similar to how PROP pools work. However, the reward is divided among the most recent N shares received, for some value of N; older shares are discarded. This approach keeps the value of shares constant, and removes any benefit to miners for pool hopping.

Rosenfeld [9] reviews the different designs of mining pools, including possible attacks and vulnerabilities.

3   Naive Pie-Splitting Model

The idea of rewarding miners for finding lesser proofs may be an obvious one, especially since it is the basic mechanism used in most mining pools. However, integrating this idea into the blockchain itself must be done with care, or else incentives may cause undesirable behavior on the blockchain.

An obvious approach, which we have dubbed the pie-splitting model, would be to divide the rewards amongst the block producer and all miners who discover near misses. A miner who finds a near miss can write a *near-miss transaction* to provide proof of the near miss and gain the reward.

However, this model has two problems: firstly, the incentive for miners to keep searching for a valid proof diminishes over the time spent to mine a block; secondly, the block producer has an incentive to discard the near-miss transactions of other miners.

For the first problem, we simply note that the number of near misses found over time will increase. Therefore, the reward for a block will be divided among more and more miners, decreasing the value for every miner. When the expected value drops to a sufficiently low level, miners will begin to use their power to mine on other blockchains, or might decide to stop work altogether. This problem is analogous to the pool-hopping attack [9] that has plagued PROP mining pools.

We next discuss the incentives for miners. For simplicity, we assume that rewards are shared evenly between the block producer and all miners with a valid near-miss transaction in the block. (We could separate the reward for the block producer and the near-miss transactions, but if the block producer has also found a near miss, this change would offer no improvement).

When a miner receives a near-miss transaction from another miner, they have two choices. If the miner accepts the near-miss transaction, and then later finds a valid proof-of work, their reward is halved. Assuming that miners are rational actors, they will therefore discard all near-miss transactions of other miners.

Game theoretically, the sharing of a set-sized award is similar to a prisoner's dilemma game in which the Nash equilibrium is for all miners to defect (i.e. throw away other near misses) (Watson [11]). The left-hand matrix in Table 1 provides an example of payoffs in such a game (note that relative magnitudes, not actual levels, are important): Where K is "Keep other's near misses", T is "Throw out other's near misses", miner Alice receives the first payment X in the set (X, Y) and miner Bob receives Y. If Bob plays "Keep" it is optimal - that is, it is the best response - for Alice to play "Throw" as she receives a payout of 5 vs. 4. If Bob plays "Throw" it is still optimal for Alice to play "Throw" as she receives 2 vs. 1. As the payoffs are symmetric, the equilibrium (a Nash equilibrium) is for both Alice and Bob to throw out each other's near miss transactions.

Table 1. Prisoner's Dilemma vs. Hawk-Dove payoffs

		Prisoner's Dilemma					Hawk-Dove	
		Bob					Bob	
		K	T				K	T
Alice	K	$(4,4)$	$(1,5)$		Alice	K	$(4,4)$	$(1,5)$
	T	$(5,1)$	$(2,2)$			T	$(5,1)$	$(0,0)$

The reasoning behind the relative payouts is that if both Alice and Bob play "Keep" (they cooperate) the blockchain network is healthy and thus the token is worth more, but they have to split the mining reward. If one of them plays "Throw" (a defect strategy) and the other plays "Keep", the value of the coin may drop as miners and users believe that mining is less "fair." However, the defector is still better off, otherwise there would be no reason to defect. If both miners defect and throw out each other's near misses, the token would likely drop further in value as miners stop participating. The theoretical result is that the system ends up at the sub-optimal outcome with both miners throwing out each other's near misses.

In the real world, this would be an iterated prisoner's dilemma game in which multiple players engage over a series of rounds. Thus there are several strategies that can lead to the "always cooperate" equilibrium, the most well recognized being "tit-for-tat": if Alice believes that Bob previously discarded her near-miss transaction, Alice could retaliate by discarding Bob's next near-miss transaction. On the other hand, if Alice believes that Bob previously included her near-miss transaction, Alice would reciprocate by including Bob's next near-miss transaction. While this approach can help ensure the proper behavior of miners, the equilibrium can be unstable as it is not sub-game perfect–that is, a player's actions are not always rational at an individual round of play.

However, if every miner cheats, the value of the coin may go to zero, in which case the payoff structure might be closer to a Hawk-Dove model. In this model, the payoff matrix would be one in which it is always optimal at the individual level to play the opposite of the other player, as in a scenario in which some miners playing "Keep" and some playing "Throw" still provides a better payoff than all miners throwing each other's near misses, or when an individual miner plays different strategies at different times. (See the right-hand side matrix in Table 1 for an example). In this case $(2, 5)$ and $(5, 2)$ are both possible equilibria given the same reasoning as used for finding the equilibrium in the prisoners' dilemma game.

We can take this a step further and assume not just two miners but a population of miners and analyze the game using evolutionary game theory as described by Friedman and Sinervo [5]. We assume that Alice doesn't play against Bob, but against the entire population of miners, and that she believes that any given member of the population will play "Keep" with probability p and therefore "Throw" with probability $1 - p$. Thus, her expected payoffs from playing either strategy is:

$$Keep = 4p + 1(1 - p) = 3p + 1 \tag{1}$$

$$Throw = 5p + 0(1 - p) = 5p \tag{2}$$

The expected payoff difference between the two strategies is therefore

$$Keep - Toss = (3p + 1) - 5p = -2p + 1 \tag{3}$$

Using a phase diagram with the horizontal axis representing the percentage of miners playing the "Keep" strategy and the vertical axis representing the expected payoff difference for Alice of playing "Keep" versus "Throw", we can get some idea of the stability of possible equilibria. In Fig. 1, the dynamics are such that in equilibrium some miners play "Keep" and some play "Throw." Thus, as long as the payoff to keeping other miners' near miss transactions is higher than the payoff to the individual miner when all miners' throw out all other miners' transactions, there can be an equilibrium in which some miners behave and others don't.

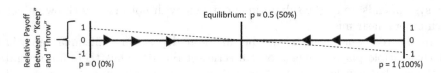

Fig. 1. Phase diagram

4   SharedWealth Model

We now review our solution for rewarding miners in a manner that avoids the problems of the pie-splitting model discussed in Sect. 3.

In many cryptocurrencies, miners are rewarded through two mechanisms: newly minted coins authorized by the protocol, and transaction fees specified by clients. We first review how newly minted coins can be allocated in our model before showing how our design can be expanded to include transaction fees.

4.1 Sharing Coinbase Rewards

Newly minted coins often account for the lion's share of rewards for miners. The protocol dictates the conditions for when these coins can be minted, who gets them, and how many coins are created. For instance, whenever a Bitcoin miner produces a block, they are allowed to create one special coinbase transaction that rewards them with new bitcoins.

In Bitcoin and many other cryptocurrencies, new coins are minted to reward the block producer. The SharedWealth model follows a similar strategy, except that the newly minted coins are not created on a per-block basis, but rather on a per-share basis, where a *share* is either a block proof or a near-miss proof.

A simple example helps to illustrate this point. If a protocol follows the classic Bitcoin approach, it might create 12.5 new coins every block to give to the miner who finds the proof-of-work. We will target this same rate of inflation.

In this example, we configure the threshold for a near miss such that we expect 4 near misses in the time that it takes to find the proof. 2.5 coins will be minted and given to the block producer, as well as 2.5 coins to each miner who finds a near miss. Thus, the expected amount of new coins created per block is expected to be 12.5 coins.

However, the *actual* number of near misses might vary. Assume that we have three miners: Alice, Bob, and Charlie. In round 1, Alice finds a block proof. In that block, she includes a near-miss transaction for herself, as well as two near-miss transactions each for Bob and Charlie. Therefore, 15 coins are created in this block.

We expect that Bob and Charlie include a transaction fee in their near-miss transaction. We also note that the number of coins minted for Alice is not affected if she decides to discard the other miners' near misses. As a rational actor, we can assume that Alice will therefore include these transactions in her block in order to gain the additional transaction fees.

In the following round, Bob finds a proof-of-work more quickly, creating a new block with only two near-miss transactions – one each for Alice and Charlie. Therefore, only 7.5 coins are minted.

Note that this design is likely to pay out more rewards when the mining community requires longer than average to find a proof, which is exactly what we want. By giving out more rewards when the proof takes longer to find, we avoid the economic problems that lead to pool hopping or *coin hopping* attacks (a term coined by Carlsten et al. [2]).

4.2   Sharing Transaction Fees

By design, Bitcoin will phase out coinbase transactions over time, instead incentivizing miner operation by transaction fees alone [8]. Therefore, it is important to consider how these fees can be divided up to reward near misses.

In the pie-splitting model, transaction fees are simply divided up among the block producer and all near-miss miners. However, as shown in Sect. 3, this approach creates a misalignment of incentives.

Instead, we adopt a *burn-and-mint strategy*. In short, we total the amount of transaction fees to determine how many additional coins to mint per proof. The coins offered as transaction fees are then "burned"; that is, the coins may be considered discarded and no longer spendable.

Returning to our previous example, let's assume that Charlie discovers a proof for a block that includes one near-miss transaction each for Alice, Bob, and Charlie. The payout per proof is 2.5 coins, as before.

In addition, we assume that a total of 0.5 coins are offered as transaction fees. As with the coinbase rewards, we divide the reward according to the expected number of proofs found. Therefore, each proof is now worth an additional 0.1 coins. Since only 4 proofs were found, one share of the transaction fees is lost and disappears from the system.

As with the pie-splitting scenario, we can also analyze SharedWealth from a game-theoretic perspective. However, given the SharedWealth mining protocol, the payoffs and the dynamics change. Consider the payoff matrix in Table 2. With SharedWealth, the payoff for defecting and playing "Throw" against someone playing "Keep" is capped by the fact that the payouts are the same regardless of the number of miners who find near misses. However, there is still a negative impact on the token value if some miners throw out near miss transactions based on the same reasoning as with the pie-splitting analysis. Therefore, miners are worse off playing "Throw" rather than "Keep" against someone else playing "Keep." Thus the optimal strategy is for both Alice and Bob to play "Keep."

Table 2. SharedWealth payoffs

		Bob	
		K	T
Alice	K	$(5,5)$	$(2,4)$
	T	$(4,2)$	$(1,1)$

This is also true from an evolutionary perspective The expected payoffs from playing either strategy are:

$$Keep = 5p + 2(1 - p) = 3p + 2 \tag{4}$$

$$Throw = 4p + 1(1 - p) = 3p + 1 \tag{5}$$

$$Keep - Toss = (3p + 2) - (3p + 1) = 1 \qquad (6)$$

Playing "Keep" always provides a relatively higher payoff than playing "Throw" regardless of what percentage of the mining population is playing one strategy versus the other. Thus, over time, everyone should play the "Keep" strategy, which provides the highest net benefit to all miners at both an individual and population level.

5 Improving Throughput and Reducing Storage

An obvious limitation of our system is that the near-miss transactions could fill up the blockchain, both reducing the throughput of *exchange transactions*[2] and increasing the storage requirements for miners.

In this section, we review an approach to partially address these issues. In short, when a miner finds a valid proof, it takes on the role of a pool operator, coordinating the effort of the miners to find the next block and tracking who should receive rewards for their shares.

Our approach takes inspiration from previous work on Bitcoin-NG [3]. We first review that protocol to provide better insight to our own approach.

Bitcoin-NG seeks to increase Bitcoin's transaction throughput. After a miner finds a block with a valid proof-of-work (dubbed a *key block*), it becomes the leader who may continue to produce additional *microblocks* that contain valid transactions, but which do not have a valid proof-of-work. If the leader is caught double spending, a *poison transaction* can provide evidence to other miners, who will then invalidate the leader's transaction fee rewards gleaned from the microblocks. The leader remains in power until another miner finds a key block and takes control.

Any transaction fee rewards are divided between the leader and its successor with a 40%/60% split. This divide both encourages the leader to produce microblocks and encourages other miners to attempt to build off of the latest microblock.

In our design, key blocks are reserved for exchange transactions. In this manner, the throughput of exchange transactions for the system remains unchanged.

In order to validate a near-miss transaction, the near-miss miner's entire block would need to be stored, including its set of transactions. Storing this amount of blocks would quickly become intractable as we increase the number of near misses.

Instead, by letting the leader select the transactions for the following block, all miners will be working to find a proof that matches that transaction set. While the additional header information would still need to be tracked, the exchange transactions themselves need to be stored only once per key block.

A downside of our approach is that the leader becomes a single point of failure; addressing related concerns remains for future work. However, we note that in the worst case, our protocol reverts to Bitcoin.

---

[2] That is, transactions that exchange coins between two clients.

We walk through the steps of our process below:

1. A miner finds a valid block proof. It publishes:
   - The proof
   - A signed list of valid exchange transactions for the subsequent block. These transactions should not be connected to the proof for the current block; otherwise, that set of transactions would need to be stored for every near miss.
2. Miners race to find a proof for the block of transactions specified by the leader. During this period, miners also collect (but do not need to publish) exchange transactions for the next block.
3. When a miner finds a near miss, it reports it to the current leader with a near-miss transaction. This transaction may include a transaction fee.
4. When a miner finds a valid block proof, it becomes the new leader. The old leader and the new leader split any transaction fees included in this block. The process starts again.

In Bitcoin-NG, the new leader and the old leader split the transaction fees with a 40/60 split. The authors' analysis [3] assumes that an individual miner's mining power would be bounded by 1/4 of the total network's mining power. From this assumption, they conclude that the current leader should gain at least 37% of the rewards, or else it might try to withhold transactions in the hopes of gaining the full value of the transaction fees in the next block. Similarly, a miner might be tempted to mine on the previous key block, and thus take the full value of transaction fee rewards that were included in subsequent microblocks. From this assumption, the authors determine that the current leader should gain at most 43% of the rewards. Therefore, assigning 40% of microblock transaction fee rewards to the current leader is within the required bounds.

In SharedWealth, near-miss transactions are invalid after a new leader is elected and selects a new transaction set. Therefore, a powerful miner has no incentive to exclude any transactions regardless of the division used. We nonetheless use Bitcoin-NG's 40/60 split to allow ourselves the flexibility of including exchange transactions in microblocks.

5.1 Including Exchange Transactions in Microblocks

In our current design, microblocks are reserved solely for near-miss transactions. Our hope is that this design would allow future optimizations to crunch down the storage of microblocks further.

However, we note that we could design our protocol to allow for exchange transactions in microblocks, and thereby increase the overall throughput of the system. Which design is preferable might depend on whether the throughput of transactions or the blockchain storage requirements were a higher priority.

6   Related Work

Miller et al. [7] introduce the idea of nonoutsourceable puzzles. In essence, the mining protocol is redesigned in such a way that a miner must have the private key for the coinbase reward in order to mine a block. With this design, a miner who finds a valid share can report it to the pool operator in order to claim its reward; however, when the miner finds a valid block proof, it can take the entire block reward for itself. Assuming miners act rationally, mining pools would not be able to effectively collaborate.

The authors suggest a tiered reward structure, where the lowest "consolation prize" is frequent, designed to even payouts for miners; the medium-value prize ensures blocks arrive regularly, given out at the same rate as bitcoin currently gives out rewards; finally, the "jackpot" gives out large payouts frequently. The goal of the jackpot is to make miners in a pool suspicious of each other and of hosted mining. Statistical analysis might be able to detect when a miner is not reporting sufficient medium-value rewards, but a miner that only takes the jackpots would be much more difficult to detect.

Carlsten et al. [2] review the importance of coinbase rewards for the stability of Bitcoin, and discuss new challenges that will appear as transaction fees become the central motivating factor for miners. They note that the variability of rewards will cause rational miners to fork off of older, more profitable blocks rather than the most recent block, leading to an increase of forks and a greater risk of "51%" attacks. Interestingly, this variability of rewards for a block remains even when the transaction fee rewards are constant, due to the variability of how long it takes a network to find a block proof. Our work might help to address this concern – on blocks where it takes longer to find a valid proof, it is probable that more near misses are discovered, and hence the value may be "evened-out"; we plan to pursue this direction in future research.

Luu et al. SmartPool [6] avoids the problems of a corruptible pool operator by replacing an operator with a smart contract. While this approach seems particularly enticing for languages like Ethereum with support for rich smart contracts, the authors show how their design could manage proofs for a different blockchain; for example, a pool with an Ethereum smart contract could be used to manage a Bitcoin mining pool.

Zamyatin et al. [12] describe Flux, a backwards compatible mining protocol (called a *velvet fork*) for the Bitcoin network. In their design, miners can be divided into legacy miners following Bitcoin's mining protocol and Flux miners who follow additional rules to build sub-blocks, essentially building a blockchain between two blocks in the Bitcoin blockchain. When the Flux miners produce a block in the Bitcoin blockchain, the transaction rewards are divided amongst all miners who produced sub-blocks. Sub-blocks require a lower proof-of-work, thereby serving to divide rewards more broadly in a similar manner to our protocol. One advantage of Flux is its backwards-compatibility with Bitcoin. In contrast, our approach alters the incentive structure, requiring a hard-fork of miners to adopt the SharedWealth model.

Szalachowski et al. [10] present StrongChain, which follows a similar approach to our design. Miners broadcast *weak proofs*, which match to the previous block with a lesser proof of work. Miners gain rewards for weak proofs, and the work required to find the weak header is added to the subsequent block. (This approach helps to defend against *selfish mining attacks* [4], where miners attempt to take more than their fair share of the mining rewards by not publishing blocks.) The authors discuss *spiteful mining*, where a block producer might discard other miners' weak proofs; as with SharedWealth, they modify the reward structure so that this approach does not benefit the spiteful miner.

7   Conclusion and Future Work

In this paper, we have shown how the SharedWealth model helps to reward miners more regularly, and thus reduce the incentives to form a mining pool. Critically, we restructure the incentives so that accepting a *near miss* from a miner does not disadvantage the miner who finds the block proof. We show transaction fees may be handled with the burn-and-mint strategy.

The differences between the SharedWealth model and the pie-splitting model are analogous to the differences between a PPS mining pool and a PROP mining pool. Like PROP mining pools, the pie-splitting model suffers from a dilution of the value of shares when it takes longer to find a valid block proof. In both PPS mining pools and the SharedWealth model, this problem is addressed by keeping the value of shares for a block constant.

An obvious question is whether a design similar to PPLNS pools would be useful. SharedWealth is not backwards compatible with Bitcoin, since we alter the reward structure in a way that would produce invalid blocks according to Bitcoin's protocol. We envision that designing our system to use an approach similar to PPLNS pools might allow us to maintain the existing reward structure, and hence be able to make a backwards compatible variant of the SharedWealth model. We intend to pursue this direction in future work.

References

1. Back, A.: Hashcash - a denial of service counter-measure (2002)
2. Carlsten, M., Kalodner, H.A., Weinberg, S.M., Narayanan, A.: On the instability of bitcoin without the block reward. In: Conference on Computer and Communications Security (SIGSAC), pp. 154–167. ACM (2016)
3. Eyal, I., Gencer, A.E., Sirer, E.G., van Renesse, R.: Bitcoin-NG: a scalable blockchain protocol. In: Symposium on Networked Systems Design and Implementation (NSDI), pp. 45–59. USENIX Association (2016)
4. Eyal, I., Sirer, E.G.: Majority is not enough: bitcoin mining is vulnerable. Commun. ACM **61**(7), 95–102 (2018)
5. Friedman, D., Sinervo, B.: Evolutionary Games in Natural, Social, and Virtual Worlds. Oxford University Press, New York (2016)
6. Luu, L., Velner, Y., Teutsch, J., Saxena, P.: Smartpool: practical decentralized pooled mining. In: USENIX Security, pp. 1409–1426. USENIX Association (2017)

7. Miller, A., Kosba, A.E., Katz, J., Shi, E.: Nonoutsourceable scratch-off puzzles to discourage bitcoin mining coalitions. In: Conference on Computer and Communications Security (SIGSAC), pp. 680–691. ACM (2015)
8. Nakamoto, S.: Bitcoin: a peer-to-peer electronic cash system (2009)
9. Rosenfeld, M.: Analysis of bitcoin pooled mining reward systems. arXiv preprint arXiv:1112.4980 (2011)
10. Szalachowski, P., Reijsbergen, D., Homoliak, I., Sun, S.: Strongchain: transparent and collaborative proof-of-work consensus. In: 28th USENIX Security Symposium, USENIX Security 2019, Santa Clara, CA, USA, August 14–16, 2019, pp. 819–836. USENIX Association (2019)
11. Watson, J.: Strategy: An Introduction to Game Theory, vol. 139. WW Norton, New York (2002)
12. Zamyatin, A., Stifter, N., Schindler, P., Weippl, E.R., Knottenbelt, W.J.: Flux: revisiting near blocks for proof-of-work blockchains. IACR Cryptol. ePrint Arch. **2018**, 415 (2018)

Modeling Smart Contracts with Probabilistic Logic Programming

Damiano Azzolini[1]([⊠]), Fabrizio Riguzzi[2], and Evelina Lamma[1]

[1] Dipartimento di Ingegneria, University of Ferrara, Via Saragat 1,
44122 Ferrara, Italy
{damiano.azzolini,evelina.lamma}@unife.it
[2] Dipartimento di Matematica e Informatica, University of Ferrara, Via Saragat 1,
44122 Ferrara, Italy
fabrizio.riguzzi@unife.it

Abstract. Smart contracts are computer programs that run in a distributed network, the blockchain. These contracts are used to regulate the interaction among parties in a fully decentralized way without the need of a trusted authority and, once deployed, are immutable. The immutability property requires that the programs should be deeply analyzed and tested, in order to ensure that they behave as expected and to avoid bugs and errors. In this paper, we present a method to translate smart contracts into probabilistic logic programs that can be used to analyse expected values of several smart contract's utility parameters and to get a quantitative idea on how smart contracts variables changes over time. Finally, we applied this method to study three real smart contracts deployed on the Ethereum blockchain.

Keywords: Blockchain · Probabilistic Logic Programming · Smart contracts.

1 Introduction

The idea of the blockchain model dates back to 1990 [12] as a method to secure timestamping digital documents, but the interest around this technology grew only after the success of the paper by Nakamoto in 2008 [18]. Afterwards, the term *smart contract*, used to identify programs written in a quasi-Turing-complete programming language that run in a blockchain environment, gained traction, thanks to the possibility of enforcing contracts between two or more parties without the need of a central authority.

Smart contracts and blockchain technology are relevant to several research fields. In distributed systems the researchers study the interaction among peers in a fully decentralized environment with an unreliable network and create methods to ensure a decentralized consensus, even in case of dishonest parties. Formal methods are used to verify the behavior of smart contracts. Cryptography encompasses methods to ensure that all the participants can see the same data and

W. Abramowicz and G. Klein (Eds.): BIS 2020 Workshops, LNBIP 394, pp. 86–98, 2020.
https://doi.org/10.1007/978-3-030-61146-0_7

that data have not been tampered with. Game theory and decision theory are used to model the behavior of the interacting parties to maximize the expected profit.

The execution of a smart contract is deterministic, i.e., it always yields the same result. However, because users can interact with it at their will, a probabilistic analysis is needed to predict its behavior. From an issuer's perspective, he may be interested in how many people interact with the contract and how profit values evolve. Similarly, from a user's perspective, he can be interested in the expected reward by interacting with a blockchain based game, such as gambling.

In this paper, we propose a method to translate smart contracts into probabilistic logic programs in order to compute the expected values of several smart contract's utility parameters. Furthermore, our approach can also be used to identify some coding errors. Several tools exist for identifying bugs but almost all of them provide a static analysis without computing a quantitative amount of the possible monetary loss. Moreover, translating a smart contract into a probabilistic logic language allows interaction with it in a practical way, without the need of a specialized checking tool for every smart contract language.

The paper is organized as follows: in Sect. 2 we briefly present blockchain technology and smart contracts. In Sect. 3 we introduce Probabilistic Logic Programming. In Sect. 4 we propose a method to translate smart contracts into probabilistic logic programs. In Sect. 5 we analyse three smart contracts that model real world applications and Sect. 6 concludes the paper.

2   Blockchain and Smart Contracts

Smart contracts were initially proposed in 1994 [26] as computer protocols to facilitate a self-enforcing agreements between two parties. Smart contracts received an increased attention only after the interest around blockchain technologies exploded, following the publication of [18]. Currently, the term smart contract is always used in the context of a blockchain environment.

In a nutshell, a blockchain is a decentralized, distributed, append only (usually) public ledger maintained by a set of peers, that records transactions between accounts. All the transactions are organized in a chain of blocks (hence the origin of the term blockchain) linked together using hash functions that guarantee the consistency and the immutability of the stored data. Each honest (full) node that follows a blockchain protocol stores his own updated copy of the ledger. In order to guarantee consistency of the data, i.e., that all the peers see the same copy of the ledger, a consensus must be reached: this is done by adopting a so-called *consensus algorithms*, such as Proof-of-Work (PoW).

Blockchain is the underlying technology of several decentralized platforms, such as Bitcoin and Ethereum. Ethereum is a decentralized transaction-based state machine [31] that executes smart contracts written in a quasi-Turing-complete bytecode language. Programmers usually develop smart contracts using a programming language called Solidity which is translated, by a compiler, into bytecode for execution. In this paper, when we write smart contract we mean a smart contract deployed on Ethereum. However, our method can be extended to general smart contracts of which source code is available. In case of Ethereum, source code is not stored in the blockchain (only a hash of the code is stored) but there are several platforms that allows developers to upload the code of a contract and verify it, such as Etherscan[1]. This website checks if the compilation of the uploaded code matches the bytecode stored into the blockchain. If so, the smart contract is considered verified and the source code is made public.

3   Probabilistic Logic Programming

A wide variety of domains can be represented using Probabilistic Logic Programming (PLP) languages under the distribution semantics [22,25]. A program in a language adopting the distribution semantics defines a probability distribution over normal logic programs called *instances* or *worlds*. Each normal program is assumed to have a total well-founded model [27]. Then, the distribution is extended to queries and the probability of a query is obtained by marginalizing the joint distribution of the query and the programs. A PLP language under the distribution semantics with a general syntax is that of *Logic Programs with Annotated Disjunctions* (LPADs) [29]. In the following part we present the semantics of LPADs for the case of no function symbols, if function symbols are allowed see [21].

Heads of clauses in LPADs are disjunctions in which each atom is annotated with a probability. Consider an LPAD T with n clauses: $T = \{C_1, \ldots, C_n\}$. Each clause C_i takes the form: $h_{i1} : \Pi_{i1}; \ldots; h_{iv_i} : \Pi_{iv_i} :- b_{i1}, \ldots, b_{iu_i}$, where $h_{i1}, \ldots, h_{iv_i}$ are logical atoms, $b_{i1}, \ldots, b_{iu_i}$ are logical literals and $\Pi_{i1}, \ldots, \Pi_{iv_i}$ are real numbers in the interval $[0, 1]$ that sum to 1. $b_{i1}, \ldots, b_{iu_i}$ is indicated with $body(C_i)$. Note that, if $v_i = 1$ the clause corresponds to a non-disjunctive clause. We also allow clauses where $\sum_{k=1}^{v_i} \Pi_{ik} < 1$: in this case the head of the annotated disjunctive clause implicitly contains an extra atom *null* that does not appear in the body of any clause and whose annotation is $1 - \sum_{k=1}^{v_i} \Pi_{ik}$. We define *substitution* θ a function mapping variables to terms. Usually θ has the form $\theta = \{X_1/t_1, \ldots, X_k/t_k\}$ meaning that each variable X_i is substituted by the term t_i. Applying a substitution θ to a LPAD T means replacing all the occurrences of each variable X_j in ϕ by the corresponding term t_j. For an exhaustive treatment of PLPs see [22].

Given an LPAD T, the main task is to perform *inference*. There are two types of inference: exact inference and approximate inference. Both exact and approximate inference are implemented into the suite *cplint* [1,23].

3.1   Exact Inference

The main goal of exact inference is to solve tasks in an exact way, without approximation. Various approaches have been presented for performing inference

---

[1] https://etherscan.io/.

on LPADs, such as PITA [24]. Starting from an LPAD, PITA performs inference using knowledge compilation [10] to Binary Decision Diagrams (BDD).

The exact inference task is in general #P-complete [16] so it is not tractable for certain domains. In these cases, *approximate* inference is needed. We will not use exact inference in this paper and so we will not cover this topic further in detail.

3.2 Approximate Inference

In cplint, approximate inference is performed using Monte Carlo algorithms [5,20]. Using these algorithms, the possible worlds are sampled and the query is tested in the samples. The estimated probability of the query is then given by the fraction of the sampled worlds where the query succeeds.

Consider a simple example were a coin is toss with uncertainty on its fairness. Our goal is to find the probability that it lands head or tail. A probabilistic logic program to model this scenario may be:

```
heads(Coin):0.5; tails(Coin):0.5 :-
    toss(Coin),not(biased(Coin)).
heads(Coin):0.6; tails(Coin):0.4 :-
    toss(Coin),biased(Coin).
fair(Coin):0.9; biased(Coin):0.1.
toss(Coin).
```

The program states: if we toss a coin that is not biased, then it lands heads or tails with the same probability 0.5. If we toss a coin that is biased, then it lands heads with probability 0.6 and tails with probability 0.4. We express our uncertainty on the bias of the coin supposing that it is fair with probability 0.9 and biased with probability 0.1. Finally, we say that the coin is certainly tossed.

To compute the probability that the coin lands head, using the module MCINTYRE from cplint. For example, we can sample heads(Coin) a certain number of times and compute the probability that it's biased using ?-mc_sample(heads(coin),1000,P). It's also possible to compute expectations by sampling using mc_expectation/4 to get, for instance, the number of consecutive toss landing head. cplint also allows the definition of probability densities using A:Density:- Body. For instance, g(X): gaussian(X,0, 1) states that argument X of g(X) follows a Gaussian distribution with mean 0 and variance 1.

4   Modelling Smart Contracts with Probabilistic Logic Programming

Probabilistic Logic Programming [11] has been successfully applied in many different fields where probability has a key role, such as natural language processing, link prediction in social networks, model checking and also Bitcoin protocol [3,4,19]. While smart contracts are, by definition, deterministic (i.e., the execution of a smart contract's function must always yield the same result), the interaction among users of the same smart contract can be seen as probabilistic because the involved parties are not under the control of the issuer.

From a smart contract issuer's perspective, he can be interested in the expected value of certain profit variables, such as the collected fees (which can be a fixed fraction of the user input value), the expected number of tokens sold in a certain amount of time or the tokens distribution after several transactions among users. From the opposite perspective, the user's perspective, he could be interested in the expected reward from participating in a smart contract game, such as gambling, which is, at the moment of writing, among the smart contract categories with most interactions (according to DappRadar[2]).

These values can be computed by translating smart contracts into probabilistic logic programs. This process is composed of two steps: the smart contract is translated into a Logic Program and then probabilistic facts are added to turn it into a PLP. The translation of a smart contract function into a logic programming predicate is straightforward. Here we use SWI-Prolog [30] implementation of the programming language Prolog. Every function can be translated into a predicate with the same name and the same number of arguments. Moreover, if needed in the logic flows of the function, the members of the globally available msg object, such as msg.value or msg.sender, can be added to the arguments list of the predicate. Since a Prolog programs does not allow the **return** keyword, all the values that are returned from a Solidity function must be added as further arguments of the predicate.

Listing 1.1 shows a simple example of a smart contract written in Solidity simulating a bank. The function **constructor** is executed only at the creation of the contract: it sets the owner of the contract and issues 1000 tokens to the creator. Each user has the possibility to call **transfer**. This function accepts two parameters, the address of the receiver and the amount the user wants to transfer to the receiver. First, it verifies that the user has enough funds to perform the operation and that the sender is different from the receiver. Both conditions are evaluated using **require**, a built-in function that checks the condition passed as input and throws an exception if it is not met. Then, the balances of both parties are updated accordingly.

To simulate the storage reserved to a smart contract, it is possible to add two further arguments to the predicate arguments list, one for the input list and one for the output list: using a user-defined predicate called find/3, we retrieve the balance of the sender (identified with Sender) from BalanceList and we store it into BalanceSender. Similarly, we retrieve the balance of the receiver. Then, we perform the checks corresponding to **require** in Listing 1.1. Finally, we update the balances and update the list (generating another list with the old balances replaced with the new one for both parties, since Prolog lists do not allow modifications) using another user-defined predicate called update/3.

Once the contract is translated into SWI-Prolog code, we can add some probabilistic facts. For instance, we may suppose that the transferred amount from a user to another user (this transfer can be seen also as placing a bet, by

---

[2] https://dappradar.com/.

```
contract simpleBank {
    address owner;
    mapping(address => uint) balances;

    constructor() public {
        owner = msg.sender;
        balances[msg.sender] = 1000;
    }

    function transfer(address receiver, uint amt) public {
        require(balances[msg.sender] >= amt);
        require(msg.sender != receiver);
        balances[msg.sender] -= amt;
        balances[receiver] += amt;
    }
}
```

Listing 1.1. Example of Solidity smart contract.

transferring funds to the address where the smart contract is stored) is uniformly distributed between 0.5 and 2 Ether. This uncertainty can be expressed with amount(A):uniform(A,0.5,2.0). The complete code is shown Listing 1.2.

```
amount(A):uniform(A,0.5,2.0).
transfer(Receiver,Amt,Sender,BalanceList,NewBalanceList):-
    find(Sender,BalanceList,BalanceSender),
    find(Receiver,BalanceList,BalanceReceiver),
    amount(Amt),
    BalanceSender >= Amt,
    Sender \= Receiver,
    NewBalanceS is BalanceSender - Amt,
    NewBalanceR is BalanceReceiver + Amt,
    update(BalanceList,Sender,NewBalanceS,NewBalanceList1),
    update(BalanceList1,Receiver,NewBalanceR,NewBalanceList).
```

Listing 1.2. Example of smart contract translated into a probabilistic logic program.

The parameter Amt is now sampled from the uniform distribution specified above. Finally, to query the program to get, for instance, the expected transferred value, we can use mc_expectation/4. The following section shows how we applied this method to three smart contracts deployed and publicly accessible on the Ethereum mainnet. The usage of a PLP language makes it possible to test the smart contract without deploying it into a test net. Moreover, a logical language for smart contracts makes the contract simpler and easier to debug.

5   Experiments

To conduct our experiments we modelled three smart contracts written in Solidity taken from Etherscan. In the experiments, we analyzed a smart contract for transferring tokens, one for a Ponzi scheme and one of a gambling game. All the experiments are conducted on a cluster[3] with Intel® Xeon® E5-2630v3 running at 2.40 GHz. The execution time is computed using the built-in SWI-Prolog predicate `statistics/2` and the memory usage is the value `maxresident` computed using GNU Time[4]. For each experiment, we used the predicate `mc_expectation/4`, available in cplint, with 1000 samples.

5.1   Transfer

In this example we model a scenario where N users trade (burn or transfer) a certain amount of tokens. The functions `burn` and `transfer` are taken from the code stored at the address `0xB8c77482e45F1F44dE1745F52C74426C631bDD52` on the Ethereum mainnet. Each user starts with 100 tokens. We want to know, for example, how many transfers are needed to produce a situation where a single user has more than 180 tokens. Those two values have been chosen just to demonstrate the process. Transfers among users are done in a random way with the transferred amount uniformly distributed between 1 and 10. Moreover, we include a small probability (5%) that, instead of trading tokens, the user will burn tokens. In Table 1 we show the relation between number of users, execution time of the experiments, memory usage and expected number of transactions. As expected, the number of transfers needed to create a situation in which a user has more than a certain number (180 for this experiment) of tokens increases as the number of users increases.

Table 1. Details for the transfer experiment (Subsect. 5.1).

# of users	Time (s)	Memory (Mb)	Expected value
5	1.643	52.744	192.724
25	8.717	102.924	421.516
50	23.431	155.636	661.514
75	44.172	202.428	874.234
100	71.367	246.136	1071.335
125	101.48	285.208	1249.186
150	140.116	327.488	1441.242

Consider now the transfer function shown in Listing 1.1. The second line checks that the sender of the tokens is different from the receiver. However, in some real-world examples[5], due to coding errors, this check is not present, causing the generation of unexpected extra tokens. The method proposed in this

---

[3] http://www.fe.infn.it/coka/doku.php?id=start.

[4] https://www.gnu.org/software/time/.

[5] https://gist.github.com/loiluu/0363070e1bada977f6192c8e78348438.

paper is also suitable to spot bugs and coding errors of this type. For modelling this situation, we run the previous experiment but this time we compute the expected value of all circulating tokens, represented as the sum of the balances of all the participants. In every run, the program chooses two random (possibly the same) users from the user's list and performs a transfer of tokens between them. As expected, the number of circulating tokens exceeds the initial amount. The results are presented in Table 2.

Table 2. Details for the transfer experiment (Subsect. 5.1) with bug.

# of users	Time (s)	Memory (Mb)	Initial amount	Final amount
5	0.755	33.324	500	627.262
25	4.967	95.996	2500	2592.483
50	12.834	154.196	5000	5077.37
75	23.828	204.036	7500	7568.241
100	37.663	253.204	10000	10064.181
125	53.451	294.352	125000	12561.246
150	72.883	344.072	150000	15058.114

5.2 Ponzi Scheme

The boost of blockchain adoption in the last years caused a massive number of smart contracts being issued. In [7] and [28] the authors identify and analyze a huge number of Ponzi schemes deployed in the Bitcoin blockchain. In [6] the analysis was extended to the Ethereum blockchain.

A Ponzi scheme is a financial fraud that promises a high return of the investment: the profit increases as long as the number of people involved in the schema increases. However, this type of business model, and in particular pyramid schemes, a variant of Ponzi scheme, quickly become unsustainable due to the need of constant increase in participants to be profitable. Probability models in these cases are fundamental for analyzing the expected reward (payoff) and avoid being cheated.

For this experiment, we collect the code for a well-known pyramid schema called Rubixi, stored at 0xe82719202e5965Cf5D9B6673B7503a3b92DE20be. This code is often used to show a critical vulnerability of smart contracts which allow anyone to become the owner of the contract and withdraw the collected fees [2]. In this experiment we ignore this problem as it is out of the scope of this example. The logic of Rubixi is simple: a user can send some Ether, at least 1, to the contract through the *fallback* function. When receiving of the amount, the contract collects the Ether, adds the new participant to the participants list and redistributes the accumulated value to the other participants if certain conditions are met. Several considerations can be done for this contract. For instance: what is the amount of collected fees in a certain amount of time?

What is the number of participants we have to wait for receiving a payment? We modelled the contract in order to answer the second question supposing that users will send an amount of tokens to the contract uniformly distributed between 0.9 and 2, since we consider also a situation where a user sends an amount less than the minimum required. In this case the amount is only added to the collected fees but the participant is not considered. Our results are shown in Table 3.

Table 3. Details for the Ponzi scheme experiment (Subsect. 5.2). The last column relates the number of users to the amount of reward distributed. For instance, the fifth user entering the scheme has to wait 13 additional people to enter the schema before getting paid.

# of users	Time (s)	Memory (Mb)	# of users to wait
5	0.159	13.384	13.783
50	1.997	37.232	69.814
100	6.111	59.384	111.138
150	12.681	85.768	152.868
200	22.051	113.436	194.760
300	50.852	174.420	278.393
400	92.934	225.832	361.172

5.3   Gambling

According to the DappRadar website, in April 2020, 5 out of 10 most used dApps (web applications with the logic implemented on a smart contract) are gambling games. In this section we analyse a smart contract implementing a gambling platform stored at the address 0x999999C60566e0a78DF17F71886333E1dACE0BAE of the Ethereum mainnet. It allows a player to bet on several games such as dice, roulette or poker. The outcome is computed considering several payout masks. Randomization is obtained using a commit value, externally generated, provided as input to the bet, combined with other values. The experiments were conducted simulating a player betting on the outcome of a single die an amount distributed with a Poisson distribution with several values of mean, and transaction fees uniformly distributed between 0.07 and 0.2 *Finney* (1 Finney = 10^{-3} Ether), according to the data taken from BitInfoCharts[6]. The results obtained are shown in Fig. 1 and Table 4. As expected, the expected payout decreases in relation to the bet amount and the number of trials.

---

[6] https://bitinfocharts.com/ethereum/.

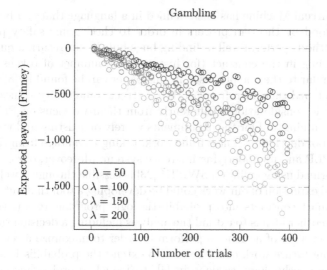

Fig. 1. Graph showing the expected payout of consecutive number of trials. λ represents the mean of the Poisson distribution.

Table 4. Resource usage for the Gambling experiment (Subsect. 5.3) with $\lambda = 150$.

# of trials	Time (s)	Memory (Mb)
5	0.517	9.344
50	4.890	11.016
100	8.757	18.924
150	12.938	28.22
200	17.439	27.076
300	27.011	143.896
400	37.437	201.996

6   Conclusions

In this paper we show that Probabilistic Logic Programming (PLP) can be applied to quantitatively model the behavior of smart contracts. We proposed an approach to translate smart contracts into probabilistic logic programs, independently from the contract language used, and we modelled three smart contracts taken from the Ethereum mainnet (i.e., Transfer, Ponzi Scheme and Gambling) in PLP.

Most of the literature about smart contracts is focused on vulnerability analysis and bug detection. There is a lot of work in the literature about automatic verification of smart contracts written using Solidity. For example, in [17] the authors used static analysis tools to automatically find bugs. Another approach can be found in [15] where symbolic model checking has been used. In [13] the

Ethereum Virtual Machine has been defined in a language that can be compiled and understood by theorem provers in order to check some safety properties. All these methods perform well in finding bugs but do not return a quantitative impact of a bug in the contract (for instance, the number of tokens lost). An approach similar to the one proposed in this paper can be found in [8] where the authors developed their own framework to extract some utility values based on game theoretic considerations. Differently from them we focus on Probabilistic Logic Programming and for the experiments we rely on existing and well-studied tools instead of developing a new framework, taking advantage from the expressiveness of PLP and its underlying inference system. Moreover, our experiments can be performed using cplint on SWISH[7] [23] accessible through a web browser.

A natural extension of our work could be developing a tool that automatically translates smart contracts into probabilistic logic programs to analyse further examples. Another idea is to extend the analysis including a decision theory approach to select a set of actions to perform in order to maximize a reward value. An interesting future work could be also to extend the probabilistic analysis to the blockchain technology, as done in [4], to model several scenarios, such as congestion of the network. Another future work is the design of a logic-based smart contract language or architecture, as suggested in [9] and [14] or probabilistic logic-based smart contract to directly allow probabilistic computation to be performed on the blockchain.

References

1. Alberti, M., Cota, G., Riguzzi, F., Zese, R.: Probabilistic logical inference on the web. In: Adorni, G., Cagnoni, S., Gori, M., Maratea, M. (eds.) AI*IA 2016 Advances in Artificial Intelligence. Lecture Notes in Computer Science, vol. 10037, pp. 351–363. Springer, Berlin (2016)
2. Atzei, N., Bartoletti, M., Cimoli, T.: A survey of attacks on Ethereum smart contracts (SoK). In: Maffei, M., Ryan, M. (eds.) POST 2017. LNCS, vol. 10204, pp. 164–186. Springer, Heidelberg (2017). https://doi.org/10.1007/978-3-662-54455-6_8
3. Azzolini, D., Riguzzi, F., Lamma, E.: Studying transaction fees in the Bitcoin blockchain with probabilistic logic programming. Information 10(11), 335 (2019)
4. Azzolini, D., Riguzzi, F., Lamma, E., Bellodi, E., Zese, R.: Modeling bitcoin protocols with probabilistic logic programming. In: Bellodi, E., Schrijvers, T. (eds.) Proceedings of the 5th International Workshop on Probabilistic Logic Programming, PLP 2018, Co-located with the 28th International Conference on Inductive Logic Programming (ILP 2018), Ferrara, Italy, 1 September 2018, CEUR Workshop Proceedings, vol. 2219, pp. 49–61. CEUR-WS.org (2018)
5. Azzolini, D., Riguzzi, F., Masotti, F., Lamma, E.: A comparison of MCMC sampling for probabilistic logic programming. In: Alviano, M., Greco, G., Scarcello, F. (eds.) AI*IA 2019. LNCS (LNAI), vol. 11946, pp. 18–29. Springer, Cham (2019). https://doi.org/10.1007/978-3-030-35166-3_2

---

[7] http://cplint.eu/.

6. Bartoletti, M., Carta, S., Cimoli, T., Saia, R.: Dissecting Ponzi schemes on Ethereum: identification, analysis, and impact. arXiv preprint arXiv:1703.03779 (2017)
7. Bartoletti, M., Pes, B., Serusi, S.: Data mining for detecting Bitcoin Ponzi schemes. In: Crypto Valley Conference on Blockchain Technology, CVCBT 2018, Zug, Switzerland, 20–22 June 2018, pp. 75–84. IEEE (2018)
8. Chatterjee, K., Goharshady, A.K., Velner, Y.: Quantitative analysis of smart contracts. In: Ahmed, A. (ed.) ESOP 2018. LNCS, vol. 10801, pp. 739–767. Springer, Cham (2018). https://doi.org/10.1007/978-3-319-89884-1_26
9. Ciatto, G., Calegari, R., Mariani, S., Denti, E., Omicini, A.: From the blockchain to logic programming and back: research perspectives. In: Cossentino, M., Sabatucci, L., Seidita, V. (eds.) Proceedings of the 19th Workshop from Objects to Agents, Palermo, Italy, 28–29 June 2018, CEUR Workshop Proceedings, vol. 2215, pp. 69–74. CEUR-WS.org (2018)
10. Darwiche, A., Marquis, P.: A knowledge compilation map. J. Artif. Intell. Res. **17**, 229–264 (2002)
11. De Raedt, L., Kimmig, A.: Probabilistic (logic) programming concepts. Mach. Learn. **100**(1), 5–47 (2015). https://doi.org/10.1007/s10994-015-5494-z
12. Haber, S., Stornetta, W.S.: How to time-stamp a digital document. In: Menezes, A.J., Vanstone, S.A. (eds.) CRYPTO 1990. LNCS, vol. 537, pp. 437–455. Springer, Heidelberg (1991). https://doi.org/10.1007/3-540-38424-3_32
13. Hirai, Y.: Defining the Ethereum virtual machine for interactive theorem provers. In: Brenner, M., et al. (eds.) FC 2017. LNCS, vol. 10323, pp. 520–535. Springer, Cham (2017). https://doi.org/10.1007/978-3-319-70278-0_33
14. Idelberger, F., Governatori, G., Riveret, R., Sartor, G.: Evaluation of logic-based smart contracts for blockchain systems. In: Alferes, J.J.J., Bertossi, L., Governatori, G., Fodor, P., Roman, D. (eds.) RuleML 2016. LNCS, vol. 9718, pp. 167–183. Springer, Cham (2016). https://doi.org/10.1007/978-3-319-42019-6_11
15. Kalra, S., Goel, S., Dhawan, M., Sharma, S.: ZEUS: analyzing safety of smart contracts. In: 25th Annual Network and Distributed System Security Symposium, NDSS 2018, San Diego, California, USA, 18–21 February 2018 (2018)
16. Koller, D., Friedman, N.: Probabilistic Graphical Models: Principles and Techniques. Adaptive Computation and Machine Learning. MIT Press, Cambridge (2009)
17. Luu, L., Chu, D.H., Olickel, H., Saxena, P., Hobor, A.: Making smart contracts smarter. In: Proceedings of the 2016 ACM SIGSAC Conference on Computer and Communications Security, pp. 254–269. ACM (2016)
18. Nakamoto, S.: Bitcoin: A peer-to-peer electronic cash system (2008)
19. Fadja, A.N., Riguzzi, F.: Probabilistic logic programming in action. In: Holzinger, A., Goebel, R., Ferri, M., Palade, V. (eds.) Towards Integrative Machine Learning and Knowledge Extraction. LNCS (LNAI), vol. 10344, pp. 89–116. Springer, Cham (2017). https://doi.org/10.1007/978-3-319-69775-8_5
20. Riguzzi, F.: MCINTYRE: a Monte Carlo system for probabilistic logic programming. Fund. Inform. **124**(4), 521–541 (2013)
21. Riguzzi, F.: The distribution semantics for normal programs with function symbols. Int. J. Approximate Reasoning **77**, 1–19 (2016)
22. Riguzzi, F.: Foundations of Probabilistic Logic Programming. River Publishers, Gistrup (2018)
23. Riguzzi, F., Bellodi, E., Lamma, E., Zese, R., Cota, G.: Probabilistic logic programming on the web. Softw. Pract. Experience **46**(10), 1381–1396 (2016)

24. Riguzzi, F., Swift, T.: The PITA system: tabling and answer subsumption for reasoning under uncertainty. Theor. Pract. Logic Program. **11**(4–5), 433–449 (2011)
25. Sato, T.: A statistical learning method for logic programs with distribution semantics. In: Sterling, L. (ed.) ICLP 1995, pp. 715–729. MIT Press (1995)
26. Szabo, N.: Smart contracts (1994)
27. Van Gelder, A., Ross, K.A., Schlipf, J.S.: The well-founded semantics for general logic programs. J. ACM **38**(3), 620–650 (1991)
28. Vasek, M., Moore, T.: Analyzing the Bitcoin Ponzi scheme ecosystem. In: Zohar, A., et al. (eds.) FC 2018. LNCS, vol. 10958, pp. 101–112. Springer, Heidelberg (2019). https://doi.org/10.1007/978-3-662-58820-8_8
29. Vennekens, J., Verbaeten, S., Bruynooghe, M.: Logic programs with annotated disjunctions. In: Demoen, B., Lifschitz, V. (eds.) ICLP 2004. LNCS, vol. 3132, pp. 431–445. Springer, Heidelberg (2004). https://doi.org/10.1007/978-3-540-27775-0_30
30. Wielemaker, J., Schrijvers, T., Triska, M., Lager, T.: SWI-Prolog. Theor. Pract. Logic Program. **12**(1–2), 67–96 (2012)
31. Wood, G.: Ethereum: A secure decentralised generalised transaction ledger. Ethereum Proj. Yellow Pap. **151**, 1–32 (2014)

Vulnerabilities and Excess Gas Consumption Analysis Within Ethereum-Based Smart Contracts for Electricity Market

Paulius Danielius[1] ⓘ, Piotr Stolarski[2]([✉]) ⓘ, and Saulius Masteika[1]

[1] Vilnius University Kaunas Faculty, Muitines St. 8, 44280 Kaunas, EU, Lithuania
{paulius.danielius,saulius.masteika}@knf.vu.lt
[2] Poznań University of Economics and Business, al. Niepodległości 10, 61-875 Poznań, Poland
piotr.stolarski@ue.poznan.pl

Abstract. Blockchain architecture can be used to implement highly decentralized electricity markets. The usage of smart contracts technology allows for fast, seamless, and reliable transaction settlement. Smart contract resistance to vulnerabilities is the key factor for functioning and ensuring the credibility of the power infrastructure.

In this paper, an analysis of some specific risks for smart contracting on Ethereum blockchain electricity markets is presented. A test set of 60 smart contracts is assembled and checked for vulnerabilities using three publicly available detection tools. Based on the results the model smart contract is developed and experimental simulations are conducted for estimating gas expenditure of basic operations characteristic to electricity auction smart contracts. The conclusions from the simulations lead to reassuring of a better quality of the smart contract's code.

Keywords: Ethereum · Electricity market · Smart contract · Vulnerability

1 Introduction

Issues related to the security of information systems are particularly significant for business environments due to the importance and value of resources controlled by these systems. They are also an important field of research for the scientific community. While the security of traditional information systems is a fairly well-developed matter, the emergence of new technologies, the use of which has not been fully tested, creates a further need for in-depth research and analysis. In particular, the presented research is focused on the Ethereum blockchain and smart contracts technologies employed to manage energy infrastructure.

In a recent couple of years, i.e. since the popularization of the idea of smart contracts, a whole series of works on the security of this type of software were produced. More and more potential vulnerabilities and threats are also known. The importance of this topic has increased when the first successful attacks started [1]. Those attacks made the public aware that the significant real value is at stake.

© Springer Nature Switzerland AG 2020
W. Abramowicz and G. Klein (Eds.): BIS 2020 Workshops, LNBIP 394, pp. 99–110, 2020.
https://doi.org/10.1007/978-3-030-61146-0_8

Much theoretical and practical work has been done since then to ensure increased safety. An example of practical actions include improvements in the EVM construction and changes to the Solidity language. Nevertheless, the research area remains open. New threats will certainly be discovered. This is due, for example, to the fact that the blockchain infrastructure and programming languages in which smart contracts are created are currently being intensively developed. It is also worth noting that new blockchain platforms offering the possibility of performing logical operations in conjunction with transactions, especially Turing-complete, are constantly emerging [2].

2 Research Workflow

The presented research is based on the three-factor approach. First, a critical literature review is done to identify potential threats in blockchain-based projects related to the electricity sector. Second, the evidence from the experimental analysis of smart contracts gathered and organized using third-party tools. The test set of selected contracts results from the first part of the study, their characteristics are described in detail in Sect. 5.1. Third, an analysis of excess gas consumption as certain vulnerability category of smart contracts in electricity sector applications is performed using a specially developed smart contract. The experimental setting is the effect of an analysis and conclusions drawn from the second part of the research.

The ordered steps of the study and their relation to the sections of the paper (numbers in brackets) are shown on Fig. 1.

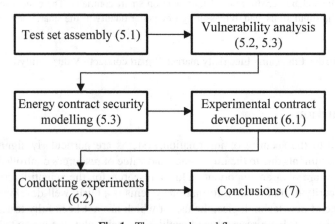

Fig. 1. The research workflow

3 Initiatives for Peer-to-Peer Electricity Markets

In the section a short summary of the idea and the most notable projects related to creation of a P2P (peer-to-peer) energy markets are given.

A smart grid is a utility grid that includes a variety of elements such as smart meters, smart appliances, multiple inputs, and output facilities with the ability to balance itself both at the micro and macro scale. The future of a utility grid involves the process of creating a hybrid with the original infrastructure. It means that the basic grid is enriched with the means of exchanging information between grid nodes, as well as the nodes themselves must have at least the functionality to monitor their internal state.

Early ideas of connecting the smart grids with blockchain infrastructure have emerged in 2016 and were developed a year later [3, 4]. Some small projects have been carried out so far, but scholars omit the impact of blockchain architecture on grid security. In our opinion, the research on the implication of blockchain security characteristics applied to the smart grid solutions is an important challenge.

A very extensive analysis of energy sector blockchain projects has been presented in [5]. Two related solutions are offered in [6] that are named Pando and Exergy. Pando enables utilities and retailers to introduce an energy marketplace. All the members of the community can trade energy. The exchange allows achieving of consumers' goals as well as maximizes community energy utility. The application is simply deployable, highly extensible, and flexible.

Another offering [7] of the blockchain-based open-source and customizable energy exchange. The company develops the D3E energy exchange engine which may be used by communities and other platforms. The exchange is designed with an emphasis on supporting environmental protection and green energy production.

3.1 Threats

There exist also a resemblance regarding the safety of energy applications is with the Internet of Things. Some of the home appliances that allow measuring or controlling power consumption are treated as specific IoT components. Two types of issues are crucial both for IoT and smart grids which are security and privacy issues [8].

4 Related Works

Currently, only a very limited number of works is devoted to the security of electricity sector-related blockchain applications. The early papers that deal with the issue of smart contracts security are dated back in the year 2016. Then, the first attempts to formally verify the soundness of the code were proposed [9]. Additionally, a survey on Ethrerum attacks has been conducted [10]. It was an important milestone as on the one hand, it raised the problem of the growing number of security incidents related to the network and on the other hand, it showed that the whole blockchain construction deals with the whole typology of potential vulnerabilities.

In 2017 another incident after DAO Hack took place. It was the Parity wallet security-hole. It was analyzed among others by [11]. The gas cost analysis of running smart contracts is made in [12]. The authors identify 7 gas-costly patterns and group them into 2 categories. They also propose and develop GASPER, a new tool for automatically locating gas-costly patterns by analyzing smart contracts' bytecodes. Their results show that some of the patterns are widespread within the smart contracts population. The issue of consuming gas is vital as it leads most of the developers to the introduction of the code neglecting the quality and security requirements.

In [13] blockchain is leveraged to a platform that facilitates trustworthy data provenance collection, verification, and management. The system utilizes smart contracts and is secure as long as the majority of the participants are honest. The last condition is the general requirement for any blockchain setting. The paper of [14] presents a novel way to permit miners and validators to execute smart contracts in parallel. It shows that a speedup of $1.33\times$ can be obtained for miners and $1.69\times$ for validators with just three concurrent threads. This kind of research introduces major innovation to smart contract engines. On the other hand, the decentralized and parallelized platforms are to be even more vulnerable. Especially, on the account of more complicated code and unpredictable effects.

The article [15] is a systematic study of the main topics addressed in conducted researches (metanalysis) that are related to smart contracts. It shows that there are four main streams of topics. These key issues are codifying, security, privacy, and performance issues. The texts that dealt with these groups of problems consisted of 67% of analyzed papers. This result is vital as it highlights the crucial aspects of smart contract safety.

The problem of information privacy in the context of blockchain and smart contracts is also tackled in the [16] text.

5   Vulnerabilities Analysis Results

In this section the analysis of available Solidity source code of projects created to work within the energy sector is presented. The general characteristics of the test set is given. The specific results of vulnerabilities detected are presented as well.

5.1   Test Set and Tools

The test set included 6 applications together with 60 smart contracts. These are all the thematically related projects identified on the basis of related works and projects review that were available as open-source. The overall number of lines of code is 4 978 and the source lines of code (SLOC) equivalent totals 4 121. All the source codes were generated between January 2018 and November 2019. The details about the test set items are presented in Table 1.

The work of [17] presented the survey of existing tools for supporting the development of secure smart contracts and analysis of smart contracts. As there are quite a number of different tools with various capabilities, the tools we have chosen were based on public availability, and the ability to detect differentiated security issues. They were positively assessed and have advantages that include a rich library of known vulnerabilities, threats categorization, as well as an associated knowledge base.

5.2   SmartCheck

In the preliminary research we acquired all analysis tools described in the literature. We rejected these that were outdated or out-of-order. Out of the three working tools the SmartCheck provides the most advanced results.

Table 1. Test set characteristics

Application Name	Latest version date	Number of files	Number of contracts	External libraries
Carbos	08/04/2019	6	6	0
SunContract	27/11/2019	2	8	0
Grid+	20/12/2018	7	7	0
GoGreenContract	19/02/2019	1	2	0
WePowerNetwork	28/01/2018	19	22	1
HivePower	18/05/2018	17	15	2
TOTAL		52	60	3

The SmartCheck [18] tool has been used to extensively test the smart contracts for potential threats. The tool is capable of indicating an impressive list of known vulnerabilities from numerous categories. It also warns about improper programming practices resulting in poor code quality. It is available online which makes the analysis a straightforward task. This tool is described in detail in [19].

The statistics obtained from the performed analysis are collected in Table 2. The table also shows the severity levels of the detected code malpractices. The total errors column shows the number of problems identified by the tool. The errors per SLOC is the result of our own calculations based on the total errors values. The severity categorization comes from the analysis tool with little downgrade adjustment of less critical malpractices.

Table 2. SmartCheck analysis results

Application Name	Total errors	Errors per SLOC	Severity 1	Severity 2	Severity 3
Carbos	22	0,06	21	1	0
SunContract	156	0,43	154	2	0
Grid+	216	0,38	216	0	0
GoGreenContract	13	0,10	13	0	0
WePowerNetwork	208	0,11	203	4	1
HivePower	90	0,11	89	0	1
TOTAL	705		696	7	2

In Fig. 2 the relations between the number of code lines and the number of detected malpractices are presented. Markers on the graph represent singled code files. The data reflects the information presented in Table 2. It is shown with more details because in the figure the files are the errors aggregation level instead of the whole projects. The graph reveals that frequently there is a visible linear relation between the length of the contract code and potential programmer mistakes. The inclinations of the lines are different for particular applications. It may reflect different backgrounds, experiences, and levels of programmers as well as expose their coding habits and style. These habits and style approaches may not always be best suited for the Solidity and the way of recommended contract construction.

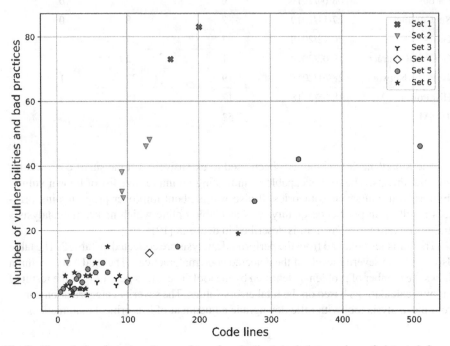

Fig. 2. The relation between the number of code lines and the number of detected flaws (SmartCheck)

The major part of detected issues, marked as low severity level (1), is related to bad coding practices and is easily remedied by following SmartCheck knowledge base recommendations.

The high (3) and middle (2) severity level vulnerability categories, as well as low-level issues (per SmartCheck) which are related to increased consumption of gas, as this is an important aspect for efficient electricity trading mechanism implementation, are presented in Fig. 3.

5.3   Other Tools and Conclusions from the Empirical Contracts Testing

Maian [20] is another publicly available tool that may be used to analyze the security of Solidity contracts. Unlike the SmartCheck it is not accessible online. Moreover, it seems that it is no longer maintained. It is able to recognize three types of code vulnerabilities. The three types of threats are related to the potential behavior of the contracts in the event of some extreme situations. These types are greediness, suicidality, and prodigality.

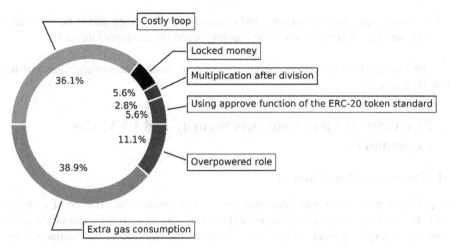

Fig. 3. Frequency of high-middle severity and low severity level vulnerabilities with increased gas consumption as detected per SmartCheck

As mentioned, the tool is no longer actively developed. Nevertheless, its code is open-sourced. As part of our study, we managed to fix the tool code and arrange the inconsistencies related to the changes in the Solidity as well as Ethereum Virtual Machine (EVM) revisions. Moreover, we extended the functionality to automatically detect the Solidity version in order to use the appropriate compiler (solc).

The results of the tool are uniform yet disappointing. Throughout the analysis of the whole set mentioned in Sect. 5.1 the behavior of only one contract has been identified as greedy (Grid +). No other vulnerabilities were discovered.

A similar situation is with the Oyente [21] tool. The trials with this tool brought no significant results.

We managed to make several important insights resulting from the analysis of results obtained during the process described in Sect. 5.2. The main points include:

- the catalog of statistically most significant vulnerabilities evolves due to common knowledge, EVM improvements, and programmers' maturity.
- As a consequence, the severity of the security vulnerabilities and programming errors change in time due to such factors as programmers' consciousness of the threats, general skills, and knowledge level as well as the development stage of the Ethereum network.
- Aside from solving the major vulnerability issues (marked as severity levels 2 and 3), which were not frequent cases, the other most relevant cases for improvement are with code using loops with arrays or mapping types and in consequence resulting in the unpredictable expenditure of gas.
- Contracts with loops do have another aspect of risk to consider, and this is especially important for P2P electricity trading in auctions: running such contracts potentially might cost huge amounts of gas, however, transactions with high gas expenditure are less likely to be picked by miners and included in a block, and this certainly could hinder smooth operation of an auction.

- In consequence, the provisioning and reception of the energy might be perturbed which may be dangerous both to the consumers and the infrastructure itself.

The identified aspects allow the construction of another experiment that will be described in Sect. 6.

6 Electricity Trading Contracts Security and EVM Gas Consumption

6.1 Experiment Description

A special smart contract was developed for the experiment. The creation of a model smart contract is a rational step as the experiment was to give the explanation for a general category of typical energy sector smart contracts. None of the contracts from the test set exemplified the expected properties of the whole category. Additionally, working on a specifically designed source code allows for greater flexibility in conducting experiments. The designed smart contract has several features to resemble the simplest operations of electricity market auction. This means looping through records with energy amounts and its prices provided for trading and updating energy amount values. Cryptocurrency transactions were not considered.

Although loops are generally advised to be avoided in smart contract code altogether, practical auction implementation is hardly achievable without using them. Besides, such design allows measuring gas consumption with a certain amount of iterations and estimate the influence of recommended fixes for two vulnerabilities "extra gas consumption" and "costly loop" on gas cost. This, in turn, allows comparing obtained values with average transaction gas expenditure with the aim to estimate how many participant records could be operated upon with reasonable gas consumption.

The average gas cost per transaction was calculated based on *bitinfo.com* real-time data (Table 3) and block gas limit on the Ethereum network which is very close to 10 000 000.

Table 3. Average gas expenditure per transaction on the Ethereum network (bitinfocharts.com, 2020-04-28).

Transactions avg. per hour	Blocks avg. per hour	Transactions avg. per block	Gas avg. per transaction
35 505	271	131	76 327

The following data structures were used for keeping market participant data: array of participant addresses which will be used to iterate through all participant records and mapping from addresses to *struct* type elements with energy amount and energy price values. Test data was generated for a number of participant records using integer numbers as a basis for conversion to address type and for *amount* and *price* values.

6.2   Obtained Results

Within the experiment, 2 types of test were conducted:

1. Estimating gas expenditure of reading operations, where 3 values are read for each participant in each loop iteration: one value from address array and two values, *amount* and *price*, from corresponding *struct* elements.
2. Same as above with the addition of updating *amount* value. Iteration count was chosen for reaching gas expenditure close to *average transaction gas* expenditure. A transaction basis fee of 21 000 gas is included in each measurement.

The test results are presented in Fig. 4 and Fig. 5.

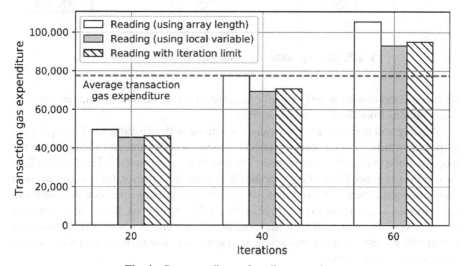

Fig. 4. Gas expenditure of reading operations

Figure 4 Displays the gas expenditure of reading operations. Three variants of stopping the loop were used:

1. based on array length value – bad practice and "extra gas consumption" category by *SmartCheck*,
2. based on local variable value to which array length value was copied (recommendation of how to improve over the first variant),
3. iteration limit set through function parameter as additional loop end condition; limiting loop iterations to known number is the recommendation of how to safeguard against the "costly loop" vulnerability category by *SmartCheck*.

It is evident from the obtained results, that checking of local variable value in the loop condition does decrease gas expenditure as compared to checking array length value, and saving accumulates with more iterations. Checking for iteration limit as an

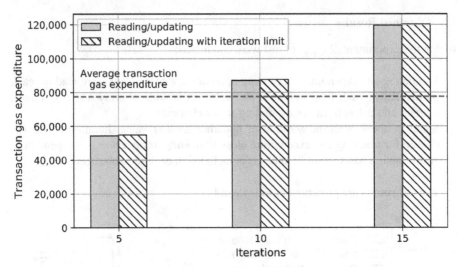

Fig. 5. Gas expenditure of reading and updating operations

additional loop condition only slightly increases gas expenditure as compared to less costly options from the first and the second.

Reading and updating records' values was done with two variants of stopping the loop: known number of iterations by checking local variable value and adding additional loop condition to limit iteration number. The results are represented in Fig. 5. The difference in the gas expenditure of both variants is negligible. However, for reaching average transaction gas expenditure, less than 10 reading and updating operation iterations can be performed as compared to performing only reading operations. In case there is a need for more operations in each loop iteration, the gas expenditure would be accordingly higher which means average transaction gas expenditure would be reached with an even lower number of iterations.

7 Conclusions

Checking for smart contract vulnerabilities and risks is not a trivial task, because there are few publicly available tools and most of them have very limited or very specific capabilities.

In this work, we explored three tools, SmartCheck, Oyente, and Mayan. By testing 60 smart contracts of 6 projects dedicated to the electricity sector, 706 vulnerability issues, and coding malpractices were found, 705 of them were detected by SmartCheck, 1 by Mayan, while Oyente bore no results. In addition, SmartCheck provided a three-level severity classification of flaws. As the major part of issues were low level, easily fixable, and were more as bad coding practice than the actual threat they were not included for further consideration. The results allowed creating a model contract source code that was used to estimate gas usage of two cases from low severity level (as categorized by SmartCheck), which concern high gas expenditure.

Developers should be aware of gas usage when designing smart contracts for Ethereum platform with special care when writing the code, where using loops is inevitable, and follow the recommended practice, because not having control of loop iteration number may result in gas overuse.

Using smart contracts for power auction trading is limited on the Ethereum blockchain network in terms of a feasible number of participating members. From experiment results, we can conclude, that electricity auction would be more suitable for trading between a small number of agents (~10) than between prosumers, whose number would be certainly far greater, in which case transaction gas expenditure can reach levels which may not be attractive for miners to include in their blocks. Consequently, auction performance may not be in line with expectancy, and some more costly in terms of gas usage transactions may stall leading to unpredictable negative effects in the grid. Another important aspect is that the number of independent electricity providers for EU counties is rising due to the liberalization of the electricity market. For example, in Lithuania, it is 4 (households) and 11 (B2B) operators [22]. The numbers correspond well to the mentioned above number of agents.

Once the Ethereum blockchain will change the consensus protocol from Proof of Work to Proof of Stake, the effective number of electricity market members could change significantly, and new evaluation will be needed.

References

1. Santos, F., Kostakis, V.: The DAO: a million dollar lesson in blockchain governance. In: School of Business and Governance, Ragnar Nurkse Department of Innovation and Governance (2018). https://digi.lib.ttu.ee/i/?9460
2. Neo: Neo Cryptocurrency (2017). https://neo.org/
3. Mannaro, K., Pinna, A., Marchesi, M.: Crypto-trading: Blockchain-oriented energy market. In: 2017 AEIT International Annual Conference, p. 1–5 (2017)
4. Mengelkamp, E., Notheisen, B., Beer, C., Dauer, D., Weinhardt, C.: A blockchain-based smart grid: towards sustainable local energy markets. Comput. Sci. Res. Dev. **33**, 207–214 (2018)
5. Pichler, M., Meisel, M., Goranovic, A., Leonhartsberger, K., Lettner, G., Chasparis, G., Vallant, H., Marksteiner, S., Bieser, H.: Decentralized energy networks based on blockchain: background, overview and concept discussion. In: Abramowicz, W., Paschke, A. (eds.) BIS 2018. LNBIP, vol. 339, pp. 244–257. Springer, Cham (2019). https://doi.org/10.1007/978-3-030-04849-5_22
6. Lo3energy: Lo3energy (2020). https://lo3energy.com/
7. Grid Singularity, Grid Singularity (2020). https://gridsingularity.com/
8. Eze, K.G., Akujuobi, C.M., Sadiku, M.N.O., Chouikha, M., Alam, S.: Internet of things and blockchain integration: use cases and implementation challenges. In: International Conference on Business Information Systems, p. 287–298. Springer, Heidelberg (2019)
9. Bhargavan, K.: et al.: Formal verification of smart contracts: short paper. In: Proceedings of the 2016 ACM Workshop on Programming Languages and Analysis for Security, p. 91–96 (2016)
10. Atzei, N., Bartoletti, M., Cimoli, T.: A survey of attacks on Ethereum smart contracts," IACR Cryptology ePrint Archive, vol. 2016, p. 1007 (2016)
11. Qureshi, H.: A hacker stole 31 M of Ether-how it happened, and what it means for Ethereum. In: Freecodecamp.org (2017)

12. Chen, T., Li, X., Luo, X., Zhang, X.: Under-optimized smart contracts devour your money. In: 2017 IEEE 24th International Conference on Software Analysis, Evolution and Reengineering (SANER), p. 442–446 (2017)

13. Ramachandran, A., Kantarcioglu, D.: Using blockchain and smart contracts for secure data provenance management. arXiv preprint arXiv:1709.10000 (2017)

14. Dickerson, T., Gazzillo, P., Herlihy, M., Koskinen, E.: Adding concurrency to smart contracts. In: Proceedings of the ACM Symposium on Principles of Distributed Computing, p. 303–312 (2017)

15. Alharby, M., van Moorsel, A.: A systematic mapping study on current research topics in smart contracts. Int. J. Comput. Sci. Inf. Technol. **9**, 151–164 (2017)

16. McCorry, P., Shahandashti, S.F., Hao, F.: A smart contract for boardroom voting with maximum voter privacy. In: International Conference on Financial Cryptography and Data Security, pp. 357–375 (2017)

17. Salzer, G., Di Angelo, A.: Survey of tools for analyzing ethereum smart contracts gernot salzer. In: 2019 IEEE International Conference on Decentralized Applications and Infrastructures (DAPPCON), pp. 69–78 (2019)

18. SmartCheck, SmartCheck (2020). https://tool.smartdec.net/

19. Tikhomirov, S., Voskre-senskaya, E., Ivanitskiy, I., Takhaviev, R., Marchenko, E., Alexandrov, E.: SmartCheck: static analysis of ethereum smart contracts. In: 2018 IEEE/ACM 1st International Work-shop on Emerging Trends in Software Engineering for Blockchain (WETSEB), pp. 9–16 (2018)

20. Nikolic, I., Kolluri, A., Sergey, I., Saxena, P., Hobor, A.: Finding the greedy, prodigal, and suicidal contracts at scale. In: Proceedings of the 34th Annual Computer Security Applications Conference, pp. 653–663 (2018)

21. Luu, L., Chu, D.-H., Olickel, H., Saxena, P., Hobor, A.: Making Smart Contracts Smarter. IACR Cryptology ePrint Archive **2016**, 633 (2016)

22. Eso, Lithuania Power Market Outlook, http://www.eso.lt/ (2020)

Central Banks Digital Currency: Issuing and Integration Scenarios in the Monetary and Payment System

Dmitry Kochergin[1] and Victor Dostov[1,2]

[1] Saint-Petersburg State University, Universitetskaya Emb. 7-9, 199034 St Petersburg, Russia
`leonova@npaed.ru`
[2] Russian Electronic Money and Remittance Association, Orlikov Per. 5/2, 107078 Moscow, Russia

Abstract. The paper reviews different approaches to implementation of the central bank digital currencies (CBDC). Retail as well as wholesale CBDC are reviewed. Implications for the domestic and cross-border usage are analyzed. Based on the theoretical assumptions and practical cases available so far, different implementation models are classified and graded. The analysis shows that different CBDC implementation models for might require different level of central bank involvement and various impact on monetary policy. It also indicates that although significant adaptation of governance and technical procedures might be required, retail and wholesale CBDC may increase efficiency of the current payment systems, ensure higher transparency and availability of payments. However, for jurisdictions with limited payment infrastructure the costs of implementing CBDC might be prohibitive.

Keywords: CBDC · Central bank digital currency · Payments · Correspondent banking · Monetary policy

1 Introduction

In recent years, digital technologies are increasingly used in financial markets around the world. The emergence of virtual currencies, stablecoins, digital tokens that act as payment medium, equity or debt instruments based on the distributed ledger technology (distributed ledger technology - DLT) created a new class of assets: crypto-assets or digital financial assets (virtual assets). This made monetary regulators face two major problems. First, they need to develop an optimal regulatory approach to these new digital assets; secondly, there is an aspiration to use DLT by the central banks, to maintain control over the money supply and also improve the efficiency and integrity of the monetary and payment systems.

Currently, there are many ways how central banks can use DLT: from optimizing settlements on the securities market to issuing bonds and managing their circulation; however, one of the key areas of DLT implementation is issuing digital currencies (central

© Springer Nature Switzerland AG 2020
W. Abramowicz and G. Klein (Eds.): BIS 2020 Workshops, LNBIP 394, pp. 111–119, 2020.
https://doi.org/10.1007/978-3-030-61146-0_9

bank digital currencies - CBDC) and using them for national payment systems and cross-border payments.

CBDCs are interesting not only because of the new ways to improve the effectiveness of existing monetary and payment systems, but because of the concerns of central banks and international financial institutions such as International Monetary Fund, Bank for international settlements and others regarding the stability of the national monetary systems and the future of central bank money.

That is why it is crucial to assess relevance of issuing central bank digital currencies, understand their forms, and economic implications. The purpose of this article is to determine the possible forms of central bank digital currencies and identify scenarios of their integration in modern monetary and payment system. This paper is structured as follows: first, we classify and identify key features of digital currencies, secondly we explore how CBDC can be integrated in the modern monetary system and their economic effects, thirdly, we make a comparative analysis of different CBDC models.

2 Interpretation of Digital Currencies and Their Classification

There is no universally accepted definition of central banks digital currencies due to the different concepts they are based on. In a general sense, central bank digital currency can be defined as electronic obligation of a central bank, expressed in the national accounts unit and used as a medium of exchange and a mean of storing value. Central bank digital currencies should be viewed as a new form of central bank money different from cash or money on reserve and settlement accounts at the central bank. Digital currencies are also different from deposits in credit institutions, although both are stored in electronic form. Moreover, although the same technologies can be used, CBDCs are different from virtual currencies.

Central bank digital currencies have a lot more in common with cash or reserve balances of central bank than bank deposits or virtual currencies. The key features are: 1) status of issuer as a centralized regulator and last-line lender; 2) the ability of an issuer to regulate procedures and conditions of issuing digital currency; 3) role of an issuer in controlling volume of digital currency. Thus, digital currencies should be identified as a new form of central banks money, which may take middle position between other forms of central banks money, because digital currency can be as universally accepted as cash and, at the same time, be stored in electronic form as money balances on reserve and settlement accounts in central bank.

Currently, there are two basic approaches to issuing central bank digital currency:

- for retail general payments – (R-CBDC);
- for wholesale payments – (W-CBDC).

Technologically, digital currencies can be token-based or account-based. A key difference between them is the form of authentication required for initiating a transaction. Token-based CBDC would require that the payee verifies the validity of a token. This should prevent parties from falsification and double spending. On the other hand, account-based CBDC require identification and authentication of the account holder identity. Account-based CBDC system needs to be protected from personal data theft, because access to these data would allow attackers to obtain unrestricted access to funds as well.

3 Key Characteristics of Digital Currencies

Issuing technology. Most virtual currencies are issued and transferred over open blockchain. Such tokens are issued in a peer-to-peer networks that have no clearly identifiable issuer. New units of virtual currency are usually issued through mining, forging, or other method. However, in case of retail central bank digital currencies, private blockchain is more preferable, as this would allow central bank to control process of issuing and manage money supply. Therefore, central bank will have to choose technology that would balance ability to control monetary supply, maintaining an acceptable level of anonymity and ensuring a low transactional cost for its users.

Method of storing currency. In most cases, when digital currencies are issued for retail payments in form of digital tokens, personal electronic wallets act as a storage device. However, funds can also be stored in the accounts opened directly in a central bank. In case of wholesale digital currencies, they can be stored as tokens in accounts opened by the banks in a central bank.

Degree of anonymity. Token-based CBDCs can provide different degrees of anonymity similar to virtual currencies. Key challenge is defining acceptable level of anonymity that would balance risks of money laundering, terrorist financing and user privacy, which is not an easy task [2, pp. 126–148].

Settlement mechanism. Cash is transferred directly from one person to another, the settlement is final when cash changes hands. Funds stored in accounts at a central bank are transferred via central bank which acts as an intermediary, transfer is final only after the funds are transferred to the payee account. In this regard, retail central bank digital currency will most likely, also have to provide for direct transfer analogous to cash. The only difference is that such digital currency will be transferred using electronic value storage devices. In case of wholesale central bank digital currencies, they will most likely be transferred through direct intermediation of central bank or commercial banks.

Method of integration in monetary system. Digital currencies for retail payments can be implemented in a monetary system in three ways. First, as a replacement for cash (digital currencies will replace cash as a legal means of payment). Secondly, as an addition to cash while cash remains in circulation (competition with non-cash payment systems). Third, as a monetary form having parallel circulation on par with cash (competition with deposits in commercial banks) [8, pp. 157–158]. Wholesale digital currencies settlements can be integrated with the existing payment system in two ways. Either in parallel with traditional cross-border RTGS systems, based on SWIFT and CLS currency settlements or as a substitute for traditional settlements based on correspondent bank accounts.

Access to funds. Currently, access to traditional forms of central bank money, excluding cash, is limited to the operational hours of central bank payment system and RTGS systems. In case of retail digital currency central banks will need to provide 24/7/365 access to such funds, while for wholesale digital currency, access can be provided at certain hours (for example, during hours of operation of wholesale payment systems).

Interest payments. By analogy with other forms of central bank money, technically, central banks can provide for payment of interest both on digital tokens and on funds in accounts. A positive interest rate (for example, applying a fixed percentage below the key rate or differentiating interest depending on the amount of funds in account) on central bank digital currency can help build the demand for digital currency, especially at the stage of their initial integration into monetary system, and stimulate competition with deposits of credit institutions [3, p. 6].

4 Scenarios for Integrating R - CBDC into a Current Monetary System

Based on the analysis of different approaches to issuing digital currencies, three main ways to integrate R-CBDC into modern monetary and payment systems can be identified: 1) as substitute for cash; 2) as supplement to cash; 3) in parallel with cash (Table 1).

Table 1. Digital currency integration scenarios for retail payments

The method of integration	Description of integration scenario	Advantages of digital currency	Impact on the monetary system	Impact on the central bank monetary policy
1. Replacement of cash (competition and substitution of cash in circulation)	Transition from cash to central bank digital currency	Ease of usage and possible anonymity in payments	CBDC is included in M0 aggregate	Insignificant
2. Addition to cash (competition with payment systems)	Outflow of funds from current accounts to central bank digital currency	Ease of usage and increased stability of payment systems	Possible effect on the structure of components in the M1 aggregate	Significant: increased role of central bank in payment systems market
3. Parallel circulation with cash (competition with deposits in commercial banks)	Outflow of funds from deposits to central bank digital currency	Ease of usage when paying, as well as the possible accrual of interest	Possible impact on both the structure and volume of the components in the M1 and M2 aggregates	Significant: change in liabilities of central bank and commercial banks

Source: compiled by the authors on: [6, p. 16]; [7, p. 94]

As seen from Table 1, depending on the form and method of integrating digital currencies into monetary system, digital currencies can have different effects on the activities of a central bank. In case of cash replacement, effect on central bank monetary policy will not be significant, as this will only result in replacement of general component of M0. A more significant effect will be observed when digital currencies will be issued in addition to cash or have parallel circulation. This is the case of e-yuan which is being piloted in some Chinese regions since May 2020: based on the reports so far, this is a token-based R-CBDC that will be distributed via large banks. E-yuan will also support offline transactions, thus acting as a cash substitute in retail transactions [5]. However, technical features of the project might change during the pilot, therefore, it is too early to make any conclusions on e-yuan effects on the national monetary system.

5   W-CBDC Integration Scenarios into a Modern Payment System

W-CBDC integration scenarios into a modern payment system are the most complicated from a technological point of view. While digital currencies for retail payments are designed primarily to replace cash settlements, wholesale digital currencies can potentially replace traditional mechanisms of cross-border settlements through correspondent accounts as well as RTGS-systems based on SWIFT messages and currency CLS settlements (Table 2).

Table 2. W-CBDC features: governance structure

1. Governance structure	
Model 1. W-CBDC	• Central governance body will be required to manage the W-CBDC system
Model 2. W-CBDC	• Management challenges will be proportionate to the number of participants and coordination costs
Model 3. W-CBDC (U-W-CBDC)	• Two governance bodies will be required: first, for the U-W-CBDC conversion and for W-CBDC; • Coordination between all the stakeholders will be required

Sources: based on: [4, 9].

There are 5 main functional problems of cross-border payments that can steer to implementation of W-CBDC: 1) different operating hours of RTGS systems and banks in different jurisdictions and time zones; 2) relying on numerous intermediaries (with corresponding costs) for cross-border payments and settlements; 3) lack of unified, agreed payment standards (technical and operational) and regulatory requirements in different jurisdictions; 4) lack of a standardized notification features about status of payments in general payment messaging network used by banks; 5) problems associated with outdated payment infrastructure in networks, central banks and commercial banks.

Introduction of W-CBDC may solve problems of a centralized settlement infrastructure based on RTGS and currency CLS settlements [1, p. 53–65]. W-CBDC systems can make wholesale payments more accessible and traceable, and integration of new messaging standards in digital currency system can lead to large-scale changes in settlement systems. However, these problems cannot be solved by W-CBDC alone. As mentioned in research papers, W-CBDC is likely to be implemented un parallel with traditional mechanisms of RTGS-based interstate settlements. Hence, there are three general models of W-CBDC systems, each of which uses token-based form of central bank obligations to carry out wholesale interbank payments and settlement operations at international level (Table 3).

Table 3. W-CBDC features: role of the central banks

1. Role of the central banks	
Model 1. W-CBDC	• Central banks will keep their position by providing settlement services within the jurisdiction • Scope of central bank powers will need to include setting the rules and criteria for the financial institutions willing to use W-CBDC • As the ecosystem grows, central banks might need to perform customer due diligence for the institutions that want to join the system, thus leading to potential regulatory risks
Model 2. W-CBDC	• Central banks may lose some control over W-CBDC, as compared to traditional forms of money, because W-CBDC can be used in multiple jurisdictions • Scope of central bank powers will need to include setting the rules and criteria for the financial institutions willing to use W-CBDC • As the ecosystem grows, central banks might need to perform customer due diligence for the institutions that want to join the system, thus leading to potential regulatory risks • If central banks can trace the issuer of the W-CBDC, they can also assess the volume of payments made within their jurisdictions
Model 3. W-CBDC (U-W-CBDC)	• Since the price of U-W-CBDC depends on the currency basket, central banks may need to be more active in managing exchange rates • Scope of central bank powers will need to include setting the rules and criteria for the financial institutions willing to use W-CBDC • Central banks will need to look into the management and oversight over the U-W-CBDC exchange house • This model provides central banks with more information to analyze the transactions made in a particular jurisdiction

Sources: based on: [4, 9].

5.1 Model 1. System with Non-convertible W-CBDC

Each central bank issues its own digital currency W-CBDC. Digital currencies are issued to participating banks in exchange for funds in national currency. Central banks A and B enter into an agreement that enables participating banks from the same jurisdiction to maintain a W-CBDC account (digital wallet) in a central bank of other jurisdiction denominated in currency of that jurisdiction. Other intermediary banks (for example, bank C1) may also support W-CBDC digital wallets in each jurisdiction. It will be like token-based version of correspondent banking, where usage of such tokens will allow central banks to keep full control over money in their jurisdiction (Table 4).

5.2 Model 2. System with Convertible W-CBDC

Central banks A and B sign an agreement that allows participating banks in both countries to exchange W-CBDCs issued by central banks of both countries (i.e. W-CBDC tokens issued by central bank A (W-CBDC-A) can be transferred to the banks in country B, and W-CBDC tokens, issued by central bank B (W-CBDC-B) can keep banks in country

Table 4. W-CBDC features: effects on monetary policy and financial markets

1. Effects on monetary policy and financial markets	
Model 1. W-CBDC	• The role of the central bank will expand, as now they need to manage the supply in W-CBDC, just as they manage the supply in cash and non-cash monies • Intermediary banks that conduct swap transactions in W-CBDC between the payers and the payees may establish exchange rates that will make the market makers
Model 2. W-CBDC *Model 3. (U-W-CBDC)*	The role of the central bank will expand, as now they need to manage the supply in W-CBDC, just as they manage the supply in cash and non-cash monies

Sources: based on: [4, 9].

A). Each participating bank supports W-CBDC accounts (or digital wallets) for different currencies in a central bank of its jurisdiction to allow receiving digital currencies and payments in various W-CBDCs in cross-border transactions with other banks.

Conversion of CBDC can be implemented through digital currencies exchange market. This W-CBDC model can operate 24/7 in parallel with RTGS for interstate and cross-border transactions.

5.3   Model 3. W-CBDC with Universal Digital Currency

Some jurisdictions or central banks agree to issue universal digital currency (U-W-CBDC). This universal currency can be backed by the currency basket issued by the respective central banks. Universal digital currency can be issued by the special purpose exchange house, established specifically to issue and buy this universal currency. Conversion of U-W-CBDC will be carried at the exchange rate established by the participating banks.

Third model seems to be most appropriate to address the challenges in cross-border settlements mentioned above. However, large-scale technological, governance and financial changes will be required to implement it. Due to the friction between the participating countries as well as the scale of changes, implementation of U-W-CBDC might take a lot of time.

Table 5 shows scale of changes that might be associated with different types of W-CBDC.

Table 5. W-CBDC features: scale of changes

1. Scale of changes	
Model 1. W-CBDC	Implementation of this model is more complex than any modification of the RTGS system, as new W-CBDC will need to be created. However, less resources will be required than in Models 2 and 3, as digital currency is used only for domestic transactions
Model 2. W-CBDC	This model is more complex than Model 1, as it will require creation of new system for W-CBDC exchange on the international levels. Central bank will need to track domestic W-CBDC transactions, cross-border issuing and redemption
Model 3. W-CBDC (U-W-CBDC)	This model is the most resource-consuming, as not only domestic changes will be required but also creation of the exchange house for U-W-CBDC as well as involvement of multiple jurisdictions

Sources: based on: [4, 9].

6 Conclusions

Based on the analysis, the following conclusions can be made.

1) Introduction of CBDCs may lead to the emergence of new forms of money, different from cash or cashless monies. CBDCs can be used for retail and wholesale payments. They can be token-based or account-based. The principal features of the CBDC are: issuing method, storage method, degree of anonymity, settlement mechanism, approach to integration into monetary system, availability of funds, accrual of interest.

2) Introduction of digital currencies both for retail and wholesale payments may have some benefits for payment and clearing systems. CBDC may provide individuals with high liquidity, low-risk, convenient and universally accepted mean of payment (R-CBDC), for the corporate users W-CBDC may provide for faster, transparent and cheap transfers. However, some central banks, including in South Korea, Japan and others indicate potential shrinkage of liquidity in a money and stock market, as well as cyber-risks associated with this innovation. That is why implementation decisions should take into account comparative value of CBDCs: for R-CBDC with instant payment systems, for W-CBDC – with RTGS and correspondent systems.

3) We have also found that the effects of the CBDCs on the monetary and payment system depend significantly on the approach chosen for their integration. For R-CBDC the biggest effect can be achieved in case of parallel circulation with traditional cash: here the demand for cash may subside, as well as transmission effects of the central bank policies. Introduction of CBDCs may also increase competition between the banks.

4) In case of W-CBDC, the most significant effect can be achieved by substituting current RTGS and correspondent cross-border payment systems with the universal CBDC. We believe that W-CBDC can provide regular 24/7 service, minimize number of intermediaries, modify payment infrastructure and allow for more coherent

set of payment standards. However, costs for the countries with limited payment infrastructure might be prohibitive.

5) In the beginning of 2020 around 80 banks were considering CBDCs. Yet, there is no operational CBDC system still. From what can be seen, e-Yuan (DCEP) in China and e-krona in Sweden are the most advanced R-CBDC systems to date, while CAD coin (Canada), e-SGD (Singapore) are good examples of interoperable cross-border W-CBDC.

6) At this point it does not seem likely that CBDC will completely substitute current settlement systems. Therefore, the scenario where two payment systems coexist and support each other seems more realistic.

References

1. Bech, M., Faruqui, U., Shirakami, T.: Payments without borders. BIS Q. Rev. March, 53–65 (2020)
2. Camera, G.: A perspective on electronic alternatives to traditional currencies. Sveriges Riksbank Econ. Rev. **1**, 126–148 (2017)
3. Coeuré, B., Loh, J.: Central Bank Digital Currencies. Committee on Payments and Market Infrastructures BIS Report, p. 34 (2018)
4. Cross-border Interbank Payments and Settlements: Emerging Opportunities for Digital Transformation. Bank of England, Bank of Canada, Monetary Authority of Singapore (November 2018). https://www.mas.gov.sg/-/media/MAS/ProjectUbin/Cross-Border-Int erbank-Payments-and-Settlements.pdf. Accessed 20 Mar 2020
5. DCEP Whitepaper Allryptowhitepapers (2020). https://www.allcryptowhitepapers.com/dcep-whitepaper. Accessed 20 Mar 2020
6. Kiselev, A.: Whether there is a future for digital currencies of central banks. Analytical note of the Bank of Russia, vol. 23 (2019). https://cbr.ru/content/document/file/71328/analytic_note_1 90418_dip.pdf. Accessed 03 Mar 2020
7. Kochergin, D., Yangirova, A.: Central bank digital currencies: key characteristics and directions of influence on the monetary and payment systems. Financ.: Theory Pract. **4**(112), 80–98 (2019)
8. Kochergin, D., Yangirova, A.: Digital currencies as a new form of central banks money. ECO **10**(544), 148–171 (2019)
9. STELLA - Synchronised Cross-Border Payments. European central bank, bank of Japan (June 2019). https://www.ecb.europa.eu/paym/intro/publications/pdf/ecb.miptopical190604.en.pdf. Accessed 18 Mar 2020

A Blacklisting Smart Contract

Byron Kruger and Wai Sze Leung(✉)

University of Johannesburg, PO Box 524, Johannesburg, South Africa
216010311@student.uj.ac.za, wsleung@uj.ac.za

Abstract. The traditional financial system has the ability to facilitate forensic investigations into crimes relating to financial fraud. The same cannot be said about the blockchain-based financial system that has come to exist over the past few years. This financial system does not tolerate third parties impeding transactions and so presents the problem of not being able to facilitate investigations. The concept of a blacklist is considered as a potential solution to this issue, specifically, a blockchain-centric blacklist. This paper describes the implementation and evaluation of such a blacklist as a potential solution for facilitating investigations involving blockchain-based financial systems.

Keywords: Blockchain · Blacklist · Decentralization · Financial crime

1 Introduction

With the advent of Bitcoin and the technology it utilizes, namely blockchain, the world has seen a shift in the way finance could be conducted across the Internet. Blockchain technology allows for the transaction of money (in the form of a cryptocurrency such as Bitcoin) from one peer to another without any third party governing such transactions. Cryptocurrencies (and the blockchain in general) are considered to be a fully-inclusive financial infrastructure, accessible to all. With the belief that approximately 10% of the global gross domestic product will likely be stored on the blockchain by 2027 [1], it is to be expected that transactions (such as cashless payments) supported by blockchain technology will achieve omnipresence.

The inclusive nature of blockchain represents an evolutionary step that provides organizations with vast potential opportunities. While such a technology offers an open infrastructure to conduct business, it is unfortunately not without negative aspects as it can also empower bad actors to perform malicious activities, due to the little to no authoritative figures (such as governments) present to oversee matters on a blockchain. This problem leads to the research question, namely whether a blacklist of blockchain addresses can be implemented in order to prevent the transactions of illegally-acquired cryptocurrencies in a decentralized way.

This research benefitted, in part, from support from the Faculty of Science at the University of Johannesburg.

W. Abramowicz and G. Klein (Eds.): BIS 2020 Workshops, LNBIP 394, pp. 120–131, 2020.
https://doi.org/10.1007/978-3-030-61146-0_10

This paper seeks to examine the challenges that the blockchain pose and proposes a solution for when investigators and law enforcement are tasked with conducting investigations that involve criminal activities conducted on the blockchain. The content of this paper is as follows: Sect. 2 will discuss the background of blockchain technology and the hurdles that have come to exist when conducting forensic investigations. Section 3 explores existing literature on the topic, leading to an outline of the model of the proposed solution, in Sect. 4. Section 5 discusses the specific implementation details of the proposed solution, Sect. 6 goes over the testing procedure in order to determine whether the solution indeed solves the problem and Sect. 7 will discuss the results obtained from testing. Finally the paper finishes with a concluding section in Sect. 8.

2   Background

A blockchain is a sequence of data in the form of blocks and is stored on all network nodes. This sequence provides the timeline of transactions in a chain-like form. All blocks are protected from modification by means of cryptography. More specifically, each block contains a timestamp and the hash value of the previous block, allowing for verification by the nodes in the peer-to-peer network. The appending of new blocks can only be done if a consensus is reached among the network nodes about the correct state of the blockchain. This ensures that all network nodes have the same correct copy of the blockchain; this is at the heart of what gives blockchain technology value [2].

Anyone wishing to utilize the blockchain and cryptocurrencies can do so via software like DApps (distributed applications), exchanges and wallet software. A distributed application is merely software run on top of a blockchain. Like most conventional web applications, they utilize services such as back-end software (smart contracts which are programs residing on the blockchain and govern business logic), front-end software, data storage and message communication. DApps can be implemented in different degrees of decentralization. Utilizing as much of the blockchain as possible will make a given DApp more decentralized.

Exchanges are the gateway between traditional currencies and cryptocurrencies as they facilitate the trading of traditional currencies for cryptocurrencies and vice versa. Exchanges come in two main forms, namely centralized exchanges and decentralized exchanges. Centralized exchanges are traditional web applications that host numerous cryptocurrency wallets and facilitate trades between users. Decentralized exchanges, on the other hand, only use traditional web technologies for UX/UI components of a given exchange and all other components are accessed from the blockchain.

Wallets allow for the storage of cryptocurrencies either by using online wallets (connected to the Internet), desktop wallets or cold storage wallets (completely offline wallets and considered to be the most secure way to store cryptocurrencies).

It is imperative to distinguish pure, public and permissionless blockchains from pseudo-blockchains (also known as private and permissioned blockchains).

The focus of this paper is on permissionless blockchains, since this type of blockchain boasts the most innovation, makes finance all inclusive and consequently poses the unique challenge to forensic investigators and law enforcement. A permissionless blockchain is a truly peer-to-peer network, in the sense that participating nodes are all equal, there are no unfair advantages and that all nodes share the cost of providing network services in return for a mutual benefit [3]. This equality and openness to participate is a characteristic only inherent in permissionless blockchains and is referred to as decentralization. The higher the level of decentralization the more difficult it is to govern a blockchain according to one or a small amount of individuals. On the other hand, permissioned blockchains are ones managed by several organizations, each of which manage and maintain one or more network nodes. On these blockchains, participants join the network by means of authorization and form a shareholder alliance with the mission to cooperatively oversee the operations of the blockchain. Only these organizations are allowed to read, write, send and record transactions. These blockchains also have the ability to support supervision [4]. In essence, permissioned blockchains do not provide any advantages over conventional databases.

In traditional finance, financial institutions, which serve as a middle-man between parties involved in a transaction, would cooperate with investigations into allegations of financial fraud. A bank, for example, can freeze an account suspected of illicit activities (such as money laundering) [5]. Similarly, permissioned blockchains also have the ability to cooperate with investigations into financial fraud and are able to seize transactions if requested by an authority.

Permissionless blockchains, on the other hand, pose the unique challenge of not being able to collaborate with these investigations due to its high level of decentralization. This is because it has five characteristics that make it permissionless: it is open, meaning that it is accessible to anyone without authorization; it is borderless – it could be accessed and utilized globally; it is neutral, in the sense that the blockchain (being a financial system) can facilitate transactions without needing to know who the sender or receiver is or for what reason a payment is made; it is censorship-resistant, so that it is impossible to prohibit payments from and to a party in the event that such a prohibition is desired, and finally, it is public, so that everything that happens on the blockchain is disclosed to everyone on the network, allowing it to be fully verifiable [6]. Of all five characteristics, a focus will be placed on neutrality and censorship-resistance.

3   Literature Review

Three instances of a blockchain will be mentioned in this paper, namely The Ripple-, Ethereum- and Bitcoin blockchains.

An instance of a permissioned blockchain is the Ripple blockchain (its Ripple token is a cryptocurrency called XRP) which does not have the decentralized nature of the Bitcoin blockchain. Only a select few "trusted" nodes or validators operate on the Ripple network and the network is mainly owned by a private company with the same name. Since gateways can be banks, money service businesses, exchanges and any other financial institutions, blockchains with links to gateways will possess the ability to seize funds upon investigation into suspicious activity and by the request of law enforcement or regulators [7].

Further to this, the use of centralized exchanges can also be seen as installing an authority in place capable of seizing funds during investigations. In such settings, users' cryptocurrencies are held in wallets hosted on the exchange, giving the exchange the power to block access to the wallet's contents. Malicious actors who are more informed, will therefore know not to store stolen funds on such platforms, opting rather to store it in a more secure, decentralized form which they have control over.

As such, this paper will focus only on the purely decentralized implementations of blockchain such as decentralized exchanges, truly distributed applications, and secure wallets.

A possible solution to the identified problem exists: by implementing the concept of a blacklist of addresses. On a blockchain, the role of an address is to have unique identifying information describing either the source or destination of a transaction. BitcoinWhosWho.com offers the usage of their API that can append addresses to a known blacklist of addresses. However, it was found that this list is simply a list in software and can be queried from websites, but is not enforceable on the blockchain.

This paper extends the solution by considering the development of a tool which addresses the aforementioned shortcoming. It is a fact that no party can prevent specific addresses from sending or receiving Bitcoin and although cryptocurrency exchanges can and indeed prohibit transactions associated with malicious addresses, the Bitcoin protocol remains resistant towards censorship [3]. This is due to how the Bitcoin blockchain works: it operates across several layers much like the Internet's protocol stack. It has an application layer (where DApps, exchanges and wallets reside) that sits on top of other layers; most importantly, the consensus protocol layer. The blacklist of addresses mentioned is only respected on the application layer, but on the consensus protocol layer, no addresses are ever blacklisted and this layer will ultimately allow blacklisted addresses to perform transactions. This phenomenon is the result of the Bitcoin blockchain's consensus algorithm that enforces only changes to the Bitcoin protocol be made if it is agreed upon by 51% or more of participating network nodes.

In order to make the proposed solution fully enforceable on the blockchain, the blacklist would have to be saved and used on the consensus layer. This is a process that the Ethereum blockchain community had conducted in the past. The "DAO" smart contract was created on the Ethereum blockchain that facilitated a crowd-funding and raised approximately $150 million. This contract contained a bug that allowed a malicious user to steal roughly $60 million. The Ethereum community decided to effectively "roll back" the Ethereum blockchain and made the hacker's funds un-spendable [8]. This is a fullproof way in which hackers could be prohibited from running off with stolen funds. However, this decision and follow-through made by the Ethereum community received a lot of criticism and ultimately could support the case that the Ethereum blockchain is in effect a glorified database like the Ripple Blockchain – this is because, having such a powerful party (here, the Ethereum development community) strips away the purity of permissionless blockchains and consequently turns it into a permissioned one. If this were to be a widely accepted solution, blockchain technology would cease to provide any unique features in the global financial system. The Bitcoin blockchain has, in its entire time of being online, never been altered in a way that supports censorship resistance.

Transactions that occur on the Bitcoin blockchain are completely public and available to anyone for inspection. This will help facilitate forensic investigations. However, users can make use of services known as mixers or tumblers to make their transactions completely anonymous. Mixing services allow for the removal of an audit trail of transactions associated with an address [9]. As with centralized exchanges, there also exists centralized mixers that introduce counter-party risk. these could facilitate forensic investigations since logs could be kept that track the mapping of a "dirty" address to a "clean" one [10]. It is however the more decentralized implementations of mixers that pose a challenge to forensic investigation, but these peer-to-peer forms of mixing see their application in wallet software [10], making it possible to implement the blacklist. On the Ethereum blockchain, mixing occurs mostly via smart contracts, again making it trivial to implement a blacklist.

4   Solution Model

The proposed solution will not be a conventional software implementation but one that is deployed in the form of a smart contract which is enforceable and respected by all interfaces to the blockchain (DApps, wallets, exchanges, etc.). A blacklist will serve as a check to see whether a certain address should be allowed to use a blockchain application. It will allow for anyone connected to the blockchain to append any other address to the blacklist. This functionality of allowing any address to append to the blacklist is to ensure that supposed malicious addresses are immediately appended to the blacklist and that no reliance is made on a supposed victim of such an attack, as this might result in the address being appended too late.

After addresses are appended to the blacklist, the blacklisted address can request that the address be removed. After this request is made, a global democratic vote ensues. Any address connected to the blockchain will have the ability to vote whether a blacklisted address should remain or be removed from the blacklist – this happens within a certain time frame after the request for removal has been made. If a blacklisted address receives more positive votes than negative, the address is removed from the blacklist, but will remain otherwise. All blockchain applications that interface with the blacklist contract will make queries to the blacklist before permitting any sensitive (able to acquire or spend cryptocurrency) functionality. This serves as a way to impede supposed criminals from spending their maliciously acquired funds. This functionality of the blacklist can be seen in Fig. 1.

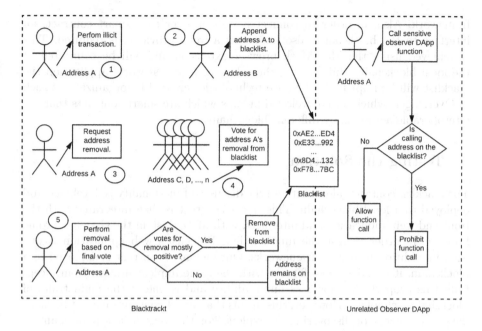

Fig. 1. High-level overview of blacklist functionality

5   Implementation Details

The Financial Action Task Force (FATF) is an independent and inter-governmental entity that enforces laws against illicit activities within the global financial system [11]. One can look at the standards that they have set for guidance as to how the blacklist should be implemented. Having said that, the blacklist functionality will align with these protocols and procedures in terms of freezing and unfreezing of funds.

The Blacklist's implementation in the form of a smart contract (a program that is run on the blockchain) makes it blockchain-centric and consequently enforceable on the application layer of the blockchain protocol. To realize this, a smart contract language, namely Solidity, is utilized. For testing and demonstration purposes (to avoid costs of real money incurred from working with smart contracts), a local Ethereum blockchain (using Ganache) is deployed.

The blacklist smart contract will involve the following functionality: the ability to append an address to the blacklist via any address, a means to request the removal of such an address appended to the blacklist via only the blacklisted address, the ability to vote for such a removal, either for or against via any address (except the blacklisted address), and finally the capability to remove the address from the blacklist which will only be successful upon having majority positive votes.

In order to make the blacklist usable to law enforcements and global voters, it is integrated into a front-end web application using the React JavaScript web framework. Also, in order to connect a web browser to a smart contract, an interface to the blockchain (also known as a web3 provider) is required. This is achieved using the MetaMask browser extension and will be connected to the local blockchain while testing the solution. In a real world application the blacklist will be implemented to use technologies created by organizations such as Overledger, which have developed mApps which are smart contracts that are interoperable across many different blockchains.

6   Testing the Solution

The blacklist contract, along with all the discussed functionality is developed and deployed to a local blockchain. This smart contract is then integrated with the front-end web application and interfaced with MetaMask in the Google Chrome browser. In order to test the functionality, Blacklist's web application is utilized in addition to testing frameworks. Once a smart contract is deployed to a blockchain, it remains there forever with the same implementation from inception; this property is known as immutability and is one of the main traits of blockchain technology. However, this means that smart contracts cannot be modified in the event of discovering an exploit. For this reason it is paramount to ensure that smart contracts are composed correctly and without any bugs. In Ethereum and the object-orientated Solidity language, this can be achieved by using the Truffle suite's testing frameworks – this was carried out in the development of the prototype.

```
2 ▾ contract Blacklist {
3        // Blacklist contract interface here
4   }
5
6 ▾ contract Quiz {
7        Blacklist bl;
8
9 ▾      constructor (address blAddress) public {
10            bl = Blacklist(blAddress); // Initializes this smart contract's
11                                       // instance of the blacklist contract.
12        }
13        // ...
14  }
```

Fig. 2. Calling contract subscribing to the blacklist contract

When a smart contract is deployed to the blockchain, it will receive an address where it resides on the blockchain. For each different instance of a given smart contract, a different address is given to that smart contract. For the purpose of the blacklist contract, it is only ever allowed to have one instance, ergo one address. Thus, this address is the one used in all other interfaces to the blockchain, including DApps, wallets, exchanges, etc. as the reference to the one and only blacklist contract.

In order to further understand the proposed solution, a software design pattern is described. The observer design pattern is realized when an object (the subject) maintains a list of other objects (observers) that depend on it and automatically notifies these observer objects of any state changes [12]. This is similar to the interaction between the blacklist contract and any smart contract that chooses to subscribe (observe) to the blacklist contract. Subscribing smart contracts are notified of any state changes albeit not automatically and only upon request by the observer smart contracts.

The exclusivity of only having one blacklist instance is achieved in the constructor of other contracts (observer contracts). In the observer contracts, a contract-wide variable (the blacklist contract) is initialized and set, upon deploying observer contracts. This is done by passing to the constructor of observer contracts the address of the blacklist (which can be seen in Fig. 2 in which subscribing to the blacklist makes the contract an observer). This gives the observer contract the ability to reference the blacklist contract for blacklist query functionality. This process was successfully carried out with another completely unrelated DApp (the observer contract – a DApp that incentivizes the studying of theoretical concepts).

```
45 ▾    function isAddrOnBlacklist(address queryAddress) public returns (bool){
46          return bl.getBlackListed(queryAddress); // queries the blacklist
47       }
48
49 ▾    function answerQuestion(bytes memory answer, uint id) public {
50          bool isOnBlacklist = isAddrOnBlacklist(msg.sender); // get address's blacklist status
51          require(isOnBlacklist == false); // only allow non-blacklisted addresses to continue
52          // ...
53       }
```

Fig. 3. Observer DApp querying the blacklist contract

In order to further test the blacklist contract, the following process occurred: blacklist queries were inserted into the caller contract at a sensitive function (it has the ability to pay out cryptocurrency to addresses that can correctly answer a question.) This can be seen in Fig. 3 where msg.sender represents the address making the call and answerQuestion() is the sensitive function and within this function the address is queried (on line 50) to the blacklist and its status stored in isOnBlacklist. This variable is then used on line 51 to determine whether the remaining part of answerQuestion() is allowed to be executed. This prevents addresses from calling these sensitive functions if they have been blacklisted. These lines of code are what is implemented on all contracts that want to enforce the blacklist in addition to the previously discussed code, in calling contracts' constructors.

7   Results and Discussions

The blacklist exhibits precisely the functionalities desired: one can successfully append any address to the blacklist and query an address, then receive a negative or positive response relating to the status of an address's membership to the blacklist.

The sensitive function mentioned earlier was called via an address that is not on the blacklist and the contract successfully executed the function. This same address was then appended to the blacklist using the blacklist front-end. In Fig. 4 one can see the affirmation of the address's membership to the blacklist. After this, the newly blacklisted address made the same call to the sensitive DApp function and it successfully prohibited such interaction as can be seen in Fig. 5 in which a reason is given for the failure of a transaction on the blockchain, namely that the calling address of the observer contract is blacklisted.

Fig. 4. Front-end displaying status of address

Fig. 5. Observer DApp rejects black-listed address

These results show that addresses could be prohibited from interacting with functions from any observer smart contracts of the blacklist contract, making it possible for effective forensic investigations into wallets belonging to such addresses if they find their way onto the blacklist due to criminal activity. This is due to the fact that suspected criminals would be prohibited from "getting rid" of their ill-gotten funds during forensic investigation and thus, provides a window of opportunity for investigators to determine the truth about a crime.

Two potential issues that arise from this solution relate to the potential abuse by users – parties are able to append to the blacklist at will while devious criminals may instantiate many wallets to control the democratic voting system for having addresses removed or remain on the blacklist. Potential solutions to address both issues are discussed later.

It is worth noting that this proposed solution will be immediately applicable in the event of cryptocurreny exchange hacks, cryptocurrency fraud and exit scams and DApp exploits, since these attacks are, from inception, on the blockchain. This is important since voters can, with confidence, be sure about a potential suspect if the details of a hack are disclosed. This removes a lot of subjectivity from the facts about a crime. For instance, in May of 2019, one of the largest cryptocurrency exchanges were hacked and the details of this hack were disclosed shortly after. The criminal had moved the funds to their own personal address and till today one can still see all the proceeds of that hack residing at this address. One can indeed be considerably certain that this address is that of the hacker, leaving no room for subjectivity in deciding (voting) whether the address must remain on the blacklist or not. The same could not be said for physical-world crimes such as the sale of illegal drugs. These types of crimes will require trusted and secure interfaces between smart contracts and the real world. These are known as oracles and is a major research area in itself in the domain of blockchain technology.

Finally, as was already discovered, the blacklist implemented is only ever respected by the higher layers of the blockchain protocol, and the lower consensus layers will ultimately permit any and all transactions irrespective of their membership to the blacklist.

8   Conclusion and Future Research

As previously mentioned, skilled criminals may attempt to acquire as many voters as possible to manipulate the voting process in their favour. Work done by [13] and [14] which focus on conducting votes using blockchain technology, have looked at such concerns.

One goal, for example, is to prohibit double voting by a single user. [14] proposes requiring a voter to register with IDs and email addresses which are then associated to a voter's Ethereum address. Similarly, [13] proposed a smart-city-orientated system that generates IDs in combination with blockchain addresses for people. This concept of identification and authentication in the voting process can be applied to the case of voting for a blacklisted address's removal. When put in place by authorities to achieve identification and authentication, such processes may prevent a criminal from generating additional wallets for skewing votes.

Another promising solution is the concept of prediction markets, specifically the Augur platform, where people can hypothesize over the results of a future event and get rewarded for predicting the correct result and penalized otherwise. This allows anyone to establish a market of an arbitrary event, trade on said market and be rewarded if correct about a prediction [15]. This platform implements a concept known as wisdom of the crowd, which dictates that the mean of a large set of predictions will most likely tend to the truth instead of the majority of the individual predictions, since there exists a diverse set of expertise [16]. Here, forensic investigators, security experts, law enforcement and even general citizens act as voters that place money at stake when voting on an address's guilt to a supposed crime. The voters that voted for the majority outcome get rewarded and the others penalized, making parties think twice before potentially risking money when casting illicit votes.

In both cases, further research into these concepts are required to reveal whether they can alleviate the shortcomings of the blacklist contract.

Since the lower layers of the blockchain protocol permits any and all transactions, the blacklist contract serves as a mechanism to deter criminals from spending funds rather than making it impossible. Potentially, oracles – mechanisms that facilitate the acquiring of data for a blockchain not inherent to the blockchain, may address this issue. However, oracles need to exist and operate in a decentralized nature [17]. In this domain, [18] and [19] have implemented algorithms that conduct transaction analysis in order to identify illicit transactions and it is this type of concept that can successfully serve as an oracle given that it is thoroughly tested for being decentralized.

The premise of permissionless blockchains is a system where money cannot be confiscated and this philosophy is well established and respected by the nodes supporting the most popular blockchain (the Bitcoin blockchain). This is due to its high level of decentralization. Chainlink is a project aiming to create oracles allowing smart contracts to interface with the real world; they bring forth different approaches to achieve oracles in a decentralized manner including: relying on many different independent sources of real world data instead of just one and implementing reputation systems in order to prevent abuse by means of enforcing repercussions [20]. Here, forensic investigators and other global voters (and possibly intelligent computing devices) act as oracles who will receive repercussions if they receive incorrect findings and subsequently incorrectly influencing votes. Artificial intelligence can also be employed albeit in a decentralized way.

Further research into oracles and its application of entering criminal addresses onto the blockchain can perhaps lead to a solution where the consensus layer respects the blacklist in a way that does not compromise decentralization. If the broader blockchain community and network nodes do not criticize such a solution, it will be supported by blockchain network nodes and accepted throughout.

With law enforcement and criminals competing with each other (law enforcement establishing new ways to apprehend criminals and criminals adapting their tactics to make this more difficult), it is important for law enforcement to be the one with the competitive advantage in order to be more effective as crime fighters. The blacklist contract proposes one manner in which law enforcement can transcend conventional, physical forms of crime fighting to keep up with the playing ground that will be created by the advent of this new financial infrastructure.

References

1. World Economic Forum: Deep Shift. Technology Tipping Points and Societal Impact (2015)
2. Tönnissen, S., Teuteberg, F.: Using blockchain technology for business processes in purchasing - concept and case study-based evidence. In: Abramowicz, W., Paschke, A. (eds.) Business Information Systems, pp. 253–264. BIS (2018)
3. Antonopoulos, A.M.: Mastering Bitcoin, 1st edn. O'Reilly, Sebastopol (2014)
4. Wang, R., He, J., Liu, C., Li, Q., Tsai, W., Deng, E.: A privacy-aware PKI system based on permissioned blockchains. In: 2018 IEEE 9th International Conference on Software Engineering and Service Science (ICSESS), pp. 928–931. IEEE (2018)
5. Reasons banks can freeze your account. http://investopedia.com/articles/markets-economy/082316/3-reasons-banks-can-freeze-your-account.asp. Accessed 14 Aug 2019
6. The five pillars of open blockchains. https://www.youtube.com/watch?v=qlAhXo-d-64. Accessed 14 Aug 2019

7. Jani, S.: An overview of ripple technology & its comparison with Bitcoin technology (2018)
8. Atzei, N., Bartoletti, M., Cimoli, T.: A survey of attacks on Ethereum smart contracts (SoK). In: Maffei, M., Ryan, M. (eds.) POST 2017. LNCS, vol. 10204, pp. 164–186. Springer, Heidelberg (2017). https://doi.org/10.1007/978-3-662-54455-6_8
9. Umair, S., Masoom, A., Sherali, Z., Abid, K.: Privacy aware IOTA ledger: decentralized mixing and unlinkable IOTA transactions. Comput. Netw. **148**(2019), 361–372 (2019)
10. Maduakor, F., Schwenk, J., Grothe, M.: Anonymous Bitcoin transactions. University in Bochum, Germany, pp. 19–22 (2017)
11. FATF (2013–2019), Methodology for Assessing Compliance with the FATF Recommendations and the Effectiveness of AML/CFT Systems, updated October 2019. FATF, Paris, France
12. Learning JavaScript design patterns. https://addyosmani.com/resources/essential jsdesignpatterns/book/. Accessed 4 Dec 2019
13. Longzhi, Y., Noe, E., Neil, E.: Privacy and security aspects of E-government in smart cities. In: Smart Cities Cybersecurity and Privacy, pp. 89–102 (2019). Chapter 7
14. Dagher, G.G., Praneeth B.M., Matea M., Jordan, M.: BroncoVote: secure voting system using Ethereum's blockchain. In: Proceedings of the 4th International Conference on Information Systems Security and Privacy (ICISSP), pp. 96–107 (2018)
15. Peterson, J., Krug, J., Zoltu, M., Williams, A.K., Alexander S.: Augur: a decentralized Oracle and prediction market platform, 1 November 2019
16. On prediction markets and blockchains. https://medium.com/swlh/on-prediction-markets-and-blockchain-48037d12039d. Accessed 9 Dec 2019
17. Antonopoulos, A.M.: Mastering Ethereum: Building Smart Contracts and DApps, 1st edn. O'Reilly Media, Incorporated (2018)
18. Wu, Y., Luo, A., Xu, D.: Forensic analysis of Bitcoin transactions. In: 2019 IEEE International Conference on Intelligence and Security Informatics (ISI), pp. 167–169. IEEE (2019)
19. Hong, S., Kim, H.: Analysis of Bitcoin exchange using relationship of transactions and addresses. In: Proceedings of the 21st International Conference on Advanced Communication Technology (ICACT 2019), pp. 67–70. IEEE (2019)
20. Ellis, S., Juels, A., Nazarov, S.: Chainlink: a decentralized Oracle network, 7 September 2017

Analysing and Predicting the Adoption of Anonymous Transactions in Cryptocurrencies

Radosław Michalski(✉)

Faculty of Computer Science and Management, Department of Computational Intelligence, Wrocław University of Science and Technology, Wrocław, Poland
radoslaw.michalski@pwr.edu.pl

Abstract. With the appearance of cryptocurrencies more than ten years back, many have been thinking that it is rather a short-term novelty that would only interest few enthusiasts and die shortly after. The history, however, has shown that not only cryptocurrencies itself are alive, but also the blockchain technology started to be applied in a variety of domains. This success required the governments to take action with respect to implementing regulations in a number of areas and, as a result of that, apart from implementing Know Your Customer (KYC) procedure, many exchanges delisted the cryptocurrencies that support anonymous transactions. In this work, we focus on the analysis of one of the cryptocurrencies offering anonymous transactions, Zcash. We are interested in answering the question of how the adoption of anonymous transactions in Zcash changes over time and whether the decisions of exchanges and new wallets in the ecosystem impacted it anyhow. Secondly, we investigate whether is it possible to predict if certain addresses that have been previously making only transparent transactions will be performing anonymous ones. The results indicate that the adoption of shielded transactions is independent of external events and it is possible to predict the involvement of an address in semi-shielded transactions.

Keywords: Cryptocurrencies · Blockchain · Anonymity · Network science · Supervised machine learning

1 Introduction

After reaching critical mass, Bitcoin, an initiative that has been only known to a few [15], started to be considered as a disruptive technology [3]. This cryptocurrency and its forks became used not only by technological nerds, but also multiple end users considered it as a potential way of making and accepting payments or money transfers. Moreover, when the exchange rates of many cryptocurrencies reached their peak in 2018 [7] and multiple individuals have been considering investing in them, the word *Bitcoin* became known worldwide, including the regulators. As it is not questionable that to become a fully-fledged user of a

© Springer Nature Switzerland AG 2020
W. Abramowicz and G. Klein (Eds.): BIS 2020 Workshops, LNBIP 394, pp. 132–144, 2020.
https://doi.org/10.1007/978-3-030-61146-0_11

cryptocurrency is not as easy as holding and exchanging cash or making money transfers, some areas became easily accessible to less-advanced users, e.g. cryptocurrency exchanges. As an entry point that does not require any technical skills, cryptocurrency exchanges could potentially play a role of popularizing other uses of cryptocurrencies than just investing in these. This is why apart from introducing the regulations on how to treat cryptocurrencies in the area of taxation, some countries started to regulate more strictly the exchanges as well [18]. An immediate result of this action was that multiple exchanges started to implement the Know Your Customer procedures asking their customers to identify themselves [16]. Yet, in order to demonstrate the transparency of exchanges, for the cryptocurrencies that do support anonymous transactions, such as Monero, Zcash or DASH, the exchanges simply started delisting them from 2019. Apart from that, some other exchanges that still permit trading of these cryptocoin, do not accept shielded (hidden) addresses as destinations of the asset transfers. This potentially poses a risk for the liquidity of the market for these cryptocoins, but the long-term outcomes of these decisions are still to come.

In this work we perform a two-fold analysis of the adoption of anonymous transactions for Zcash cryptocurrency. Zcash belongs to the cryptocurrencies that support anonymous (shielded) transactions as an option. Firstly, we have been interested in knowing how the adoption of shielded transactions looked over time with special emphasis on recent year where the decisions of exchanges could be possibly observed. Secondly, we are interested in finding out whether is it possible to predict if certain entities will become a side of semi-shielded transactions. To do so, we did develop a set of features to be computed for Zcash addresses and performed supervised learning classification to find out whether is it possible to predict whether certain addresses that so far have been involved only in transparent transactions will become part of semi-shielded ones. These features consist from two areas: the history of transactions, but we also look at the blockchain as at a complex network and we compute the network measures of addresses [2].

The work is structured as follows. In the next section we shortly describe the Zcash cryptocurrency and four types of transactions that are possible to be performed. The next section is describing the related work in the area of adoption of anonymous transactions as well as means of deanonymising certain aspects of cryptocurrencies. In Sect. 5 we look at how the adoption of anonymous transactions in Zcash looks over time. Next, Subsect. 6.1 describes the experimental setting for predicting the adoption of anonymous transactions by Zcash users and the subsequent sections presents the results. Lastly, Sect. 7 summarizes the paper with proposing future work directions.

2   Zcash Cryptocurrency

The Zcash cryptocurrency appeared in 2016 [9]. Albeit is shares some properties with Bitcoin, from the very beginning it was aiming the users that are interested in protecting their anonymity when making transactions. To offer this feature, Zcash is using so-called zk-SNARK that is the zero-knowledge succinct non-interactive argument of knowledge [20]. Thanks to that, contrary to many cryptocurrencies, Zcash offers the ability to acquire a special type of addresses, z-addresses that are then used for anonymous transactions. This feature is offered in parallel to the typical transparent t-addresses that are used where anonymity does not need to preserved.

The existence of these two types of addresses results with four possible types of transactions in the Zcash ecosystem:

- transparent to transparent – a transaction involving only t-addresses.
- transparent to shielded – a transaction from a t-address to a z-address.
- shielded to transparent – a transaction from a z-address to a t-address(es).
- shielded to shielded – a transaction involving only z-addresses.

The intermediate transaction types, transparent to shielded and shielded to transparent, are needed in order to have the communication between two layers in Zcash blockchain: transparent and anonymous Zcash pools. Otherwise, it would be impossible to transfer the assets between these two.

Thanks to supporting two types of addresses, Zcash could attract both types of uses: the ones that would like to perform transparent transactions and the ones that have been more interested in protecting their privacy. The feature of providing the ability to conduct anonymous transactions is the foundation of this cryptocurrency and its removal seems to be not considered as an option at all.

However, it must be also underlined, that albeit Zcash protocol itself is able to support two types of addresses from the very beginning, this cannot be said regarding its ecosystem. Those interested in acquiring z-addresses, until recently needed to use full node, namely to download the whole blockchain and to work in command line using zcash cli. The first wallet supporting z-transactions, ZecWallet FullNode, appeared at the end of 2018 and its lite sibling that does not require downloading full blockchain, was made public at the end of 2019. Another lite wallet, Shielded Guarda suited for mobile phones, was released in mid-2019. In total, out of 16 wallets, only four, including the zcash client itself, support z-addresses and two of them require downloading the whole blockchain. Moreover, as it was already mentioned, the exchanges have been either delisting the Zcash and other privacy-preserving cryptocurrencies from their markets or forcing to use t-addresses. Hypothetically, these decisions could also affect the adoption of Zcash shielded addresses. However, these decisions seem to not affect the community and its founders and the development of Zcash protocol is progressing fast. It also gained a lot of attention from others as one of the most sophisticated cryptocurrencies regarding the privacy options, so the perspectives for Zcash seem optimistic.

Recent developments in Zcash ecosystem, especially related to wallets introduction and the decisions of some exchanges, also suggest that last year is an interesting period for studying how the adoption of shielded transactions. This analysis - done from the beginning of Zcash but also looking in detail at last year will be performed later, in Sect. 5. However, it must be noted, that due to the nature of Zcash, for shielded-only transactions, the findings will be limited, as both sides of the transaction remain anonymous. Nevertheless, some aspects of shielded transactions still can be quantified. Another aspects of analyses is answer the question whether some t-addresses become a transparent side of shielded transactions and, if so, whether this can be predicted by using supervised learning applied to the features built upon the transaction history of addresses as well as the complex network measures for these addresses.

3   Related Work

The research on blockchain is taking many different directions, as described in [22]. Some researchers are looking at the security of blockchain-based systems. This is mainly due to the fact that in many cases, the well-known 51% majority attack still poses a risk to the integrity of the whole ecosystem [5,6,13]. Another important research direction in terms of security is the fork problem leading to the emergence of a rivalling sibling of a given blockchain [17]. These two risks are mainly the cases for systems with open participation. However, for some industrial applications, the risk is still not negligible.

In this work we investigate some areas related to the anonymity of nodes - firstly by looking at the adoption of shielded transactions in Zcash over time and then by studying whether is it possible to predict if certain address will be part of semi-shielded transactions. A family of this studies tackles the problem at different levels, e.g., by identifying IP addresses of nodes [11], by studying transaction motifs analysis [10] or transaction history summarisation [14]. This direction has been summarized in [1].

Due to the innovative anonymity mechanisms embedded in Zcash, it is often being studied by researchers interested in privacy. One of the most extensive analysis of privacy in Zcash has been presented in [12]. Here, authors studied the properties of shielded pool such as its volume and then proposed a number of heuristics in order to deanonymise the character of nodes in Zcash blockchain. Another interesting study has been performed in [19] where authors did investigate whether is it possible to find some links between shielded and transparent addresses, especially by looking at the amount and timing of these transactions.

Our work tackles at the problem of privacy in Zcash from another perspective. Here, we are interested in fining out whether is it possible to discover whether a certain t-address will be part of a semi-shielded transaction, i.e. a transaction where on one side there is a t-address and on another side is a z-address. In order to do so, we developed a set of features of nodes that combine features related to the history of transactions, such as frequency or recency of transactions, but also look at network properties of addresses, i.e. with how many other addresses the transactions have been made or how central the node is in a network built upon blocks. These features have been the input to the classifiers that have been learning a model for classifying unknown new instances whether an address will become a side of a semi-shielded transaction. This part of research belongs to the domain of machine learning.

Machine learning is an art of programming where computers can learn from given data without being precisely programmed. Two most essential types of machine learning methods are supervised learning and unsupervised learning. In this work, we focused on the first one in which an algorithm is learning from data and is provided with the correct solution (labels). In this work we investigate two algorithms: Decision Trees and Support Vector Machines.

Decision Tree algorithm is a process of sequential decisions, which starts from a *root* and after making a decision based on the given rule it spreads out creating new *branch* and lead down to left or right node [21]. At the end of this structure are *leafs*, which contain predicted classes for decision paths that lead from root to the corresponding leaf. Decision Tree algorithm looks for a rule that decreases impurity measure the most.

SVM, similar to Logistic Regression, is solving binary classification problem which can be extended for multiclass classification [4]. However it is not a probabilistic model. For N number of features, it divides N-dimensional space with hyperplanes that looks for maximum distance between datapoints from the training set. New objects are classified based on which side of hyperplanes they are placed. Contrary to Linear Regression this method is effective for high dimensional data.

4 Data

4.1 The Dru Platform

Usually, researchers asking any blockchain research questions have been independently working on gathering data and further analyzing it. In the case of Bitcoin, apart from several online tools that typically impose limits on their API, such as blockchain.info[1] or blockexplorer.com[2], the solutions space is scarce. One can also find libBitcoin-database[3] that offers in-memory storage of data that is troublesome due to blockchain size. This lack of widespread available solutions led to the development of the *Dru* platform that was built and released as an open source project[4]. The Dru platform used for gathering data for this this research is intended to support the researchers as it provides a consistent environment for acquiring data on a blockchain that can be used for further analysis - an exemplary plot generated by a single endpoint API call to Dru is presented in Fig. 1. This section can be also understood as a call for action to create a unified way of gathering data and doing research on cryptocurrencies, so interested researchers are welcomed to contribute to the development of the platform.

---

[1] https://blockchain.info, last access April 30th, 2020.
[2] https://blockexplorer.com, last access April 30th, 2020.
[3] https://github.com/libbitcoin/libbitcoin-database, last access April 30th, 2020.
[4] https://dru.readthedocs.io.

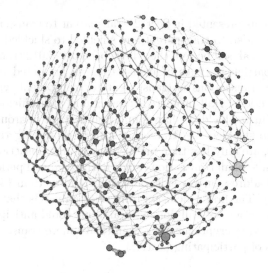

Fig. 1. The visualisation of blocks 1050–1060 of Zcash cryptocurrency. The size of a vertex is bound with its indegree, the width of edges represents the amount of cryptocurrency being transferred between nodes, colors indicate separate clusters. Data for this plot was gathered by using *get_edges* endpoint of *Dru* platform. (Color figure online)

4.2   Data Description

Data used in this study is based on Zcash blockchain. We investigated transactions from the genesis block (block 0) dated October 28th, 2016 to block 785,000 dated April 4th, 2020, so the data covered nearly three and half years. As the coinbase transactions are also sent to t-addresses, these have been counted as transparent ones.

5   Shielded Transactions Adoption

As it has been already presented in Sect. 2, only partial information about shielded transactions is available, so in this part of research we focused mainly on studying two aspects:

- the adoption rate of Zcash shielded or semi-shielded transactions by the community using Zcash expressed as the number of transactions since the appearance of this cryptocurrency,
- the analysis whether the recent appearance of wallets supporting shielded transactions and the decisions of exchanges impacted the popularity of shielded or half-shielded transactions.

To investigate the two above aspects, using the data on 785,001 blocks gathered using the Dru platform, we investigated how the adoption of particular types of transactions looked over time. These transactions could be made in one of four configurations presented in Sect. 2: transparent to transparent, transparent to shielded, shielded to transparent and shielded to shielded.

In total, in the analysed period, there have been 6,126,012 transactions made. The number of particular types of transactions is presented in Table 1. The majority of these transactions is transparent to transparent with nearly 87% share of all transactions. Half-protected transaction types, shielded to transparent and transparent to shielded, share similar percentage of around six percent. Lastly, shielded to shielded, i.e. transactions fully protecting the addresses of their participants, are only less than one percent of all transactions. Even when summing the number of all transactions made in the analysed period that at least have one shielded address on any side, only slightly more than 13% transactions are made that way. The first conclusion from this analysis is that transparent to transparent transactions are still dominant and one could anticipate that Zcash as a privacy preserving cryptocurrency would have more transactions that protect the identities of participating parties.

Table 1. The total number and ratio of particular transaction types in Zcash blockchain for blocks 0 to 785,000.

Transaction type	Number of transactions	Percentage
Transparent to transparent	5, 312, 089	86.713%
Transparent to shielded	359, 423	5.867%
Shielded to transparent	407, 115	6.646%
Shielded to shielded	47, 385	0.774%
Total	6, 126, 012	100%

However, it is also interesting to investigate how the share of particular transactions types changed over time - this is shown in Fig. 2. Here one can see that despite the smallest share among all transactions types, shielded to shielded transactions are actually the ones that note the highest rise over time. When looking at the numbers, we analysed what was the increase in terms of numbers of transactions from block 246,612 mined on January 1st, 2018 to the last block analysed. While the number of transparent to transparent transactions increased by the factor of 2.12, shielded to shielded noted the increase by the factor 6.54, more than three times more. Surprisingly, shielded to transparent and transparent to shielded transaction rose slower than transparent ones, 1.44 and 1.69, respectively. This indicates that these transaction types are used solely for transferring assets into or from shielded pool and then shielded transactions become dominant. This leads to the next conclusion that the privacy-preserving transactions are becoming more important over time and the Zcash ecosystem should prepare that more transactions will be made as shielded ones. As a consequence, entities offering services or goods should also offer this way of accepting Zcash cryptocurrency.

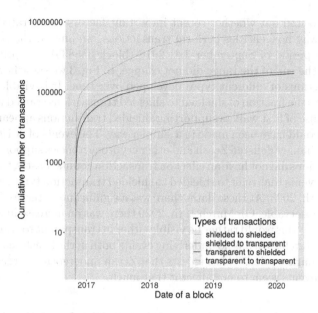

Fig. 2. Cumulative number of particular transaction types over time for Zcash cryptocurrency. The range of analysed blocks is from 0 to 750,000.

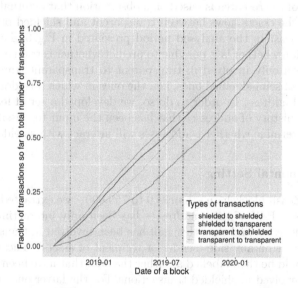

Fig. 3. The number of particular transaction types divided by total number of transactions over time for Zcash cryptocurrency for a limited number of blocks dating from September 1st, 2018 to April 4th, 2020. Vertical lines refer to the following events: orange – October 1st, 2018 – the introduction of zec-qt-wallet, blue – June 11th, 2019 – Shielded Guarda mobile wallet supports Zcash shielded transactions, black – August 26th, 2019 – Coinbase delists Zcash, and red – November 7th 2019 – ZecWallet lite client releases. (Color figure online)

In order to find out whether recent important factors enumerated in Sect. 2 changed the way how entities perform transactions, we limited the scope of the blocks to the period of September 1st, 2018 (block 385835) to April 4th, 2020 (785,000), so the end of the range did not change. In Fig. 3 we show how the number of transactions of different types developed over time. It is visible that from the end of 2018 the fraction of shielded to shielded transactions started to increase, possibly because of first wallets supporting shielded transactions appearing, so the transactions could have been made in a simpler way. The events of 2019Q3 when a number of exchanges delisted Zcash and other cryptocurrencies supporting anonymous transactions did not have an effect on transaction counts. Lastly, there are two unexplained events that refer to shielded to shielded transaction types: March 19th and March 29th, 2020. At these dates there was a significant increase of this transactions type observable. On March 19th, 2020 there was an increase in Zcash price by 25%, however this would rather be visible either in transparent to transparent or mixed types. March 29th has no underlying events both in the Zcash ecosystem and in the Zcash market price. This suggests that Zcash undergoes an organic growth and external events seem to not affect it that much.

6 Predicting Transactions with the Shielded Pool

The next part of the research is based on a observation that external events seem to not impact the assets flows between transparent and shielded pools in Zcash that much, at least in the analysed period presented in Fig. 3. This led to the research question whether is it possible to predict whether certain addresses that so far have been only involved in transparent to transparent transactions will become a part of semi-shielded ones, i.e., the ones in which a t-address interacts with a shielded address. In order to do so, we developed a set of features based on transaction history of addresses that has been the input to classifiers that are expected to determine whether t-addresses will interact with shielded addresses.

6.1 Experimental Setting

Based on the Zcash blockchain – blocks 0 to 750,000 – we extracted 2,000 transparent addresses. Half of these addresses has been only interacting with other transparent addresses and the other half has been involved in transactions with shielded addresses on any side (sender or recipient of assets). These two groups of addresses could be then labelled as either the ones that have been not involved or have been involved in shielded transactions. For the latter ones we only used the history of transactions before semi-shielded ones. Each of the addresses has been characterized by the following feature set:

- current balance.
- maximum balance.
- the number of transactions.
- the number of transactions peers.
- the average value of a transaction.

- the average value of an outgoing transaction.
- the average value of an incoming transaction.
- the number of transactions in which this address was a sender.
- the number of transactions in which this address was a recipient.
- the number of a block in which the address appeared for the first time.
- the number of a block in which the address appeared for the last time.

Next set of features has been referring to the network we built using the last transaction we found particular address in. The method of constructing the network was the following: we took the address as the root of our network. Then, around this root we built a network by looking at prior transactions in such a way that we took all the addresses interacting with this address and their neighbourhood for up twenty transactions back limiting the depth of this approach also to twenty in order to not to recreate the whole blockchain. Then for each node we generated additional statistics based on complex network approach [2]. In this case we look at a network as at an undirected unweighted network. Moreover, the procedure of creating the network led to only one connected component. The measures we computed for each of the addresses based on the network built around it are the following:

- the diameter of the network.
- the average path length.
- the degree of the address.
- the betweenness of the address.
- the closeness of the address.
- the average degree in the network.
- the maximum degree in the network.

These two sets of features have been combined and altogether with the class information (only transparent or semi-shielded) became the input to the classifiers. The training dataset consisted of 80% of 2,000 addresses and the testing one of 20%, the number of instances belonging any of two classes has been balanced.

We used two supervised machine learning approaches: Decision Trees and Support Vector Machines. The parameters of the classifiers are presented in Table 2. The performance metric for evaluation of the algorithms was F-score [8] and the baseline reference was a simple random approach that randomly assigns an address to the class and results with the F-score of 50%.

6.2   Results

The results of classifying the addresses into one of two classes demonstrate that in fact it is possible to predict to some extent whether certain addresses will become involved in the transactions with shielded addresses. Our results show the following performance of two evaluated classifiers:

- Decision Trees: 78.95%.
- Support Vector Machines: 81.44%.

Table 2. Parameters tested for each classifier for proposed set of features based on transaction list. Chosen ones that performed the best are in bold.

Parameter name	Evaluated values
Decision trees	
Class weight	Balanced, **Balanced subsample**, Uniform
Criterion	**Entropy**, Gini
SVM	
C parameter	1, 10, 100, **1000**
Kernel	**Linear**, RBF

Both of these classifiers performed better than a random approach and suggest that in fact the history of the addresses is indicative about future involvement of an address in semi-shielded transactions. Albeit the difference in the performance of the classifiers is not significant, Decision Trees due to their nature are able to say more about the most important features from the feature set that contributed to the decisions. Top five of them in the descending order of importance are: the diameter of the network, the betweenness of the address, the average degree, the number of transactions and the number of transactions peers. This indicates that both - transactions statistics and complex network measures contribute to the accuracy of the classification.

7   Summary and Future Work

In this work we looked at two aspects related to anonymous or shielded transactions in a Zcash ecosystem. Firstly, we investigated how the adoption rate of these transactions looks like compared to transparent ones. Albeit in numbers the shielded or semi-shielded transactions are still a minority, the increase factor of shielded ones is the highest among other transactions types. This leads to the conclusion that the Zcash ecosystem is systematically moving towards shielded pool independently of external events. In the second part of the work we researched whether is it possible to predict if certain transparent addresses will become involved in semi-shielded transactions in the future. By using supervised learning techniques applied to a feature set containing both: transactional information and complex network statistics we demonstrated that in fact it is possible to discover whether addresses will be part of semi-shielded transactions. As for future work we would like to extend the feature set and perform more fine-tuning of the classifiers.

Acknowledgments. The development of the Dru platform was funded by the Zcash Foundation.

References

1. Alsalami, N., Zhang, B.: SoK: a systematic study of anonymity in cryptocurrencies. In: Proceedings of the IEEE Conference on Dependable and Secure Computing (DSC 2019), pp. 1–9. IEEE (2019)
2. Barabási, A.L., et al.: Network Science. Cambridge University Press, Cambridge (2016)
3. Baur, A.W., Bühler, J., Bick, M., Bonorden, C.S.: Cryptocurrencies as a disruption? Empirical findings on user adoption and future potential of Bitcoin and Co. In: Janssen, M., et al. (eds.) I3E 2015. LNCS, vol. 9373, pp. 63–80. Springer, Cham (2015). https://doi.org/10.1007/978-3-319-25013-7_6
4. Cortes, C., Vapnik, V.: Support-vector networks. Mach. Learn. **20**(3), 273–297 (1995)
5. Dasgupta, D., Shrein, J.M., Gupta, K.D.: A survey of blockchain from security perspective. J. Bank. Financ. Technol. **3**(1), 1–17 (2018). https://doi.org/10.1007/s42786-018-00002-6
6. Dey, S.: Securing majority-attack in blockchain using machine learning and algorithmic game theory: a proof of work. In: Proceedings of the 10th Computer Science and Electronic Engineering (CEEC 2018), pp. 7–10. IEEE (2018)
7. Dyhrberg, A.H., Foley, S., Svec, J.: How investible is bitcoin? Analyzing the liquidity and transaction costs of Bitcoin markets. Econ. Lett. **171**, 140–143 (2018)
8. Goutte, C., Gaussier, E.: A probabilistic interpretation of precision, recall and F-score, with implication for evaluation. In: Losada, D.E., Fernández-Luna, J.M. (eds.) ECIR 2005. LNCS, vol. 3408, pp. 345–359. Springer, Heidelberg (2005). https://doi.org/10.1007/978-3-540-31865-1_25
9. Hopwood, D., Bowe, S., Hornby, T., Wilcox, N.: Zcash Protocol Specification. GitHub, San Francisco (2016)
10. Jourdan, M., Blandin, S., Wynter, L., Deshpande, P.: Characterizing entities in the bitcoin blockchain. In: Proceedings of the IEEE International Conference on Data Mining Workshops (ICDMW 2018), pp. 55–62. IEEE (2018)
11. Juhász, P.L., Stéger, J., Kondor, D., Vattay, G.: A Bayesian approach to identify Bitcoin users. PLoS ONE **13**(12), e0207000 (2018)
12. Kappos, G., Yousaf, H., Maller, M., Meiklejohn, S.: An empirical analysis of anonymity in Zcash. In: Proceedings of the 27th {USENIX} Security Symposium ({USENIX} Security 18), pp. 463–477 (2018)
13. Lin, I.C., Liao, T.C.: A survey of blockchain security issues and challenges. Int. J. Netw. Secur. **19**(5), 653–659 (2017)
14. Lin, Y.J., Wu, P.W., Hsu, C.H., Tu, I.P., Liao, S.W.: An evaluation of Bitcoin address classification based on transaction history summarization. In: Proceedings of the IEEE International Conference on Blockchain and Cryptocurrency (ICBC 2019), pp. 302–310. IEEE (2019)
15. Nakamoto, S.: Bitcoin: a peer-to-peer electronic cash system. Technical report, Manubot (2019)
16. Navarro, R.R.: Preventative fraud measures for cryptocurrency exchanges: mitigating the risk of cryptocurrency scams. Ph.D. thesis, Utica College (2019)
17. Neudecker, T., Hartenstein, H.: Short paper: an empirical analysis of blockchain forks in Bitcoin. In: Goldberg, I., Moore, T. (eds.) FC 2019. LNCS, vol. 11598, pp. 84–92. Springer, Cham (2019). https://doi.org/10.1007/978-3-030-32101-7_6
18. Pieters, G., Vivanco, S.: Financial regulations and price inconsistencies across Bitcoin markets. Inf. Econ. Policy **39**, 1–14 (2017)

19. Quesnelle, J.: On the linkability of Zcash transactions. arXiv preprint arXiv:1712.01210 (2017)
20. Sasson, E.B., et al.: Zerocash: decentralized anonymous payments from Bitcoin. In: Proceedings of the 2014 IEEE Symposium on Security and Privacy, pp. 459–474. IEEE (2014)
21. Utgoff, P.E.: Incremental induction of decision trees. Mach. Learn. 4(2), 161–186 (1989)
22. Yli-Huumo, J., Ko, D., Choi, S., Park, S., Smolander, K.: Where is current research on blockchain technology?-A systematic review. PLoS ONE 11(10), e0163477 (2016)

Volatility and Value at Risk of Crypto Versus Fiat Currencies

Viviane Naimy$^{(\boxtimes)}$, Johnny El Chidiac, and Rim El Khoury

Notre Dame University– Louaize, 72 Zouk Mikayel, Zouk Mosbeh, Lebanon
{vnaimy,jfchidiac,rkhoury}@ndu.edu.lb

Abstract. This paper examines the behavior of Bitcoin and Ripple compared to the three fiat currencies, EURUSD, GBPUSD and CNYUSD, by comparing their volatility and VaR during the period extending from March 01, 2016 to February 28, 2019. EWMA, GARCH (1, 1), GARCH (p, q) and EGARCH (1, 1) were used to forecast volatilities. EWMA model outperformed the rest of the models for all of the selected fiat and cryptocurrencies during the in-sample period and for EURUSD, GBPUSD and Bitcoin during the out-of-sample period. GARCH (p, q) was the optimal model for the CNYUSD and Ripple in the out-of-sample period. Bitcoin and Ripple exhibit an asymmetry in their volatility which is significantly higher than all the volatilities of the studied currencies. When estimated volatilities were compared to the implied volatility, the GARCH (1, 1), GARCH (6, 6) and EWMA were the optimal models for the EURUSD, CNYUSD and GBPUSD, respectively for the in-sample period. VaR results were accepted for the EURUSD, GBPUSD and Bitcoin at all confidence levels. For the CNYUSD, VaR measures underestimated the risk at the 99% confidence level unlike Ripple's VaR that was accepted at the 90% and 99% confidence levels. Our results suggest that Bitcoin and generally the cryptocurrencies market cannot act as alternatives to fiat currencies at the moment.

Keywords: Bitcoin · Ripple · Fiat currencies · GARCH models · Value at risk (VaR)

1 Introduction

Regardless of the huge capitalization and although cryptocurrencies are managed through advanced encryption techniques, many investors and governments are still cautious toward them as they are a possible source of uncertainty and financial instability [1]. Cryptocurrencies have been subject to many accusations such as price bubbles and money laundering [2]. Even after roughly ten years of their initiation, there still exists debates on the classification of cryptocurrencies, some describe them as currencies, while others argue that they should be classified as commodities [3].

Cryptocurrencies are characterized by high volatility and return behaviors [4, 5]. For example, Bitcoin prices increased by 4,380% from March 01, 2016 to December 16, 2017 to reach a peak of $19,497. Similarly, Ripple prices witnessed a huge growth of 42,150% from $0.008 at the start of the period on March 01, 2016 to $3.38 on January

© Springer Nature Switzerland AG 2020
W. Abramowicz and G. Klein (Eds.): BIS 2020 Workshops, LNBIP 394, pp. 145–157, 2020.
https://doi.org/10.1007/978-3-030-61146-0_12

7, 2018. Thus, the appropriate techniques to understand the behavior and volatilities of cryptocurrencies are to define them in terms of their distribution function. Given that normality of cryptocurrencies returns is uncertain and inconsistent if new information does not arrive symmetrically or if investors do not react linearly, and since most of the financial markets are characterized by informational asymmetric with volatility clustering, many complex models for estimating volatility are used such as autoregressive conditional heteroscedasticity ARCH models of Engle [6] and EWMA (Exponentially Weighted Moving Average).

This paper attempts to explore and assess several types of volatility models (EWMA, GARCH (1, 1), GARCH (p, q) and EGARCH, models) applied on cryptocurrencies (Bitcoin, Ripple) and fiat currencies (EURUSD, GBPUSD, CNYUSD) to choose the best model suitable to each instrument and check whether cryptocurrencies have the same properties of fiat currencies and if the cryptocurrencies' behavior in terms of risk is similar to traditional currencies. The winning volatility model will be incorporated to calculate the VaR (Value at Risk)[1] and compare its performance and variabilities among crypto and fiat currencies. Section 2 explores the review of literature. Section 3 exposes the procedure and methodology adopted to model the volatilities of the selected instruments. Section 4 presents the results and Sect. 5 discusses and concludes the findings.

2 Literature Review

Although asset volatility is considered as one of the most important areas for many researchers, few studies focused on the volatility of cryptocurrencies as compared to other types of assets. The below studies summarize and illustrate the most recent outputs (2015–2019) sometimes in contradiction the one with the other.

Urquhart [7] and Bariviera [8] studied Bitcoin price clustering with periods extending from May 2012 till April 2017 and from August 2011 till February 2017, respectively, and found a persistence in the volatility and evidence of price clustering. Chan et al. [9] examined the volatility of seven cryptocurrencies (Bitcoin, Dash, Litecoin, Ripple, Monero, Dogecoin and Maidsafecoin) using twelve GARCH models with a period extending from June 2014 to May 2017. The findings revealed that IGARCH (1, 1) and GJRGARCH (1, 1) models are the best in modelling the volatility associated with virtual currencies. They also pointed at the high volatility associated with cryptocurrencies making it a suitable investment for risk seeking investors. Stavroyiannis and Babalos [10] investigated the dynamic properties of Bitcoin versus other asset classes using several GARCH models with a time horizon extending from July 2013 till December 2016. The findings indicate that Bitcoin is highly volatile and violates VaR measures more than other assets, such as the S&P500 index and Gold. They also suggested that cryptocurrencies are unable to provide investors with benefits such as diversification and hedging when compared to the US market.

Naimy and Hayek [3] applied EWMA, GARCH and EGARCH (Exponential GARCH) models to examine the volatility of Bitcoin during a period extending from

---

[1] VaR is a standard risk measure which summarizes the downside risk and is defined as the maximum loss expected with a given probability over a specific period of time.

April 2013 till March 2016. The findings showed that EGARCH, which captures the leverage effect, outperformed other models. The authors also noted that the behavior of Bitcoin is different from that of fiat currencies. Gkillas and Katsiampa [11] implemented an extreme value analysis for VaR and Expected Shortfall on five major cryptocurrencies (Bitcoin, Ripple, Ethereum, Litecoin and Bitcoin Cash) to examine the tail behavior of the returns. They used data varied in sample size and ranged between July 2010 and July 2017. Surprisingly, their results revealed that Bitcoin and Litecoin were least risky while Bitcoin cash was the riskiest with the highest recorded volatility. However, the number of observations used for Bitcoin cash was limited, since the latter was still in its early stages.

Bouri et al. [12] conducted a comparative analysis on Bitcoin with other indices taking the S&P500 as a benchmark with Brent oil, Crude oil and Gold. The selected time period extended from July 2011 through December 2015. The findings revealed that the GARCH (1, 1) model was the most effective in forecasting volatility in the digital currencies. They also noted that Bitcoin can be used in a portfolio for diversification purposes. On the other hand, Kim [13] compared the volume and spread of Bitcoin versus fifteen fiat currencies such as Euro, US dollar, Chinese yuan and many others. The research was conducted in a time horizon extending from April 2014 till April 2015. The findings indicated that Bitcoin is considered as alternative to the foreign exchange market for international transactions and settlements.

Dyhrberg [14] implemented a volatility analysis on Bitcoin, Gold and the US dollar using GARCH (1, 1) and EGARCH (1, 1) models with a time frame extending from July 2010 till May 2015. The results showed the hedging capabilities of Bitcoin to the US dollar and Gold when using GARCH (1, 1). However, with the asymmetric EGARCH (1, 1) model, the author suggested the use of Bitcoin as a hybrid classification between commodities and fiat currencies. Also, Dyhrberg [15] applied T-GARCH (Threshold GARCH) to investigate if Bitcoin holds similar hedging capabilities to gold in a period extending from July 2010 till May 2015. The results showed that Bitcoin can serve as a hedging vehicle against the FTSE 100 index. In addition, Bitcoin can be used as a hedge against the US dollar in the short term. A later research by Baur et al. [16] extended Dyhrberg's study [14] on Bitcoin, Gold and the US dollar with the same time horizon. However, they concluded that Bitcoin characteristics are neither similar to Gold nor to the US dollar.

Katsiampa [4] evaluates the volatility of Bitcoin using six different GARCH models with a period extending from July 2010 till October 2016. The author revealed that Bitcoin market is highly speculative. The results showed that the AR-CGARCH (Auto Regressive Component GARCH) model outperformed other models, which in turn demonstrates the added value of using both a long-run and a short-run conditional variance. Also, Bouoiyour and Selmi [17] implemented a study on Bitcoin using several GARCH models for a period extending from December 2010 till July 2016; their findings showed that Bitcoin is extremely volatile with a conditional variance trailing an "explosive" process, where positive shocks have less effect than negative shocks. They suggested that the virtual currencies market was still immature even after it reached low levels of volatility.

Another interesting paper by Philip et al. [18] studied 224 virtual currencies using the stochastic volatility model with a time horizon extending from July 2010 till July 2017. The author noted predictable patterns with the presence of volatility clustering. Mild leverage effects were noticed on cryptocurrencies except for Ripple. Urquhart and Zhang [19] used several GARCH models on the studied variables (Bitcoin, Australian Dollar, Canadian Dollar, Swiss Franc, Euro, Great Britain Pound and the Japanese Yen), for a period extending from November 2014 till October 2017. They found that the Bitcoin can be used as a hedge against the Swiss Franc, Euro and the Great Britain Pound, and as a diversifier for the Australian Dollar, Canadian Dollar and the Japanese Yen. Furthermore, Bitcoin can be considered a safe haven during market turbulence when compared to the Canadian Dollar, Swiss Franc and Great Britain Pound but not when compared to the Australian Dollar, Euro and the Japanese Yen.

A recent study by Trucíos [20] investigated the best model in forecasting Bitcoin volatility and Value at Risk using twelve GARCH models with a time horizon extending from September 2011 till December 2017. The results showed that AVGARCH (Absolute Value GARCH) model outperformed other GARCH-type models. However, when using robust procedures, the robust GARCH outperformed all the non-robust models. Peng et al. [21] implemented a study on Bitcoin volatility using ten GARCH models, such as GARCH (1, 1), EGARCH (1, 1), GJR-GARCH (1, 1) and many others. The sampling consisted of two data sets: a daily data characterized by low frequency during a period extending from January 2016 till July 2017, and an intraday hourly data considered as high frequency ranging from January 2016 till July 2017. RMSE and MAE error metrics were used in order to validate the findings. SVR-GARCH (Support Vector Regression-GARCH) outperformed other models in both samples. This is due to its ability to cover nonlinearity and dynamics in the financial series represented by volatility clusters and leptokurtic data distribution.

Even though a lot of research have been conducted in modelling assets' volatility in different financial sectors worldwide, only few studies have been focusing on modelling the structure of crypto-currency markets' volatility. This present study therefore, attempts to contribute to the literature through comparing crypto and fiat currencies while suggesting the optimal model for each currency's volatility forecast and through testing the accuracy of VaR measures among our selected assets for the period March 01, 2016 through February 28, 2019.

3 Procedures and Methodology

3.1 Data

This study used secondary data for the daily closing prices of three years for Bitcoin/USD, Ripple/USD EUR/USD, GBP/USD and CNY/USD. The Bitcoin/USD and Ripple/USD data are downloaded from coinmarketcap.com website, whereas EUR/USD, GBPUSD and CNY/USD data are extracted from the Bloomberg platform. The in-sample period ranges from March 01, 2016 through February 28, 2018 with 489 observations and the out-of-sample period ranges from March 01, 2018 till February 28, 2019 with 243 observations. Although cryptocurrencies prices are quoted daily including the weekends, only weekday's data are used to match the closing prices for each of the selected fiat currencies. The daily volatility is estimated based on daily return and is calculated as the percentage change in the daily closing price as follows:

$$u_i = \frac{S_i - S_{i-1}}{S_{i-1}} \tag{1}$$

Where u_i is the return on the i^{it} day, S_i and S_{i-1} are respectively the price of the asset at the end of the i^{it} day and at the end of the previous day $i - 1$.

Volatility data are obtained from Bloomberg except for the in-sample period for Ripple, which is manually calculated using Merton's approach [22] as follows:

$$\sigma_n = \sqrt{\sum\nolimits_{i=1}^{n} u_i^2} \tag{2}$$

3.2 Descriptive Statistics

Several preliminary tests are conducted to examine the behavior of the two cryptocurrencies and the three fiat currencies before investigating their volatility. The stationarity of the returns is tested using the Augmented Dickey Fuller (ADF) test. All p-values are statistically significant at 1%, providing an evidence of the presence of stationary in the return series of all currencies.

3.3 Selection of the Volatility Models

We use EWMA, GARCH (1, 1), and GARCH (p, q) models since they incorporate structures that enable forecasts of the future level of variance rate to be produced relatively straightforwardly. EGARCH model is also used since it captures the leverage effect and the asymmetry of the conditional variance. The description of all the models used in this paper is depicted in Table 1.

Parameters estimation are performed for the in-sample period (March 01, 2016 till February 28, 2018), under each of the volatility models. The realized volatilities are compared to the calculated in-sample volatilities to determine the most accurate model for measuring and predicting volatility. The comparison is addressed using three error metrics, namely: Root Mean Square Error (RMSE = $\sqrt{(\sum_{t=1}^{n}(f - Y)^{\wedge}2)/n}$), Mean Absolute Error (MAE = $\frac{1}{n}\sum_{t=1}^{n}/(f - Y)/$) and Mean Absolute Percentage Error (MAPE = $\frac{100}{n}\sum_{t=1}^{n}/\left(\frac{Y-f}{Y}\right)/$), where n is the number of periods, Y is the true value and f is the prediction value. The model with the least error difference is considered the most accurate. Subsequently, the estimated in-sample parameters are used to forecast the volatility in the out-of-sample period (March 01, 2018 till February 28, 2019). The same error statistics process is performed to check whether the same optimal model for the in-sample returns applies to the out-of-sample returns. The calculated volatilities are compared to both, the realized and implied volatilities for fiat currencies, and to the realized volatility for cryptocurrencies, given that their options market is still new and immature.

Table 1. Model Specification

Model	Equation	Proposed By	Explanation
GARCH (p,q)	$\sigma_n^2 = \omega + \sum_{i=1}^{p} \beta \sigma_{n-i}^2 + \sum_{i=1}^{q} \alpha u_{n-i}^2$	Bollerslev [23]	σ_n^2 is today's variance, σ_{n-i}^2 is the variance of the return p days ago and u_{n-1}^2 is the square of the return q days ago. ω is the intercept, α refers to ARCH effect, and β refers to GARCH effect. The model is considered stable when the weights sum-up to one.
EGARCH	$\log \sigma_n^2 = \gamma v_l + \beta g(z_{n-1}) + \alpha \log \sigma_{n-1}^2$	Nelson [24]	σ_n^2 is today's variance, v_l is the long run variance rate, σ_{n-1}^2 is the variance of the previous day's return, $g(z_{n-1})$ is an explanatory variable which accounts for the leverage effect. The weights assigned to v_l, σ_{n-1}^2, and $g(z_{n-1})$ are γ, α, and β respectively. The logarithmic representation preserves a positive process, meaning that the model respects the non-negativity constraint.
EWMA	$\sigma_n^2 = \lambda \sigma_{n-1}^2 + (1 - \lambda) u_{n-1}^2$	Zangari [25]	σ_n^2 is the variance of today, σ_{n-1}^2 is the variance of the previous day, u_{n-1}^2 is the square of the previous day's return and λ is the decay factor and smoothing parameter, which ranges between 0 and 1. If $\lambda = 1$ then today's variance is entirely dependent on the most recent variance, whereas if $\lambda = 0$ then the model converges to the Random Walk model. High lambda denotes a low reaction to new market changes.

The parameters values of all models are estimated using the maximum likelihood (ML) method[2], mathematically expressed as follows:

$$f(y_1......, y_n | \theta) = \prod_{i=1}^{n} f(y_i | \theta) = L(\theta | y) \tag{3}$$

$$\ln L(\theta | y) = \sum_{i=1}^{n} lnf(y_i | \theta) \tag{4}$$

The probability density function is represented by f(y|θ), where y represents a random variable constrained on a parameters set θ. The likelihood function is known as the joint density function and is expressed in Eq. (3). The vector model parameter is denoted by θ and y is used to indicate the time series at time i. In addition, the parameters are constant, and their estimation is based on the data selection. The best selected models in the out-of-sample period are incorporated to calculate VaR of the crypto and fiat currencies. Each model's parameters are estimated twice. More specifically, the parameters are estimated for the first 100-day period (February 13, 2018 till July 16, 2018) and for the remaining 150-day period (July 17, 2018 till February 28, 2019).

For the VAR calculation and in order to create 399 scenarios for 250 days, the sample period extends from June 30, 2016 till February 28, 2019 with a total of 649 days of observations. The first sub-sample period, consisting of 400 days of observation from June 30, 2016 till February 13, 2018, is used to create 399 scenarios for the upcoming result on day 400 (February 14, 2018). This procedure is repeated over 250 days with rolling windows, each one incorporating 399 scenarios where the last day is February 28, 2019. By implementing this approach, the value of each currency and cryptocurrency under the ith scenario becomes as follows:

$$V_{ithscenario} = v_n \frac{v_{i-1} + (v_i - v_{i-1})\sigma_{n+1}/\sigma_i}{v_{i-1}} \tag{5}$$

Where: v_i is the value of the currency or cryptocurrency on day i; v_n is the value of the currency or cryptocurrency on the last day of the chosen time period; σ_i is the estimate of the daily volatility on day i; σ_{n+1} is the most recent estimate of the daily volatility.

Using the following equation, the return scenarios is calculated under each simulation trial leading to 399 return scenarios, to subtract the expected gains and losses on the first day of the out-of-sample period:

$$u_{ith\,scenario} = \frac{(V_{ith\,scenario} - v_n)}{v_n} \tag{6}$$

Where: $V_{ith\,scenario}$ is the value of the currency or cryptocurrency under the ith scenario; v_n is the value of the currency or cryptocurrency on the last day of the chosen time period.

Finally, Kupiec test [26] is used for back-testing to evaluate the accuracy of VaR measures. The Log-likelihood Ratio (LR) is denoted as follows:

$$LR = -2\ln\left[\left(1 - p\right)^{T-N} p^N\right] + 2\ln\left[\left(1 - \frac{N}{T}\right)^{T-N} \left(\frac{N}{T}\right)^N\right] \tag{7}$$

---

[2] Maximum Likelihood involves choosing values for the parameters corresponding to each model that maximize the chance of the data occurring.

Where: N is the number of exceptions; T is the number of trials; and p is the probability of failure.

The VaR model is rejected if LR is greater than 3.84. Furthermore, the probability of failure values (p) are 0.01, 0.025, 0.05 and 0.1 which parallel VaR confidence levels of 99%, 97.5%, 95% and 90% respectively. The number of trial value (T) is 250 and is constant at all confidence levels. The number of exceptions (N) will be obtained by counting the recorded exceptions where the actual loss exceeds VaR on a given day.

4 Findings

The analysis moves further to determine the best fitted model that can help in modelling the volatility of return series for the three fiat currencies and the two cryptocurrencies.

4.1 GARCH and EWMA Parameters

GARCH (1, 1), EGARCH and EWMA parameters are depicted in Tables 2, 3 and 4.

Table 2. GARCH (1, 1) Estimated Parameters

	EURUSD	CNYUSD	GBPUSD	Bitcoin	Ripple
Omega (ω)	0.00000088	0.00000076	0.00000766	0.00005507	0.00022216
ARCH component (α)	0	0.134	0.187	0.159	0.235
GARCH component (β)	0.963	0.715	0.715	0.841	0.765
LT Volatility	7.76%	3.53%	13.92%	–	–

Table 3. EGARCH (1, 1) Estimated Parameters

	EURUSD	CNYUSD	GBPUSD	Bitcoin	Ripple
Omega (ω)	−3.190063	−0.992132	−0.460894	−0.143102	−0.39817
Leverage coefficient (γ)	−0.01	0.206	0.226	0.243	0.313
ARCH component (α)	0.138	0.09	0.07	0.053	0.163
GARCH component (β)	0.7	0.919	0.951	0.971	0.914
LT Volatility	7.71%	3.52%	14.39%	135.50%	155.76%

The daily in-sample returns are used to obtain the GARCH (p, q) parameters using several combinations of the ARCH components (p) and the GARCH components (q). The ARCH components (p) are taken up to 6 and the GARCH components (q) are

Table 4. EWMA Estimated Parameters

	EURUSD	CNYUSD	GBPUSD	Bitcoin	Ripple
Lambda	0.938	0.936	0.924	0.911	0.946

taken up to 9. When estimating the GARCH (p, q) parameters for the EURUSD, the log likelihood function could not attain any higher value while manipulating the parameters. This means that the GARCH (1, 1) parameters maximized the log likelihood function for EURUSD. Also, when estimating the GARCH (p, q) parameters for the Bitcoin and Ripple, the sum of the ARCH and GARCH components was 1. This means that in the 53 combinations of the ARCH (p) and GARCH (q), the models converged to the IGARCH (p, q) model where the long run variance is unattainable.

4.2 Optimal Models: In-Sample Vs. Out-of-Sample

When comparing in-sample to out-of-sample results, Table 5 shows that EWMA is the optimal model despite the absence of a long-run average variance.

Table 5. Optimal Models: In-Sample vs. Out-Of-Sample

Optimal Selected Models Based on the Realized Volatility Comparison					
	EURUSD	CNYUSD	GBPUSD	Bitcoin	Ripple
In-Sample	EWMA	EWMA	EWMA	EWMA	EWMA
Out-Of-Sample	EWMA	GARCH (6, 2)	EWMA	EWMA	GARCH (1, 8)
Optimal Selected Models Based on the Implied Volatility Comparison					
	EURUSD	CNYUSD	GBPUSD		
In-Sample	GARCH (1, 1)	GARCH (6, 6)	EWMA		
Out-Of-Sample	EWMA	EWMA	EGARCH (1, 1)		

More specifically, the EWMA proved to be the best model in both contexts for the EURUSD, GBPUSD and Bitcoin when compared to the realized volatility. Whereas, for the CNYUSD and Ripple, the EWMA was the optimal model during the in-sample period and the GARCH (p, q) yielded better results for the out-of-sample period. However, when comparing the modelled volatilities to the implied volatility for the currencies, the results were mixed. Results show that EGARCH (1, 1) did not prove its superiority and produced the least desirable results among the selected models.

4.3 VaR Results

The optimal selected models corresponding to the out-of-sample period are integrated to calculate VaR. All parameters have been estimated for a period of 100 and 150 days respectively. As previously mentioned, based on the estimated parameters and using the returns of each currency and cryptocurrency, the daily variances are calculated 399 times for each sub-sample using EWMA for the EURUSD, GBPUSD and Bitcoin, and GARCH (p, q) for the CNYUSD and Ripple. Applying these volatilities values in Eq. (5) produced different price scenarios for each currency and cryptocurrency which are fitted into Eq. (6) to determine the return scenarios. Final results are presented in Table 6.

Table 6. VAR and Kupiec Test Outcome

	Method Applied	VaR CL	Exceptions Recorded	Allowed Exceptions	LR	95% Critical Value	Test Outcome
EUR USD	Incorporating volatility to historical simulation using **EWMA**	90%	34	[17, 35]	3.27	3.84	**Accept**
		95%	16	[7, 20]	0.95	3.84	**Accept**
		97.5%	11	[2, 11]	3.03	3.84	**Accept**
		99%	3	[0, 5]	0.09	3.84	**Accept**
CNY USD	Incorporating volatility to historical simulation using **GARCH (6, 2)**	90%	41	[17, 35]	9.73	3.84	**Reject**
		95%	22	[7, 20]	6.26	3.84	**Reject**
		97.5%	10	[2, 11]	1.96	3.84	**Accept**
		99%	2	[0, 5]	0.11	3.84	**Accept**
GBP USD	Incorporating volatility to historical simulation using **EWMA**	90%	30	[17, 35]	1.05	3.84	**Accept**
		95%	15	[7, 20]	0.5	3.84	**Accept**
		97.5%	9	[2, 1]	1.09	3.84	**Accept**
		99%	3	[0, 5]	0.09	3.84	**Accept**
Bitcoin	Incorporating volatility to historical simulation using **EWMA**	90%	26	[17, 35]	0.04	3.84	**Accept**
		95%	14	[7, 20]	0.18	3.84	**Accept**
		97.5%	7	[2, 11]	0.09	3.84	**Accept**
		99%	2	[0, 5]	0.11	3.84	**Accept**
Ripple	Incorporating volatility to historical simulation using **GARCH (1, 8)**	90%	20	[17, 35]	1.18	3.84	**Accept**

(continued)

Table 6. (*continued*)

Method Applied	VaR CL	Exceptions Recorded	Allowed Exceptions	LR	95% Critical Value	Test Outcome
	95%	0	[7, 20]	–	3.84	**Reject**
	97.5%	0	[2, 11]	–	3.84	**Reject**
	99%	0	[0, 5]	–	3.84	**Accept**

The EURUSD, GBPUSD and Bitcoin had the most accurate results at all confidence levels, with LR values less than 3.84. However, VaR results underestimated the risk of CNYUSD and were rejected at the 90% and 95% confidence levels. Interestingly, VaR results were rejected at the 95% and 97.5% for the Ripple since the recorded exceptions were less than the allowed amount of exceptions which means that the model overestimated the corresponding risk. The Ripple VaR measures were accepted at the 90% and 99% confidence levels.

5 Discussion and Conclusion

This study reveals original inferences. The first one relates to the forecasting and predictive ability of the selected volatility models when applied on Bitcoin. EWMA model outperformed other volatility models in both contexts (in-sample and out-of-sample). This contradicts the findings of many recent studies which confirmed that the GARCH models were the most effective in forecasting the volatility of Bitcoin and other virtual currencies [3, 4, 12, 14]. Such contradictions may be related to the selected time period we opted to choose and to the evolvement of cryptocurrencies behavior. The second inference relates to the volatility asymmetry of the selected cryptocurrencies. Our research concludes that Bitcoin and Ripple exhibit an asymmetry in their volatility where good news have a larger effect than bad news. In fact, the values of the leverage coefficients for the Bitcoin and Ripple are 0.24 and 0.31, respectively (Table 3), in line with Radovanov et al. [27] results.

Another implication concerns the behavior of Bitcoin when compared to fiat currencies. Bitcoin and Ripple volatility is significantly higher than all the volatilities of the fiat currencies that we examined: EURUSD, CNYUSD and GBPUSD. This corroborates the findings of Cermak [28] and Naimy and Hayek [3]. During the out-of-sample period, our results exclusively revealed that Bitcoin's optimal volatility model was the same as the EURUSD and GBPUSD. GARCH (p, q) was the winning model for the Ripple same as for the Chinese Yuan. This highlights some similarities in the volatility behavior between the studied fiat and cryptocurrencies, an observation that is new to the literature.

Also, our results oppose those of Stavroyiannis [10] who stated that Bitcoin violates VaR and other risk measures. Two possible reasons might have caused the difference in the results. First, Stavroyiannis used a different calculation methodology based on a simple calculation of the daily volatility to estimate VaR, while we used the adjusted historical simulation method after incorporating the optimal estimated volatility model. Second, we used a different (recent) time horizon compared to the Stavroyiannis' time frame with a lag of almost 18 months.

Finally, Bitcoin and generally the cryptocurrencies market cannot act as real alternatives to fiat currencies at the moment. Financial managers and investors need to be prudent when considering investing in cryptocurrencies, given their unexpected and extremely volatile behavior. Also, market participants aiming at diversifying their portfolios or seeking a risky position could consider cryptocurrencies, given their unique behavior compared to other instruments.

References

1. Ngunyi, A., Mundia, S., Omari, C.: Modelling Volatility Dynamics of Cryptocurrencies Using GARCH Models (2019). https://doi.org/10.4236/jmf.2019.94030
2. Corbet, S., Meegan, A., Larkin, C., Lucey, B., Yarovaya, L.: Exploring the dynamic relationships between cryptocurrencies and other financial assets. Econ. Lett. **165**, 28–34 (2018). https://doi.org/10.1016/j.econlet.2018.01.004
3. Naimy, V.Y., Hayek, M.R.: Modelling and predicting the Bitcoin volatility using GARCH models. Int. J. Math. Modell. Numerical Optimisat. **8**, 197–215 (2018)
4. Katsiampa, P.: Volatility estimation for Bitcoin: a comparison of GARCH models. Econ. Lett. **158**, 3–6 (2017). https://doi.org/10.1016/j.econlet.2017.06.023
5. Osterrieder, J., Lorenz, J.: A statistical risk assessment of Bitcoin and its extreme tail behavior. Ann. Finan. Econ. (AFE) **12**, 1–19 (2017)
6. Engle, R.F.: Autoregressive conditional heteroscedasticity with estimates of the variance of united kingdom inflation. Econometrica **50**, 987–1007 (1982). https://doi.org/10.2307/1912773
7. Urquhart, A.: Price clustering in Bitcoin. Econ. Lett. **159**, 145–148 (2017). https://doi.org/10.1016/j.econlet.2017.07.035
8. Bariviera, A.F.: The inefficiency of Bitcoin revisited: a dynamic approach. Econ. Lett. **161**, 1–4 (2017). https://doi.org/10.1016/j.econlet.2017.09.013
9. Chan, S., Chu, J., Nadarajah, S., Osterrieder, J.: A Statistical Analysis of Cryptocurrencies. 1–23 (2017). https://doi.org/10.3390/jrfm10020012
10. Stavroyiannis, S., Babalos, V.: Dynamic properties of the Bitcoin and the US market. Soc. Sci. Res. Netw. (2017). http://dx.doi.org/10.2139/ssrn.2966998
11. Gkillas, K., Katsiampa, P.: An application of extreme value theory to cryptocurrencies. Econ. Lett. **164**, 109–111 (2018). https://doi.org/10.1016/j.econlet.2018.01.020
12. Bouri, E., Molnár, P., Azzi, G., Roubaud, D., Hagfors, L.I.: On the hedge and safe haven properties of Bitcoin: Is it really more than a diversifier? Finance Res. Lett. **20**, 192–198 (2017). https://doi.org/10.1016/j.frl.2016.09.025
13. Kim, T.: On the transaction cost of Bitcoin. Finance Res. Lett. **23**, 300–305 (2017). https://doi.org/10.1016/j.frl.2017.07.014
14. Dyhrberg, A.H.: Bitcoin, gold and the dollar – A GARCH volatility analysis. Finance Res. Lett. **16**, 85–92 (2016). https://doi.org/10.1016/j.frl.2015.10.008
15. Dyhrberg, A.H.: Hedging capabilities of bitcoin. Is it the virtual gold? Finance Res. Lett. **16**, 139–144 (2016). https://doi.org/10.1016/j.frl.2015.10.025
16. Baur, D.G., Dimpfl, T., Kuck, K.: Bitcoin, gold and the US dollar – a replication and extension. Finance Res. Lett. **25**, 103–110 (2018). https://doi.org/10.1016/j.frl.2017.10.012
17. Bouoiyour, J., Selmi, R.: Bitcoin: a beginning of a new phase? Econ. Bull. **36**, 1430–1440 (2016)

18. Phillip, A., Chan, J.S.K., Peiris, S.: A new look at Cryptocurrencies. Econ. Lett. **163**, 6–9 (2018). https://doi.org/10.1016/j.econlet.2017.11.020
19. Urquhart, A., Zhang, H.: Is Bitcoin a hedge or safe haven for currencies? An intraday analysis. Int. Rev. Finan. Anal. **63**, 49–57 (2019). https://doi.org/10.1016/j.irfa.2019.02.009
20. Trucíos, C.: Forecasting bitcoin risk measures: a robust approach. Int. J. Forecast. **35**, 836–847 (2019). https://doi.org/10.1016/j.ijforecast.2019.01.003
21. Peng, Y., Albuquerque, P.H.M., Camboim de Sá, J.M., Padula, A.J.A., Montenegro, M.R.: The best of two worlds: forecasting high frequency volatility for cryptocurrencies and traditional currencies with Support Vector Regression. Expert Syst. Appl. **97**, 177–192 (2018). https://doi.org/10.1016/j.eswa.2017.12.004
22. Merton, R.C.: On estimating the expected return on the market: an exploratory investigation. J. Finan. Econ. **8**, 323–361 (1980). https://doi.org/10.1016/0304-405X(80)90007-0
23. Bollerslev, T.: Generalized autoregressive conditional heteroskedasticity. J. Econ. **31**, 307–327 (1986). https://doi.org/10.1016/0304-4076(86)90063-1
24. Nelson, D.: Conditional heteroskedasticity in asset returns: a new approach. Econometrica. **59**, 347–370 (1991). https://doi.org/10.2307/2938260
25. Zangari, P.: Estimating volatilities and correlations. In: RiskMetrics-Technical Document. pp. 43–66. Morgan Guaranty, New York (1994)
26. Kupiec, P.H.: Techniques for verifying the accuracy of risk measurement models. J. Derivatives. **3**, 73–84 (1995). https://doi.org/10.3905/jod.1995.407942
27. Radovanov, B., Marcikić, A., Gvozdenović, N.: A time series analysis of four major cryptocurrencies. Ser.: Econ. Organ. **15**, 271–278 (2018)
28. Cermak, V.: Can bitcoin become a viable alternative to fiat currencies? An empirical analysis of bitcoin's volatility based on a GARCH model. An empirical analysis of bitcoin's volatility based on a GARCH model. Social Science Research Network (2017). http://dx.doi.org/10.2139/ssrn.2961405
29. Stavros, S.: Value-at-risk and related measures for the Bitcoin. J. Risk Finance. **19**, 127–136 (2018). https://doi.org/10.1108/JRF-07-2017-0115

How Much Identity Management with Blockchain Would Have Saved Us? A Longitudinal Study of Identity Theft

Razieh Nokhbeh Zaeem[✉] and K. Suzanne Barber

The Center for Identity, The University of Texas at Austin, Austin, TX 78712, USA
{razieh,sbarber}@identity.utexas.edu
https://identity.utexas.edu

Abstract. The use of blockchain for identity management (IdM) has been on the rise in the past decade. We present the first work to study the actual, large-scale impact of using blockchain for identity management, particularly how it can protect Personally Identifiable Information (PII) to curb identity theft and fraud. Our insight is that if blockchain-based IdM protects PII, it can reduce the number of theft and fraud cases that take advantage of such PII. At the Center for Identity at the University of Texas at Austin, we have modeled about 6,000 cases of identity theft, and PII exploited in them. We utilize this model to investigate how three real-world blockchain-based IdM solutions (Civic, ShoCard, and Authenteq) could have reduced the identity theft loss over the past 20 years if they had been universally used. We identify which PII protected by blockchain is more critical. We also suggest new PII to include in blockchain-based IdM. Our work paves the way for the design of more effective blockchain-based IdM or any other new line of IdM for that matter.

Keywords: Identity management · Blockchain · Identity theft · Self-sovereign identity

1 Introduction

The blockchain technology has found a variety of applications beyond cryptocurrency, from insurance to the Internet of Things. One of these applications is Identity Management (IdM), the framework that identifies, authenticates and authorizes users in order to control access to resources. Many concrete blockchain-based IdM solutions exist today, ranging from successful startups to open source projects and foundations.

One of the most prominent goals of IdM, when utilized to manage Personally Identifiable Information (PII) (such as email addresses, passwords, fingerprints, and driver's licenses), is to protect PII from identity theft and fraud. Identity theft and fraud is the fraudulent acquisition and use of such PII. According to the latest statistics from the U.S. Department of Justice published in 2019 [8],

© Springer Nature Switzerland AG 2020
W. Abramowicz and G. Klein (Eds.): BIS 2020 Workshops, LNBIP 394, pp. 158–168, 2020.
https://doi.org/10.1007/978-3-030-61146-0_13

10% of all U.S. residents reported that they had been victims of identity theft in 2016 with a total loss of $17.5 billion.

Blockchain-based IdM solutions offer user control, decentralization, immutability, transparency, security, and privacy for the PII they manage–added properties that can prevent identity theft and fraud. The actual effect of these solutions on the landscape of identity theft and fraud, however, has not received the attention it deserves.

We present the first work to examine the large-scale effect of this new type of IdM through the lens of a longitudinal study of close to 6,000 cases of identity theft and fraud, which took place over the past 20 years. We estimate the potential of blockchain-based IdM in preventing identity theft and fraud, based on its ability to protect PII leveraged in identity management. Our insight is that if blockchain-based IdM secures and protects some types of PII, it can eliminate (or substantially reduce) the number of theft and fraud cases that take advantage of those types of PII. Consequently, it can prevent millions of dollars of loss that harm identity theft victims every year.

We make the following contributions:

1. We introduce the idea of extrapolating from previous identity theft and fraud cases to study the future effect of blockchain-based IdM.
2. We take advantage of our longitudinal study of the past 20 years of news-reporting on identity theft and fraud—modeled by a team of modelers at the University of Texas at Austin Center for Identity in over six years. We investigate the frequency, monetary loss, and other properties of identity theft cases involving PII that could be protected with blockchain-based IdM.

We discover that while blockchain-based IdM cannot eliminate identity theft and fraud altogether, its effective protection of PII such as Social Security Card can potentially end millions of dollars of identity theft loss. The protection blockchain-based IdM solutions extend to *some* PII is more valuable than others. Reducing the chance of exposure of these PII (Social Security Card, Healthcare ID and Driver's License) would have saved identity theft victims from the highest amount of loss in the past 20 years.

Based on our study of the identity theft cases, we recommend new PII to include in blockchain-based IdM to further mitigate the effect of identity theft. Some of the most prominent PII we suggest include: Financial Information such as Banking, Credit/Debit Card, and Taxpayer Information as well as Employer Information and ID.

Our work sheds light on actual, large-scale potential of new IdM solutions, with a focus on blockchain-based IdM. Our results enlighten developers of new IdM and their users and scientifically direct these efforts to maximize their impact. Furthermore, once new blockchain-based IdM is widely used, we can study its actual effect on the landscape of identity theft and fraud through the same methodology of examining reported identity theft cases.

2   Related Work: Identity Management with Blockchain

In this section we explain fundamentals of blockchain and blockchain-based IdM, review commercially available examples of such IdM solutions, and cover related work on the actual large-scale benefits of migrating to IdM with blockchain.

2.1   Blockchain

The term blockchain is rooted in the seminal Bitcoin white-paper [12] that introduced a novel crypto-currency (i.e., electronic cash) technology. This technology allows online transactions to take place without the need to go through a trusted financial third party. Digital signatures and a peer-to-peer network form the backbone of the Bitcoin technology. The two parties of the transaction communicate through digital signatures (i.e., public and private keys). The peer-to-peer network timestamps transactions by hashing them into a chain of blocks, forming a record of transactions. The longest chain of blocks serves as a tamper-proof ledger of all witnessed transactions. This ledger (also known as blockchain) cannot be altered without the consensus of the network majority.

2.2   Blockchain-Based Identity Management

The blockchain technology has been used for a whole host of applications [10], including identity management (IdM). IdM is the framework that identifies, authenticates, and authorizes users to access resources. Blockchain-based IdM solutions [18] adopt blockchain for identity management.

In an IdM, when the user needs to make a *claim* (i.e., assert something about one's identity, like the citizenship of a country), he/she provides an *identity proof* (i.e., some form of document that provides evidence for the claim, like a passport). Identity proofs always contain PII, i.e., any information that could be used to identify an individual. Such identity proofs should be *attested* (i.e., validated) by the relevant identity authority (e.g., the agency that issued the passport).

Blockchain-based identity solutions encrypt a user's identity, hash it, and add its attestations to the blockchain ledger. These attestations are later used in order to prove the user's identity. Two categories of blockchain-based IdM solutions exist [7]:

1. Decentralized Identity (e.g., **ShoCard**, **Authenteq**, BitID, ID.me, and IDchainZ): This identity solution is similar to conventional identity management solutions where credentials from a trusted service are used. The only difference arises is the storage of validated attestations on a distributed ledger for later validation by a third party.
2. Self-sovereign identity (e.g., **Civic**, Sovrin, uPort, and Onename.io): The user owns and controls his/her identity without heavily relying on central authorities. In essence, self-sovereign identity is very similar to how non-digital identity documents work today. Every user keeps their own identity documents in

their device. The user creates a public/private key pair and contacts identity authorities to associate and attest his/her public key with an identity proof. When the user makes a claim, the user provides his/her attested public key. The verifier accepts only the claims signed with the user's private key.

Blockchain-based IdM improves identity management in several ways. Digital signatures, one of the major components of the blockchain technology, provide authenticity of the identity proof and attestation. The peer-to-peer network, the other major component of blockchain, eliminates the need for a central repository of users' identity. Hence, blockchain can make IdM solutions decentralized, tamper-resistant, and enhance security and privacy.

2.3   Examples of Commercially Available IdM Solutions

Many foundations, companies, startups, and open source projects have utilized [9] blockchain for identity management. In this work, we have studied different blockchain-based IdM services offered by multiple companies like Authenteq [2], ShoCard [3], and Civic [1]. In all these services, the user first logs into the web/mobile app, and provides email address and phone number to create an account. The user then selects which government-issued identity documents to use for identity verification, e.g., Passport, National Identity Card, Driver's License, or Social Security Card as shown in Table 1. The user scans one of these identity verification documents and the app verifies the document against a third party. After the check, the user identity is confirmed, attested on the blockchain and can be reused.

Table 1. Identity document options in three blockchain identity management solutions.

Civic	ShoCard	Authenteq
Passport	Social Security Card (SSC)	Any Government ID
National Identity Card (NID)	Green Card (GC)	
Driver's license	Health Card/Photo ID	

2.4   Evaluation of Blockchain-Based IdM Solutions

There are very few studies evaluating blockchain-based IdM solutions. For example, Baars [4] studied ten blockchain-based identity management systems to identify their properties and compare them together. These ten solutions were Onename.io, Qiy, iDIN, eHerkenning, IRMA, PKIoverheid, Jumio, Tradle, Idensys, and uPort. Dunphy and Petitcolas [7] used the "laws of identity" framework (including user control and consent, minimal disclosure, justifiable parties, directed identity, design for a pluralism of operators, human integration, and consistent experience across contexts) to evaluate three blockchain-based IdM solutions (uPort, ShoCard, and Sovrin).

Our previous work [14] investigated different PII options given to users for authentication on current blockchain-based IdM solutions. Based on our Identity Ecosystem model [5,6,11,13,15], we evaluated these options and their risk and liability of exposure. Powered by real world data of about 6,000 identity theft and fraud stories from ITAP (Sect. 3), our model recommended some authentication choices and discouraged others.

None of these studies, however, have looked at the potential large-scale effect of this new line of IdM solutions. In this work, we investigate how the added security and privacy of blockchain-based IdM can thwart identity theft. Clearly, using blockchain would not eliminate all identity theft nor it would fully protect identity documents. There still exists a chance that the device containing the identity proof falls into the wrong hands and one poses as the identity owner. With that said, we can still estimate the potential of blockchain-based IdM in reducing the effect of identity theft since it dramatically reduces the chance of compromise for the identity proof/PII. Assuming that blockchain-based IdM protects the identity proof/PII, we approximate how much these IdM solutions would have saved identity theft victims by eliminating identity theft and fraud cases that happened because a particular identity proof/PII was stolen/misused.

3   Identity Threat and Assessment Prediction (ITAP)

In order to estimate the potential of blockchain-based IdM in reducing financial loss of identity theft, we need a database of identity theft and fraud cases that records details of such cases and spans a long period of time and a diverse set of victims. Our Identity Threat and Assessment Prediction (ITAP) project [16,17] is such a longitudinal study of identity theft and fraud cases.

ITAP is a structured database of news stories that gathers and models data about incidents of identity theft, fraud, and abuse. Through these news stories, we seek to determine the methods and resources used to carry out these crimes, the vulnerabilities exploited, and the consequences of these incidents. The ITAP model is a large and continually growing, structured repository of such information. ITAP spans national and international identity theft cases over the past 20 years (2000–2020) and currently includes 5,906 unique identity theft incidents.

In order to populate ITAP, using a wide variety of online sources, we gather the data from news stories that report on incidents involving the exposure, theft, or fraudulent use of PII such as names, social security numbers, and credit card numbers. We find candidate news stories in two ways: via an RSS feed set up through Feedly[1] and via several Google Alerts[2] based on relevant key phrases. We regularly monitor the stories thus gathered and store those that we deem appropriate for ITAP.

A team of modelers record salient information about these incidents. The recorded information about identity thieves includes:

---

[1] https://feedly.com.
[2] https://www.google.com/alerts.

1. Performers: the performers of the identity theft incident, including thief, frustrater, abuser, or non-malicious actor.
2. Inputs: PII that the performers initially obtained and used.
3. Outputs: PII that the performers created, acquired, or exposed based on the inputs.
4. Resources: The types of tools, devices, applications, and other instruments used by the performers in carrying out the incident.
5. Steps: The steps that the performers took.
6. Criminal Activities: The relationship between the performers and the victims, e.g., insider, outsider, or both.

The information collected about victims includes:

1. Age Group of Victims.
2. Gender of Victims.
3. Citizenship of Victims.
4. Education Level of Victims.
5. Annual Income of Victims.
6. Profession of Victims.
7. Organization Affected.
8. Sector of Society or Infrastructure Involved.
9. Counter Measures Taken by the Victim Organization.

Finally, the information about the incident itself includes:

1. Location of the Event, or the "Internet".
2. Date When Event Occurred.
3. Date of Article or Announcement.
4. Type of Loss Incurred, e.g., emotional, reputation, or financial.
5. Financial Loss Amount (converted to US$s).
6. Reputation Loss.
7. Emotional Distress.

4 Experimental Results

In this work, we seek to answer the following main research question: What is the potential of the blockchain-based identity management solutions (namely Civic, ShoCard, and Authenteq) in thwarting identity theft? For example, how much financial damage would have been avoided if all the ITAP identity theft and fraud cases that have a Social Security Number as input where eliminated, because the owner protected his/her Social Security Number with blockchain-based identity management? In order to answer this question, we report statistics from ITAP to answer the following research questions (RQ):

1. How many cases have at least one PII input that could be protected with blockchain-based IdM (Table 1)?
2. What is the average financial cost of those cases?

3. What are other PII that could be added to blockchain-based IdM to prevent
   (a) high-loss identity theft incidents?
   (b) frequent identity theft incidents?

We retrieved the list of PII involved in ITAP identity theft and fraud cases (i.e., all PII that thieves and fraudsters used as inputs to carry out the incidents or expose/generate other PII as well as all PII that they exposed or fraudulently generated as outputs) and manually scanned them for relevant PII according to Table 1. Table 2 lists these PII and shows which IdM solution could possibly protect them.

4.1 RQ1: How Many Cases Have at Least One Input PII that Could Be Protected with Blockchain-Based IdM?

There are a total of 5,906 identity theft and fraud cases in ITAP, and Table 3 shows how many of those cases have at least one input PII from a given category of PII. (Categories of PII are shown in Tables 1 and 2.) Note that Green-Card-related PII (e.g., Visa Details) happen to appear only as *output* in the current set of ITAP cases, and therefore have zero cases in which they are PII *inputs*. Also, there are only two cases with National ID as their input PII, none of which report their financial loss. As a result, we cannot calculate the average loss for the Green Card and National ID categories of PII. This might be due to the fact that ITAP is predominantly U.S.-oriented, where Visa and National ID contribute to a very small number of identity theft cases.

It is clear from the number of cases reported in Table 3 compared to the total number of 5,906 cases in ITAP that blockchain-based IdM solutions do not eliminate all identity theft and fraud, even if universally applied to protect the PII these solutions do cover. These IdM solutions, however, would save the identity theft and fraud victims from considerable amount of financial loss.

4.2 RQ2: What Is the Average Financial Cost of Preventable Cases?

Most of the PII categories of Table 3 pertain to an average loss amount of over $1M (not averaged per victim but per incident[3]). For example, if all PII of the Passport category (including Passport Information, Number, Expiration Date, and Country of Issue) where protected with blockchain-based IdM, 16 identity theft cases of ITAP with an average loss value of $1,252,464 would have been avoided because at least one of their input PII would not be compromised.

The Social Security Card PII category (including Social Security Number, Suffix, and Social Security Card) is the input to 567 cases with an average financial loss of over $27M. To provide some context, Social Security Number is one of the PII involved in the highest loss value cases: the average loss value of all the ITAP cases is just over $11M, while Social Security Card and Number cases have an average loss of over $27M.

---

[3] We did not calculate the average loss per victim because the number of victims is usually not reported in the identity theft and fraud news story.

Table 2. PII that could be protected with the blockchain-based IdM solutions Civic, ShoCard, and Authenteq, as found in the ITAP repository.

		PII	Category
	ShoCard	Insurance Policy Information	Private Health ID
		Health Plan Member Number	
		Health Insurance Policy Number	
		Individual Healthcare Plan	
		Health Insurance Policy Information	
		Healthcare Provider's Name	
		Medical ID Number	
		Healthcare Provider Information	
		Health Insurance ID Card Information	
		Health Plan Group Number	
		Health Insurance Company Name	
		Medicare Provider ID Number	Gov. Health ID
		Medicaid ID Number	
		Medicare ID Number	
		Medicaid Provider ID Number	
		National Health Index Number (New Zealand)	
		Social Security Number (SSN)	SSC
		SSN Suffix (Last Four Digits)	
		Social Security Number - Invalid	
		Visa Details	GC
Authenteq	Civic	Passport Information	Passport
		Passport Number	
		Passport Expiration Date	
		Passport Country of Issue	
		National Identity Number	NID
		Driver's License Number	Driver's License
		Driver's License Information	
		Stolen Driver's License Information	
		Fake Driver's License Information	
		Driver's License Photo	
		Driver's License Expiration Date	
		Government Issued ID Number	Other Gov. ID
		State Identification Number	
		ID Card Number	
		Universal ID Number	
		ID Card Information	
		Aadhaar Unique Identification Number (India)	
		Personal Record Identifier (PRI) (Canada)	
		DNI (Documento Nacional de Identidad - Spain) Information	

4.3   RQ3: What Are Other PII to Add to Blockchain-Based IdM?

Top 5% PII/identity proof documents in terms of feeding input to high-loss identity theft incidents are as follows:

- Financial Information, including Credit Card Information, and Taxpayer Information such as Electronic Filing Identification Number (EFIN)
- Social Security Number
- Employer Information and ID
- Medicare ID Number

Social Security Number and Medicare are already included in many blockchain-based IdM solutions. We recommend further addition of Credit Card Information and Taxpayer Information, as well as Employer Identification to avert high-loss identity theft and fraud.

Top 5% PII/identity proof documents frequently involved as input to incidents are as follows:

- Social Security Number
- Financial Information, e.g., Credit/Debit Card, Bank Account, and Taxpayer Information
- Driver's License Information
- Patient Medical Record

These frequently abused PII have a considerable overlap with PII involved in high-loss incidents. We reiterate that one category of PII/identity proof that we strongly recommend adding to blockchain-based IdM is Financial Information, e.g., Debit/Credit Card, Bank Account, and Taxpayer Information. In fact, the investigation of the websites of Authenteq, Civic, and ShoCard shows they are planning to include this information in their IdM solutions.

Table 3. The number of ITAP cases that have a PII protected with the blockchain-based IdM solutions Civic, ShoCard, and Authenteq, as *input*.

IdM	Authenteq				Civic			
	ShoCard							
PII Cat.	Private Health ID	Gov Health ID	Social Security Card	Green Card	Passport	National ID	Driver's license	Other Gov. ID
# Cases	28	19	567	0	16	2	86	42
Avg. Loss	$1,661,308	$4,052,047	$27,465,086	N/A	$1,252,464	N/A	$1,751,281	$492,459

5   Conclusion and Future Work

We presented the first work in the literature that investigated the real-world, large-scale impact of blockchain-based IdM on identity theft and fraud. We utilized a model of about 6,000 reported identity theft and fraud cases that occurred in the past 20 years (from 2000 to 2020), and estimated how much blockchain-based IdM solutions (Civic, ShoCard, and Authenteq) could have saved identity theft victims, if they had been used. We found that some PII are more important to protect: complete protection of Social Security Card, for example, would have eliminated about 10% of all the incidents we recorded and modeled, with an average monetary loss of over $27M per incident. We also recommended new PII to add to blockchain-based IdM solutions: most notably Financial Information, e.g., Debit/Credit Card, Bank Account, and Taxpayer Information. Our results are useful to developers of new this new line of IdM and their users. In the future, we can retrospectively evaluate the effect of the universal use of blockchain-based IdM through the same methodology, once it is widely adopted.

Acknowledgments. This work was in part funded by the Center for Identity's Strategic Partners. The complete list of Partners can be found at https://identity.utexas.edu/strategic-partners.

References

1. Civic decentralized reusable KYC services - blockchain-powered. https://www.civic.com/solutions/kyc-services/. Accessed 04 Nov 2019
2. Identity verification & KYC, authenteq. https://authenteq.com/. Accessed 04 Nov 2019
3. Shocard identity management use cases — shocard. https://shocard.com/identity-management-use-cases/. Accessed 04 Nov 2019
4. Baars, D.: Towards self-sovereign identity using blockchain technology. Master's thesis. University of Twente (2016)
5. Chang, K.C., Zaeem, R.N., Barber, K.S.: Enhancing and evaluating identity privacy and authentication strength by utilizing the identity ecosystem. In: Proceedings of the 2018 Workshop on Privacy in the Electronic Society, pp. 114–120 (2018)
6. Chang, K.C., Zaeem, R.N., Barber, K.S.: Internet of Things: securing the identity by analyzing ecosystem models of devices and organizations. In: 2018 AAAI Spring Symposium Series, pp. 111–116 (2018)
7. Dunphy, P., Petitcolas, F.A.: A first look at identity management schemes on the blockchain. IEEE Secur. Privacy **16**(4), 20–29 (2018)
8. Harrell, E.: Victims of identity theft (2016, revised in 2019)
9. Jacobovitz, O.: Blockchain for identity management. The Lynne and William Frankel Center for Computer Science Department of Computer Science. Ben-Gurion University, Beer Sheva (2016)
10. Lakhani, K.R., Iansiti, M.: The truth about blockchain. Harv. Bus. Rev. **95**, 118–127 (2017)
11. Liau, D., Zaeem, R.N., Barber, K.S.: Evaluation framework for future privacy protection systems: a dynamic identity ecosystem approach. In: 2019 17th International Conference on Privacy, Security and Trust (PST), pp. 1–3. IEEE (2019)

12. Nakamoto, S.: Bitcoin: a peer-to-peer electronic cash system (2008). Technical report, Manubot (2019)
13. Rana, R., Zaeem, R.N., Barber, K.S.: Us-centric vs. international personally identifiable information: a comparison using the UT CID identity ecosystem. In: 2018 International Carnahan Conference on Security Technology (ICCST), pp. 1–5. IEEE (2018)
14. Rana, R., Zaeem, R.N., Barber, K.S.: An assessment of blockchain identity solutions: Minimizing risk and liability of authentication. In: 2019 IEEE/WIC/ACM International Conference on Web Intelligence (WI), pp. 26–33. IEEE (2019)
15. Zaeem, R.N., Budalakoti, S., Barber, K.S., Rasheed, M., Bajaj, C.: Predicting and explaining identity risk, exposure and cost using the ecosystem of identity attributes. In: 2016 IEEE International Carnahan Conference on Security Technology (ICCST), pp. 1–8. IEEE (2016)
16. Zaeem, R.N., Manoharan, M., Yang, Y., Barber, K.S.: Modeling and analysis of identity threat behaviors through text mining of identity theft stories. Comput. Secur. **65**, 50–63 (2017)
17. Zaiss, J., Nokhbeh Zaeem, R., Barber, K.S.: Identity threat assessment and prediction. J. Consum. Aff. **53**(1), 58–70 (2019)
18. Zyskind, G., Nathan, O., et al.: Decentralizing privacy: using blockchain to protect personal data. In: 2015 IEEE Security and Privacy Workshops, pp. 180–184. IEEE (2015)

DigEx

DigEx-2020 Workshop Chairs' Message

Preface

It is our pleasure to present the post-proceeding papers of the Second International Workshop on Transforming the Digital Customer Experience (DigEx 2020), that was held in conjunction with the 23rd International Conference on Business Information Systems (BIS 2020). It was the second edition of a workshop that responded to an increasing need from the scientific and business communities to find a space to exchange ideas and knowledge in this emerging area. Even though this year DigEx took place online due to the COVID-19 pandemic, it did not stop the scientific community from participating in the workshop and exchanging ideas.

In this second edition, the workshop had seven submissions in total, and among them three were accepted for presentation after the reviewing process. Each article was reviewed by at least three reviewers. Among the accepted papers, two of them were presented during the online session. The authors were allowed to revise the papers with the reviewers' comments and the discussions of the presentation before including them in the proceedings.

This year, the Program Committee included 13 researchers representing 9 countries from three continents (America, Asia, and Europe). Their evaluation of the submitted papers, by using the highest scientific standards, enabled us to ensure the workshop quality in terms of the value of the scientific contributions, the excellence of the presentations, and the pertinence of the discussion.

We would like to thank the authors of the submitted papers and the members of the Program Committee that participated and helped in any way to promote the workshop. We are grateful to the organizers of the BIS 2020 conference, which belongs to the Poznań University of Economics and Business, Poland, and the University of Colorado, USA, for their help in the transmission and organization of the conference and associated workshops online. Without their efforts, it would have been impossible to organize this academic event.

August 2020

Oscar Avila
Virginie Goepp
Beatriz Diaz

Organization

Chairs

Oscar Avila (Chair) — Universidad de los Andes, Colombia
Virginie Goepp (Co-chair) — INSA Strasbourg, France
Beatriz Diaz (Co-chair) — Universidad Nacional de Colombia, Colombia

Program Committee

Oscar Avila — Universidad de los Andes, Colombia
Sonia Camacho — Universidad de los Andes, Colombia
Alejandro Cataldo — Universidad de Talca, Chile
Beatriz Díaz — Universidad Nacional de Colombia, Colombia
Lauri Frank — University of Jyväskylä, Finland
Virginie Goepp — INSA Strasbourg, France
Laura González — University of the Republic, Uruguay
Paul Grefen — Eindhoven University of Technology, The Netherlands
Tiina Kemppainen — University of Jyväskylä, Finland
Michael Petit — University of Namur, Belgium
Camille Salinesi — Université de Paris, France
Markus Siepermann — TU Dortmund, Germany
Mathupayas Thongmak — Thammasat Business School, Thailand

Benefits of the Technology 4.0 Used in the Supply Chain - Bibliometric Analysis and Aspects Deferring Digitization

Anna Maryniak[✉] and Yuliia Bulhakova[✉] [iD]

Poznań University of Economics and Business, 61-875 Poznań, Poland
{anna.maryniak,yuliia.bulhakova}@ue.poznan.pl

Abstract. Nowadays, Economy 4.0 is more and more often discussed in relation to communities, cities, centres of public administration, the non-profit sphere, public institutions, factories, and the service sphere. This research is at an early stage of development and there is therefore a need to formulate a scientific framework and carry out research to bring new knowledge into the discipline of management science. The paper focuses on 4.0 technologies that are used in chain management and the benefits they bring. They are an integral part of Economy 4.0. The considerations aim to define the 4.0 supply chain, to indicate the interest in particular technologies in quantitative terms, to identify the benefits of their implementation, to identify the most frequently related keywords, and to propose directions for further research.

Research has shown, among other things, that there is a need to identify factors that delay the digitalization of supply chains.

In the area of quantitative research, among other things, the productivity of leading authors and scientific centres, which most often take up the topics from SC 4.0, was presented. Bibliometric analysis in relation to generated records were used as a research method.

Keywords: Industry 4.0 · Supply chain · Digitalization

1 Introduction

Nowadays, in the context of the supply chains one can more and more often see terms such as Industry 4.0. Examples of different definitions of this term are given in the review work [1], which also contains definitions of individual technologies 4.0. However, it is not possible to give a single universal definition for all technologies, as each has its specificities. Since the supply chain 4.0 (SC 4.0) is under consideration, it is assumed that:

supply chain 4.0 is a chain in which there is a flow of goods (and accompanying information and financial flows) between vertically integrated companies creating a multidirectional network of connections in real time through universal technologies such as Internet of Things, Big Data, Cyber-Physical Systems, Cloud Computing, Cyber-Security and through technologies specifically dedicated to logistics processes such as

W. Abramowicz and G. Klein (Eds.): BIS 2020 Workshops, LNBIP 394, pp. 173–183, 2020.
https://doi.org/10.1007/978-3-030-61146-0_14

Additive Manufacturing, Augmented Reality, Autonomous Robots, Vehicles Autonomous, Real-Time Locating System, etc. This applies to both upstream and downstream flows, which in economic reality most often form a multidirectional network of links between supply chain participants. The proposed scope of SC 4.0 is difficult to confront with other proposals because, according to the authors' knowledge, such a definition does not exist in the literature.

It should be mentioned that SC 4.0 is sometimes synonymous with such terms as Intelligent Supply Chain, Digital Supply Chain, Intelligent Supply Chain. This is a research area that is still poorly recognized, although many articles highlight the usefulness of intelligent supply chain solutions. Reviews are slowly beginning to emerge, but they do not cover a wider set of technologies that are directly related to the flow of goods between contractors in the supply chain. Therefore, the aim of further reflection is to identify the most frequently discussed technologies of this type, on the topics addressed in SC 4.0, the benefits of implementing the technologies, and to define the directions for further research.

2 Research Methodology

The need to deepen the supply chain 4.0 theme was born out of the results of the pilot study [2] aimed to identify what kind of technologies are used in business practice and whether there is the intention to digitalization supply chains to a greater extent. The survey was conducted on the basis of in-depth interviews in several Polish links in the supply chain (producers of raw materials, semi-finished products, final products, and trade). In total, several entities were examined. The respondents were decision-makers whose responsibilities are related to logistics (e.g. supply, transport, storage).

Research has shown a low level of implementation of 4.0 technology in supply chains.

In the first stage, on the basis of the literature review, basic groups of technologies that are directly related to the physical flow in the supply chain were distinguished.

In the second research stage, articles under the slogan "supply chain" were generated and then the number of records was narrowed down to the items in which any of the five separate technology groups 4.0 was discussed according to the established search path: (TITLE("Additive Manufacturing" OR "3D printing" OR "Rapid prototyping" OR "Direct digital manufacturing" OR "DDM" OR "Layered manufacturing")… further by analogy).

Based on the obtained results, has been identified the research areas in which the articles are published, the most productive authors, countries and research centres specialising in this topic as well as keywords that most frequently accompany articles on technology 4.0 in supply chains. Presented also the number of publications. The search area was made up of titles and keywords.

The most common research topics were then identified using keyword analysis. The analysis was made by correlating the keyword "supply chain 4.0 with the names of all technologies (which are detailed in Sect. 4). In this part of the research, the area of searches were keywords. In total, 19500 different keywords were analysed, with five search paths combined in this case covering all technologies in total.

The Scopus database was used for the research. Scopus is a global database of scientific publications that contains information on more than 70 million works. Based on content analysis, Scopus adds appropriate keywords for each publication that authors of the publication may have omitted or made an error in writing these keywords. Therefore, more complete search results can be obtained from generally accepted and modern keywords.

In a third stage, selected examples point to the benefits of using particular technologies in the supply chain.

In the final stage, the authors indicated future research directions.

3   Technologies 4.0 Used in Supply Chains

Stage 1. In the literature on the subject, the proposals of technology classification 4.0 include from a few to a dozen or so technologies, however, it is possible to distinguish the types of technologies that are most frequently repeated [2–7]. Taking into account the research topics, the technologies that are more directly related to physical flows were selected for further analysis. However, it should be noted that other technologies are also useful for supply chain management, as indicated by many authors (Kamble, Gunasekaran i Gawankar [8] - IoT, block-chain; Makris, Hansen i Khan [9] - Big Data, cloud computing; Din i in. [10] - CPS; Perboli, Musso i Rosano [11] - Blockchain; Hatterscheid i Schluter [12] - CPS; Awwad i in. [13] - Big Data; Juhász i Bányai [14] - CPS; Li, Sijun i Cui [15] - Big Data, cloud computing; Rajput over Singh [16] - Artificial Intelligence).

Therefore, in the next stage, for empirical research, the following technology groups have been identified with strong links to the operational (physical) sphere of SC management:

A. Layered production of products using a laser or electron beams referred to as slogans: Additive Manufacturing/3D Printing/Rapid Prototyping/Direct Digital Manufacturing (DDM)/Layered Manufacturing;
B. Recording by the camera of images of the physical world and synchronizing it with   computer graphics and animations, that is: Augmented Reality/Virtualization/Visualization Technology/Virtual Reality/Digital Twin/Augmented Wearables;
C. A technical device which, to accomplish a given task, moves smoothly in a given environment and without direct operator intervention, called: Autonomous Robots/Robotics/Advanced Robotisation/Robotisation/Collaborative Robots/Cobots;
D. A computer-controlled car, usually unmanned, described under the password: Vehicles Autonomous/Wireless Vehicles/Autonomous Transport/Drones/Drone;
E. Technologies of systems locating moving objects in real-time, which primarily include: Real-Time Locating System Technologies (RTLS)/Radio-Frequency Identification (RFID)/GPS.

At the same time, the above passwords have been included in the search paths of individual technologies, which have been separated by the suffix "or".

4   Technologies 4.0 Used in Supply Chains - Statistical Analysis

Stage 2. As a result of bibliometric research (based on the methodology presented in Sect. 2) was determined that most of the literature is devoted to technologies supporting vehicle location and product identification. A particularly large number of scientific studies concern RFID (Fig. 1).

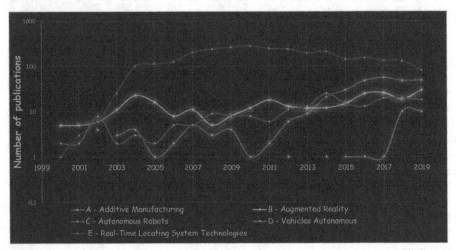

Fig. 1. Number of publications on "supply chain 4.0" per technology. Source: own research based on Scopus database.

To the core research areas (according to the Scopus classification) in which scholars write about the 4.0 technologies used in supply chains belong to Engineering; Computer Science; Business, Management, and Accounting; Decision Sciences. They account for three-quarters of all studies analyzed.

Detailed research shows that the overall percentage distribution in the main research areas for individual technologies is similar. The largest differences in the percentage distribution of research areas in relation to other technologies are found in the Autonomous Robots technology group. There is a greater share of articles in the Engineering area and a smaller share in the Decision Sciences area.

The authors come mainly from the United States, China, Great Britain, Germany and China.

Taking into account particular groups of detailed technologies, it can be concluded that, apart from the countries mentioned above, the research areas are dominated by Finland and Italy (Additive Manufacturing), Italy and India (Augmented Reality), Japan and India (Autonomous Robots), Canada and India (Vehicles autonomous), Italy and Australia (Real-time locating system technologies). The United States is the leader in terms of the number of publications published for all technologies.

Some of the most productive authors (who write most about it) in the scope of the discussed topic include: Piramuthu, S., Li, Y., Masciari, E., Patrono, L. Within the framework of the different technologies these are: [17–19] - additive manufacturing;

[20–22] - augmented reality; [23–25] - autonomous robots; [26–28] - real-time locating system technologies.

The most productive research centres are Hong Kong Polytechnic University, University of Florida and Pennsylvania State University.

As a result of the analysis of keywords assigned to the generated literature items, it has been concluded that the 4.0 theme in the context of specific technologies is most often combined with the issues of production and inventory management (Fig. 2).

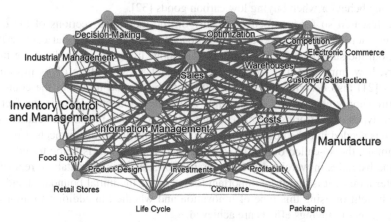

Fig. 2. Networking of keywords from Industry 4.0 literature on Industry 4.0 in supply chains Source: own research based on Scopus database. **Access to the interactive version**: http://data. lewoniewski.info/supplychain/test3.html

5   Technology 4.0 in the Supply Chain and Its Benefits

Stage 3. The literature review shows that the potential of technology 4.0 in supply chains is still underutilised. Many applications are at the conceptual stage or are implemented on a small scale in specific case studies. Most articles highlight their positive impact on supply chain management.

In the context of supply chain digitalization, one of the most popular topics today is **3D Printing**. This technology can be used in many sectors, such as aerospace, Huang, and others [17]. Empirical research indicates many advantages of 3D Printing, such as the simplification of supply chains, by omitting production stages of individual parts, the possibility to decentralize the production of authors [19] and thus shorten the delivery time of products which are manufactured closer to the consumer [17], saving energy and CO_2 emissions during the life cycle of products [29], the possibility of recycling waste from such production [29], the possibility of including prosumers in the creation of products [30].

Despite the huge potential of this technology and demonstrating that it can be a source of additional revenue [30], there is a need for more research into the economics of processing and transporting certain materials (e.g. powdered metal) and estimating the costs of such products for the environment and human health (e.g. in terms of electron radiation or disposal of printing materials).

Another technology - the so-called **Augmented Reality** - among other things, it allows the design of parts (using CAD and CAM computer systems), production simulation, and visualization of these processes [20], virtualization of products [21], risk management in the chain through the use of digital twins [31], shortening the time and cost of developing the project [20], improving the quality products [22], studying purchasing behavior when buying low carbon goods [32].

The research so far has focused mainly on the individual sections of the chain. However, more emphasis is needed on a broader range of studies that capture flows throughout the supply chain. E.g. authors [21] show, using the example of the flower industry, how important it is to visualize a product at every stage of its movement. Authors [21] also describe the virtualization of objects and things in the context of the entire supply chain using the example of the food industry. Despite the fact that different types of platforms are increasingly affordable also for smaller entities [21], research is needed into the cost-effectiveness of implementing such projects. Currently, for example, the so-called digital glove integrated with glasses is being tested, which scans the barcodes in warehouses and searches for subsets in the database replacing a computer mouse. It is therefore important to determine the initial boundary conditions for the benefit of using this type of innovation and whether, in addition to improving logistics, specific savings effects are achieved.

Another of the analyzed technologies are **Autonomous Robots** [25]. On specific examples, the authors prove that robots: are particularly helpful in situations of fast-changing product lines, with complex and labor-intensive tasks and high start-up costs of [33], in cooperation with agent technologies, help solve complex planning and scheduling problems and minimize production costs for [24] and [25], can operate continuously without intervention, assuming [33], can capture dangerous situations resulting from human error [34].

Despite the many advantages of robots, research shows that even if companies increase the automation of their logistics processes, they do not generally intend to invest in more expensive and advanced technologies [23]. Despite the proven effectiveness of the use of robots [33] is rarely used in small and medium-sized enterprises as well. In the supply chain management process (in ports, warehouses) even in larger enterprises are only partially used.

Taking into account the next technology group - **Vehicles Autonomous** - the literature underlines that by using wireless car networks, Intelligent Transport Systems (ITS) have the potential to shape the future of multimodal logistics. Among other things, it is indicated that unmanned vehicles: can be useful in humanitarian actions, in which logistics operations involving humans are severely hampered by [34], reduce energy consumption and greenhouse gas emissions, [35], can help with the inventory process, have lower operating costs, provide greater safety, shorter delivery times, and have a positive effect on sustainable development in transport and storage [36].

Recently, autonomous vehicles have proved to be particularly useful in the provision of food in city logistics and the disinfection process in hospital logistics, during a pandemic. This direction of research - so-called social logistics - is poorly described. There is a lack of extensive research studies and analyses indicating the costs connected with it.

The last of the analyzed groups is the **Technology of Systems Locating Moving Objects in Real-Time.** Technologies of this type, among, reduce labor costs, simplify business processes, shorten delivery times and reduce the costs of inventory control of [37], automate and optimize logistic tasks in the scope of grouping, routing and creating harmonograms [38], together with other technologies others such as wireless sensor networks (WSN), middleware, cloud computing, and IoT application software - they can create essential IoT technologies [39].

This is the most widely discussed group in the scientific literature. The literature presents mainly benefits resulting from the application of this technology, points to several device and application functionality. However, there is a lack of benchmarking analyses indicating cost relations before and after the implementation of this technology about the management of the whole supply chain.

In the analysed articles, simulations of case studies or traditional case studies were used as a research method. Quantitative research is only at the initial stage of development and usually concerns one selected technology. Therefore, there is a need to create a model that could be tested on a representative group of enterprises.

The considerations focus primarily on indicating the benefits of implementing individual solutions, on the description of changes in configuration and flow of goods (including in particular inventory levels).

6 Delaying Implementation of S.C. 4.0 - Future Directions of Research

Stage 4. On the basis of the conducted pilot projects interviews, it appears that public and business entities are not always interested in implementing modern technologies or are only interested in them in selected areas of their activity. According to representatives of enterprises (including those with a well-known brand), which are the main coordinators of supply chains and at the same time have great bargaining power over other participants in the supply chain, in many cases there is no justification for introducing innovative solutions.

There is therefore a need to identify factors that de facto do not have positive impact on the economic performance of supply chains.

For example, the market within which the supply chain is embedded may not require the introduction of spectacular technologies, as relying on existing solutions and coexistence in current technological relations may ensure the maintenance of a competitive advantage.

It therefore seems that there may be a deliberate "postponement of supply chain digitalization", as such solutions are not necessarily indispensable under the current market, configurations and economic conditions.

Therefore, there is a need to establish a list of factors that determine whether the implementation of digitised chains should be postponed or completely abandoned (i.e. barriers to their implementation) and prioritised, as well as to determine whether the implementation of such chains brings unambiguous economic effects. **In particular, it is worth concentrating on those factors that cause a deliberate "postponement" and are not an obvious barrier, such as (e.g. lack of competent people, lack of knowledge of technology) and focus on negative results** (Fig. 3).

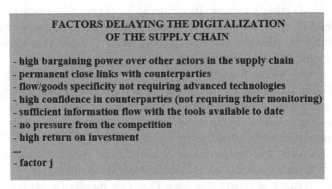

Fig. 3. Factors delaying the digitalization of the supply chain

There are generally few studies on success factors for supply chain management in the literature. Among other things, there are items on universal aspects for each chain [40], on specific industries [41], IT systems [42], and the use of specific technologies 4.0 [43]. There are also few items on factors for the implementation of various supply chains [44].

Concerning the factors hindering the implementation of 4.0 technologies, they are considered without reference to vertically integrated economic actors [8]. Sometimes only factors are indicated for the very reasons of the implementation of the digitalization chains about specific technologies [45].

Therefore, the results obtained under the proposed new research line will be difficult to confront them with those already existing and to make benchmarking analyses. However, they may become an important reference point for further analyses.

7   Restrictions on Research

Search trails contain the most popular keywords for a given technology. However, these search paths can be extended with further keywords and writing methods.

Due to the format of the article work, only sample items of literature are referred to.

Moreover, in the future research should be extended to other technologies, such as Big Data, the Internet of Things (IoT), Physical Cybersystems (CPS), Cloud Computing and CyberSecurity, and the benefits of their application should be considered multi-faceted, depending on the number of simultaneous implementations of several technologies.

Due to editorial limitations, only selected technology applications from all generated and analyzed records are cited. For this reason, it is also not possible to carry out generalization with an indication for each analyzed source.

8 Conclusions

As a result of the research work carried out, it was found, inter alia, that there is an upward trend in publications in the scope of all surveyed technologies. More than 70% of the items generated concern only four research areas. It is also likely that more research will become interdisciplinary in the future. At present, authors from the United States, China, the United Kingdom and Germany dominate. The description of technology most often refers to the sphere of production and inventory management.

The literature analysis concluded that theme 4.0 in the context of specific technologies is most often combined with production and stock management issues.

On the basis of the collected material, it can be concluded that there are numerous benefits resulting from the implementation of technology 4.0 in the areas of supply chain management. At the same time, the level of digitalization of the supply chain is not high. Therefore, research should be carried out in the future to identify factors that delay the digitalization of the supply chain. As there are no such studies in the literature, this is an interesting and future direction of research is needed.

Acknowledgements. We would like to thank dr Włodzimierz Lewoniewski for technical support for data generation for this article.

References

1. Tay, S.I., Lee, T.C., Hamid, N.Z.A., Ahmad, A.N.A.: An overview of industry 4.0: definition, components, and government initiatives. J. Adv. Res. Dyn. Control Syst. **10**(14), 1379–1387 (2018)
2. Konecka, S., Maryniak, A.: Intelligent and innovative solutions in supply chains. In: Ahram, T., Karwowski, W., Pickl, S., Taiar, R. (eds.) IHSED 2019. AISC, vol. 1026, pp. 854–859. Springer, Cham (2020). https://doi.org/10.1007/978-3-030-27928-8_129
3. Manavalan, E., Jayakrishna, K.: A review of Internet of Things (IoT) embedded sustainable supply chain for industry 4.0 requirements. Comput. Ind. Eng. **127**(1), 925–953 (2019)
4. Zhong, R.Y., Xu, X., Klotz, E.T., Newman, S.T.: Intelligent manufacturing in the context of industry 4.0. A review. Engineering **3**(5), 616–630 (2017)
5. MPI Distribution and Logistics Report, The MPI Group. https://www.youtube.com/watch?v=ZXGCRuZ_fp4&list=PL6YSN5R3uIn_iIOP-C6sF2RKOzgV0xfhs. Accessed 15 May 2017
6. Barata, J., Da Cunha, P.R., Stal, J.: Mobile supply chain management in the industry 4.0 era. An annotated bibliography and guide for future research. J. Enterp. Inf. Manag. **31**(1), 173–193 (2018)
7. McKinsey: Digital, Industry 4.0 after the initial hype Where manufacturers are finding value and how they can best capture it. McKinsey & Company. https://www.mckinsey.com/~/media/mckinsey/business%20functions/mckinsey%20digital/our%20insights/getting%20the%20most%20out%20of%20industry%204%200/mckinsey_industry_40_2016.ashx. Accessed 21 Nov 2016

8. Kamble, S.S., Gunasekaran, A., Gawankar, S.A.: Achieving sustainable performance in a data-driven agriculture supply chain. A review for research and applications. Int. J. Prod. Econ. **219**, 179–194 (2020)

9. Makris, D., Hansen, Z.N.L., Khan, O.: Adapting to supply chain 4.0: an explorative study of multinational companies. Supply Chain Forum: Int. J. **20**, 1–16 (2019)

10. Din, F.U., Henskens, F., Paul, D., Wallis, M.: Formalisation of problem and domain definition for agent oriented smart factory (AOSF). In: 2018 IEEE Region Ten Symposium (Tensymp), Sydney, Australia, pp. 265–270 (2018)

11. Perboli, G., Musso, S., Rosano, M.: Blockchain in logistics and supply chain: a Lean approach for designing real-world use cases. IEEE Access **6**, 62018–62028 (2018)

12. Hatterscheid, E., Schluter, F.: Towards a decision support approach for selecting physical objects in collaborative supply chain processes for cyber physical system-transformation. In: 2018 5th International Conference on Industrial Engineering and Applications (ICIEA) (2018)

13. Awwad, M., Kulkarni, P., Bapna, R., Marathe, A.: Big data analytics in supply chain: a literature review. In: Proceedings of the International Conference on Industrial Engineering and Operations Management, Washington DC, USA, 27–29 September 2018 (2018)

14. Juhász, J., Bányai, T.: What industry 4.0 means for just-in-sequence supply in automotive industry?. In: Jármai, K., Bolló, B. (eds.) VAE 2018. LNME, pp. 226–240. Springer, Cham (2018). https://doi.org/10.1007/978-3-319-75677-6_19

15. Li, F., Sijun, L., Cui, Y.: Logistics Planning and Its Applications for Engine Plant under "Industry 4.0". SAE Technical Paper Series (2018)

16. Rajput, S., Singh, S.P.: Connecting circular economy and industry 4.0. Int. J. Inf. Manag. **49**, 98–113 (2019)

17. Huang, S.H., Liu, P., Mokasdar, A., Hou, L.: Additive manufacturing and its societal impact: a literature review. Int. J. Adv. Manuf. Technol. **67**(5–8), 1191–1203 (2012). https://doi.org/10.1007/s00170-012-4558-5

18. Mellor, S., Hao, L., Zhang, D.: Additive manufacturing: a framework for implementation. Int. J. Prod. Econ. **149**, 194–201 (2014)

19. Khajavi, S.H., Partanen, J., Holmström, J.: Additive manufacturing in the spare parts supply chain. Comput. Ind. **65**(1), 50–63 (2014)

20. Chryssolouris, G., Mavrikios, D., Papakostas, N., Mourtzis, D., Michalos, G., Georgoulias, K.: Digital manufacturing: history, perspectives, and outlook. Proc. Inst. Mech. Eng. Part B: J. Eng. Manuf. **223**(5), 451–462 (2009)

21. Verdouw, C.N., Wolfert, J., Beulens, A.J.M., Rialland, A.: Virtualization of food supply chains with the internet of things. J. Food Eng. **176**, 128–136 (2016)

22. Lawson, G., Salanitri, D., Waterfield, B.: Future directions for the development of virtual reality within an automotive manufacturer. Appl. Ergon. **53**(Part B), 323–330 (2016)

23. Pyke, D., Robb, D., Farley, J.: Manufacturing and supply chain management in China: a survey of state-, collective-, and privately-owned enterprises. Eur. Manag. J. **18**(6), 577–589 (2000)

24. Ito, T., Mousavi Jahan Abadi, S.M.: Agent-based material handling and inventory planning in warehouse. J. Intell. Manuf. **13**(3), 201–210 (2002). https://doi.org/10.1023/A:101578682 2825

25. Giordani, S., Lujak, M., Martinelli, F.: A distributed multi-agent production planning and scheduling framework for mobile robots. Comput. Ind. Eng. **64**(1), 19–30 (2013)

26. Wu, D., Rosen, D.W., Wang, L., Schaefer, D.: Cloud-based design and manufacturing. A new paradigm in digital manufacturing and design innovation. Comput.-Aided Des. **59**, 1–14 (2015)

27. Kamel Boulos, M.N., Wilson, J.T., Clauson, K.A.: Geospatial blockchain: promises, challenges, and scenarios in health and healthcare. Int. J. Health Geogr. **17**(1), 1–10 (2018)

28. Coronado Mondragon, A.E., Lalwani, C.S., Coronado Mondragon, E.S., Coronado Mondragon, C.E., Pawar, K.S.: Intelligent transport systems in multimodal logistics: a case of role and contribution through wireless vehicular networks in a sea port location. Int. J. Prod. Econ. **137**(1), 165–175 (2012)

29. Zander, E., Gillan, M., Lambeth, R.: Recycled polyethylene terephthalate as a new FFF feedstock material. Addit. Manuf. **21**, 174–182 (2018)

30. Halassi, S., Semeijn, J., Kiratli, N.: From consumer to prosumer: a supply chain revolution in 3D printing. Int. J. Phys. Distrib. Logist. Manag. 1–18 (2018)

31. Ivanov, D., Dolgui, A., Das, A., Sokolov, B.: Digital supply chain twins: managing the ripple effect, resilience, and disruption risks by data-driven optimization, simulation, and visibility. In: Ivanov, D., Dolgui, A., Sokolov, B. (eds.) Handbook of Ripple Effects in the Supply Chain. ISORMS, vol. 276, pp. 309–332. Springer, Cham (2019). https://doi.org/10.1007/978-3-030-14302-2_15

32. Tong, W., Mu, D., Zhao, F., Mendis, G.P., Sutherland, J.W.: The impact of cap-and-trade mechanism and consumers' environmental preferences on a retailer-led supply chain. Resour. Conserv. Recycl. **142**, 88–100 (2019)

33. Guerin, K.R., Lea, C., Paxton, C., Hager, G.D.: A framework for end-user instruction of a robot assistant for manufacturing. In: 2015 IEEE International Conference on Robotics and Automation (ICRA), Seattle, WA, pp. 6167–6174 (2015)

34. Askarpour, M., Mandrioli, D., Rossi, M., Vicentini, F.: Formal model of human erroneous behavior for safety analysis in collaborative robotics. Robot. Comput.-Integr. Manuf. **57**, 465–476 (2019)

35. Rabta, B., Wankmüller, C., Reiner, G.: A drone fleet model for last-mile distribution in disaster relief operations. Int. J. Disaster Risk Reduct. **28**, 107–112 (2018)

36. Chiang, W.-C., Li, Y., Shang, J., Urban, T.L.: Impact of drone delivery on sustainability and cost. Realizing the UAV potential through vehicle routing optimization. Appl. Energy **242**, 1164–1175 (2019)

37. Han, K.H., Bae, S.M., Lee, W.: Integrated inventory management system for outdoors stocks based on small UAV and Beacon. In: Rocha, Á., Adeli, H., Reis, L.P., Costanzo, S. (eds.) WorldCIST'18 2018. AISC, vol. 746, pp. 533–541. Springer, Cham (2018). https://doi.org/10.1007/978-3-319-77712-2_50

38. Zhu, X., Mukhopadhyay, S.K., Kurata, H.: Review of RFID technology and its managerial applications in different industries. J. Eng. Tech. Manag. **29**(1), 152–167 (2012)

39. Lam, C.Y., Ip, W.H.: An integrated logistics routing and scheduling network model with RFID-GPS data for supply chain management. Wirel. Pers. Commun. **105**(3), 803–817 (2019). https://doi.org/10.1007/s11277-019-06122-6

40. Lee, I., Lee, K.: The Internet of Things (IoT): applications, investments, and challenges for enterprises. Bus. Horiz. **58**(4), 431–440 (2015)

41. Ab Talib, M.S., Abdul Hamid, A.B., Thoo, A.C.: Critical success factors of supply chain management: a literature survey and Pareto analysis. EuroMed J. Bus. **10**(2), 234–263 (2015)

42. Pilyavskyy, A., Maryniak, A., Bulhakova, Y.: Key success factors in managing supply chain – ISM analysis. In: Kolinski, A., et al. (eds.) Contemporary challenges in supply chains. WSL FORUM Conference 2019, Spatium, vol. 1, pp. 94–105 (2019)

43. Fernando, E., Surjandy, Warnars, H.L.H.S., Meyliana, Kosala, R., Abdurachman, E.: Critical success factor of information technology implementation in supply chain management: literature review. In: 2018 5th International Conference on Information Technology, Computer, and Electrical Engineering (ICITACEE), pp. 315–319 (2018)

44. Kumar, S., DeGroot, R.A., Choe, D.: Rx for smart hospital purchasing decisions. Int. J. Phys. Distrib. Logist. Manag. **38**(8), 601–615 (2008)

45. Maryniak, A.: Zarządzanie zielonym łańcuchem dostaw. UEP, Poznań (2017)

How Do Movie Preferences Correlate with e-Commerce Purchases? An Empirical Study on Amazon

Marcin Szmydt[✉]

Department of Information Systems, Poznan University of Economics and Business,
Aleja Niepodległości 10, Poznan, Poland
marcin.szmydt@ue.poznan.pl

Abstract. The following paper presents a study on the relationship between customer movie preferences and online purchases of products from different categories. The analysis was based on the dataset of 233.1 million Amazon reviews and followed the CRISP-DM methodology. The presented findings confirm that movie preferences correlate with specific product purchase preferences. For instance, customers who watch movies from the categories *documentary* and *drama* are more likely to be interested in *books* purchase, whereas people who watch *action* movies are having higher scores in *electronics*. The following paper contributes especially in directly linking movie preferences and product categories purchases. Provided analysis and generalized model should be interesting for both researchers and practitioners from the e-commerce domain.

Keywords: Recommender systems · Cross-domain · Big data · e-commerce · Amazon · CRISP-DM

1 Introduction

Recommender systems (RS) are commonly used as information search and decision support tools. They are addressing the problem of information overload by generating personalized suggestions that suit user's needs. RS are intended to help users find products or services such as books, movies, or even people, based on a different kind of information about the user or recommended item [1].

There are three major types of recommender systems: content-based (CB), collaborative filtering (CF), and hybrid recommender systems [18]. CB exploit similarities among items, e.g., recommending music of the same genre or news articles on the same topic. In contrast, CF exploits similarity and relationships among users to provide recommendations [14]. The most successful examples of CF recommender systems are employed on Amazon[1], Netflix[2], Spotify[3], and

---

[1] http://amazon.com.
[2] http://netflix.com.
[3] http://spotify.com.

© Springer Nature Switzerland AG 2020
W. Abramowicz and G. Klein (Eds.): BIS 2020 Workshops, LNBIP 394, pp. 184–196, 2020.
https://doi.org/10.1007/978-3-030-61146-0_15

Last.fm[4]. According to the Microsoft Research report, 30% of Amazon.com's pages views were from recommendations [17]. Similarly, RS used by Netflix is so effective that more than 80% of movies watched by Netflix users came through recommendations [11]. Therefore, RS are a very important element of current online businesses. However, despite the high effectiveness of RS, companies are still seeking to improve their algorithms since any improvements in this area can be very lucrative for them. Thus, research in this domain seems justified. Moreover, while most of the companies are focused on offering recommendations for items belonging to a single domain (e.g. Netflix, Spotify), there are large e-commerce sites like Amazon or eBay which often store customer purchasing behavior from multiple domains. It is beneficial for them to leverage the knowledge from one domain (source domain) to the other domain (target domain) to generate better recommendations. These kinds of recommendations are called *cross-domain recommendations* and are getting increased interest. Many papers are analysing cross-domain recommendations [3,9,22]. Yet, to the best of the author's knowledge, there are no articles nor publications with the main focus on exploring the relationship between movie preferences and online product purchases in different categories. The unprecedented success of Netflix RS is another reason to explore the predictive power of movie preferences combined with other domains. Those potential relationships might be also another suggestion on how to solve the so-called cold-start problem for the users for which movie preferences are already known. The above motivation gives the way to outline the main research question investigated in this paper, which is: *Is there a statistically significant correlation between movie preferences and online purchases in different product categories?*

2   Literature Review

2.1   Search Methodology

A comprehensive literature review has been conducted to analyze the current state of the art of research in the area of cross-domain recommender systems. Key recommendations provided by Webster and Watson in [20] has been used as a guideline to enhance the rigor of the research process. The chosen research databases consisted of Springer and IEEEExplore since they covered many papers related to Information Systems, IT, and e-commerce. Moreover, the Google Scholar search engine and Mendeley tool were used to provide relevant articles that were not found in the databases listed above. The search was conducted using the keywords and phrases: "cross-domain recommender systems"; "cross-domain collaborative filtering"; "cross-domain recommendations"; "movie preferences and product purchases"; "movie preferences and product preferences"; "cross-domain recommendations based on movies" and combining them with the domain keywords: "e-commerce", "online"; and "Amazon".

---

[4] http://last.fm.

The main goal of this literature review was to summarise existing knowledge in recommender systems with a primary focus on the cross-domain recommendations in e-commerce. The secondary goal of this literature review was to find any papers exploring potential relationships between movie preferences and product purchases in different categories.

An initial number of publications related to cross-domain recommendation systems consisted of 26,908 items (Springer) and 232 items (IEEE). Filtering to publications from the last six years (2015–2020) allowed to decrease this number to 11,743 items (Springer) and 199 items (IEEE). The list of publications was limited to the top 1000 papers recommended by the search engine. Then the title screening was conducted and 88 papers were selected. After title filtering, abstracts were analyzed and 40 positions were selected for full-text reads. Then backward and forward selection added 12 more papers. Not being able to describe all of them, the author selected and referenced 19 positions that are the most representative and interesting studies. The literature has been sorted historically with a description of their key contributions.

2.2   Related Work

The first paper worth investigating [5] covers an ontology-based customer profile model with the multi-source cross-domain nature and focused on clear semantics of customer interests. In general, the paper is using the Amazon case study dataset to highlight the basic conceptual issues of the 3G (third-generation) recommending systems model. The model seems interesting from the perspective of cross-domain recommendations. Yet, no empirical results were presented here.

In the same year, [15] proposed an approach for recommender systems based on graph metrics and linked open data. According to their evaluation based on the three tasks of the 2015 LOD-RecSys challenge[5], the results for cross-domain recommendations, outperformed collaborative filtering. As evaluation metrics, they used the recall, precision, and F-measure for the top 10 recommendations. Their graph-based approach seemed to find reasonable relations between books and movies (e.g., common genres) and leverage those for creating predictions.

Analysis and identification of a customer personality is another common approach when it comes to the new user cold-start problem in CF. Researchers in [4] tried to alleviate this problem by exploiting personality information (Big Five scores) inferred from different domains. The main contribution of their paper is a cross-domain rating prediction method. The evaluation of the method was conducted on a myPersonality database subset of 5,027,593 likes from 159,551 users on 16,303 items from three categories, namely books, movies, and music. Presented results show that personality-based methods achieved performance improvements that range from 6 to 94% for users completely new to the system while increasing the novelty of the recommended items by 3–40% concerning the non-personalized popularity baseline. Unfortunately, this research was not extended to other cross-domain recommendations.

---

[5] https://recsys.acm.org/recsys15/challenge/.

Another interesting paper [10] presents a neural network-based cross-domain RS. Their system named CCCFNet combines CF and CB filtering in a unified framework. Evaluation of their system was conducted on two rating datasets: Douban[6] and MovieLens 20M[7]. According to their experiment, CCCFNet consistently outperforms the baselines of Average Filling, Bias Matrix Factorization, CMF, MV-DNN, and SVDFeature.

Scholars in [23] also focused their efforts on providing cross-domain recommendations and alleviating the issues of a cold-start. They proposed a cross-domain recommender system, including three approaches (three auxiliary domain information fusion-based cross-domain CF algorithms), based on multi-source social big data. Evaluation of their work was based on an actual dataset in the book and music domains. According to the presented findings, proposed cross-domain approaches are providing better accuracy scores than the conventional CF-based approaches and matrix factorization. These findings seem promising, yet according to the paper, this approach is restricted to exploiting data from the auxiliary domain that is related to the target domain (e.g. books and music). Domains distantly related are not considered in this study.

A highly related study was conducted by [2]. Researchers in this paper presented the cross recommendation system based on the data from three different domains namely music, movies, and books. Their model is capable of generating one-to-many cross-domain recommendations exploiting movie domain knowledge to generate recommendations for books and music. Yet, their dataset is very small since it covers only 40 users and the analysis is not extended to other domains.

Another interesting approach is presented in [22]. The authors proposed an approach that exploits the potential of partially overlapping domains to improve cross-domain recommendations. Their method utilizes information about the entities in the two domains that partially overlap. The paper presents a cross-domain recommender system based on kernel-induced knowledge transfer, called KerKT. Presented experiments showed that KerKT had 1.13–20% better prediction accuracy compared with six benchmarks (PCC[8], FMM[9], CBT[10], PMF[11] PMFTL[12], RMGM[13]).

More recent studies seem to keep investigating the potential for improvements in the target domain recommendations based on the knowledge gathered from the source domain. In [8] authors presented a generic cross-domain recommendation framework that outlines three main phases ("domain selection

---

[6] Douban is a leading Chinese online community that allows users to record information related to multiple domains, including movies, books, music, and activities.

[7] https://grouplens.org/datasets/movielens/20m/.

[8] Person Correction Coefficient.

[9] Flexible Mixture Model.

[10] Code-Book-Transfer.

[11] Probabilistic Matrix Factorization.

[12] Probabilistic Matrix Factorization Transfer Learning.

[13] Rating-Matrix Generative Model.

criteria", "knowledge transfer", and "recommendation generation") and components essential to conduct cross-domain recommendation. Their framework was based on the analysis of 128 papers gathered from two selected by them literature review papers.

Summing up, there seems to be no universal algorithm for RS, and combining different approaches for different cases might be a good solution. Therefore, exploring relationships between customer movie preferences and product purchases in different categories might be useful for situations where those movie preferences of customers might be easily accessible. This paper contributes in directly linking movie preferences and product categories purchases. Such research is a specialization of existing methods into a specific application domain. However, the study proposes also a general conceptual model for RS based on customer preferences vector.

3   Research Methodology

The CRoss-Industry Standard Process for Data Mining (CRISP-DM) is the underlying research methodology used in this study. It is a popular methodology for increasing the success of data mining projects [21]. CRISP-DM provides a sequence of six phases helping to build and implement data mining models to support business decisions in a real environment (Fig. 1). It defines a project as a cyclic process, where several iterations can be used to allow the final result to be tuned towards the business goals.

The first phase (business understanding) is related to the analysis of the domain and identification of the goals to achieve. In this study, the research question was stated and motivated in the introduction. The second phase covers data understanding. When designing the data mining analysis, it was decided to exploit the potential of Amazon.com since it is the biggest worldwide e-commerce company that offers products from many different categories (including movies). Their website allows retrieving user reviews regarding particular products. Moreover, with each review, there is an associated "Verified Purchase" indicator that shows whether this particular customer has purchased the product on Amazon or not. By this means it is possible to collect data about customers who both purchased and reviewed products from different Amazon categories including movies (reviews after purchasing in the traditional form e.g. DVD's or after watching on Amazon Prime[14]). This way it is possible to compare customers' movie preferences and their purchasing behavior in different product categories. The data did not need to be collected specifically for this research since it was already publicly available. Details about the dataset are presented in the dedicated section of this paper. A thorough analysis of this dataset allowed to achieve the data understanding phase. The next step of CRISP-DM involves data preprocessing, which in this case, took a significant amount of time and resources since there needed to be done many transformations and data merging. Then, the correlation model was created (modeling phase) and evaluated (evaluation

---

[14] Amazon Prime (or Amazon Prime Video) is paid subscription service offered by Amazon that gives users access to additional services such as movies or TV shows.

phase) using a statistical metric (p-value). The results of the correlation analysis are presented in the next section. Finally, conclusions were drawn from the obtained results, the initial research question could be answered and suggestions for deployment were presented.

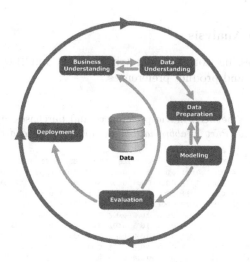

Fig. 1. Cross industry standard process for data mining. Source: [21]

4 Results

The main aim of this paper is to investigate the potential relationship between movie preferences and e-commerce product purchases in different categories. The first part of this section describes the empirical dataset used in this study, while the second part presents Spearman correlation analysis results.

4.1 Dataset Description and Pre-processing

The analysis was carried out using a dataset of 233.1 million Amazon reviews between 1998 and 2018 collected by [12] and publicly available[15]. The preprocessing stage covered combining the reviews dataset with the products' metadata. Then, there was created a dataset of customers who reviewed at least 5 movies (Verified Purchases) and reviewed at least 5 other products (Verified Purchases). Moreover, to limit the dimensionality of the analysis top 30 most popular product categories and 39 movie categories were selected (listed in the Appendix).

Structure of the final dataset covered unique customer identification number, encoded columns representing movie preferences (number of verified reviews in each category), and encoded columns representing product preferences (number

---

[15] http://jmcauley.ucsd.edu/data/amazon/.

of verified reviews in different product categories). The dataset covered 147,130 unique customers. Such large dataset is significant advantage of this paper since in [6,7] the data sets contains only 111 users, in [13] and [16] it is around 100, and in [19] only 52. The dataset covering 20 years of e-commerce data allows to extract correlation patterns that are timeless and lasts for a long time.

4.2 Correlation Analysis

After data preprocessing, correlation matrix (Tables 1 and 2) was built using the dataset with movie and product preferences.

Table 1. Correlations between movie preferences and purchases in different product categories on Amazon - Part 1 *(abbreviations explained in Appendix A)*

	PC1	PC2	PC3	PC4	PC5	PC6	PC7	PC8	PC9	PC10	PC11	PC12	PC13	PC14	PC15
MC1	0,027	0,046	0,034	0,023	0,038	0,000	-0,006	0,007	0,011	0,000	0,043	-0,015	0,021	0,022	-0,009
MC2	0,015	0,060	0,017	0,011	0,023	0,017	0,008	0,008	0,010	0,012	0,031	0,020	0,015	0,034	0,004
MC3	0,022	0,081	0,029	0,018	0,039	-0,004	-0,003	-0,001	0,011	0,008	0,060	0,036	0,010	0,112	0,003
MC4	-0,001	0,007	0,002	0,000	0,001	0,001	-0,001	0,000	0,004	0,002	0,006	-0,002	0,002	0,006	0,004
MC5	0,013	0,020	0,015	0,012	0,014	0,023	0,006	0,002	0,010	0,007	0,008	0,004	0,008	0,005	0,006
MC6	0,004	0,095	-0,008	0,002	0,025	-0,004	-0,008	0,022	0,020	-0,006	0,081	-0,035	0,018	0,104	-0,026
MC7	0,021	0,081	0,010	0,017	0,044	0,003	0,009	0,027	0,028	0,005	0,062	-0,017	0,039	0,063	-0,021
MC8	0,018	0,052	0,012	0,023	0,013	0,073	0,013	0,021	0,009	0,025	0,007	0,021	0,001	0,059	0,001
MC9	0,035	0,053	0,066	0,024	0,067	0,001	-0,014	-0,016	-0,019	0,000	0,047	0,049	0,008	0,038	-0,001
MC10	0,048	0,166	0,049	0,035	0,089	0,008	0,007	0,033	0,031	-0,001	0,126	-0,043	0,052	0,143	-0,026
MC11	0,025	0,161	-0,008	0,024	0,053	0,049	0,041	0,054	0,058	0,024	0,064	-0,015	0,064	0,149	-0,017
MC12	-0,001	0,030	-0,002	0,000	0,002	-0,006	-0,013	0,002	-0,001	-0,008	0,024	-0,019	-0,002	0,031	-0,016
MC13	0,007	-0,004	-0,018	0,010	0,010	0,018	0,030	0,040	0,051	0,031	0,013	-0,006	0,034	0,019	0,019
MC14	0,003	0,038	0,003	0,003	0,011	0,010	0,002	-0,001	0,004	0,002	0,015	0,007	0,007	0,023	-0,007
MC15	-0,002	0,007	-0,020	0,002	0,001	-0,012	-0,013	0,020	0,002	-0,004	0,015	-0,020	-0,003	0,020	-0,008
MC16	0,014	0,022	0,013	0,014	0,015	0,012	0,004	0,007	0,004	0,009	0,010	0,016	0,005	0,027	0,008
MC17	0,078	0,101	0,118	0,067	0,081	0,083	-0,005	-0,006	0,004	0,038	0,044	0,032	0,014	0,032	0,020
MC18	0,008	0,035	0,000	0,008	0,013	0,017	0,013	0,015	0,016	0,013	0,022	0,011	0,016	0,033	-0,002
MC19	-0,023	0,036	-0,025	-0,019	-0,015	-0,020	-0,015	-0,010	-0,012	-0,022	0,025	-0,019	-0,009	0,047	-0,030
MC20	-0,023	0,062	-0,046	-0,018	0,004	-0,010	0,009	0,017	0,024	-0,018	0,038	-0,021	0,026	0,106	-0,040
MC21	0,001	0,012	-0,018	0,009	-0,015	0,013	0,007	0,022	0,032	0,048	0,017	0,072	-0,007	0,010	0,020
MC22	0,008	-0,058	-0,001	0,016	-0,041	-0,001	-0,010	-0,004	-0,003	0,046	-0,027	0,150	-0,046	-0,061	0,046
MC23	-0,020	-0,032	-0,044	-0,017	-0,016	0,009	0,029	0,020	0,039	0,013	-0,019	-0,035	0,020	-0,013	0,014
MC24	0,024	0,114	0,019	0,021	0,045	0,043	0,016	0,022	0,030	0,020	0,044	0,004	0,038	0,082	-0,012
MC25	-0,022	-0,089	-0,058	-0,007	-0,054	0,009	0,029	0,026	0,054	0,057	-0,041	0,088	-0,005	-0,058	0,053
MC26	0,053	0,062	0,047	0,046	0,059	0,045	0,029	0,034	0,057	0,059	0,055	0,078	0,041	0,030	0,047
MC27	0,000	0,004	-0,003	0,003	-0,006	0,001	0,004	0,005	0,001	0,000	-0,004	-0,009	0,000	0,014	-0,006
MC28	0,002	0,063	-0,023	0,004	0,019	0,013	0,024	0,032	0,041	0,006	0,018	-0,012	0,024	0,222	-0,012
MC29	0,001	0,009	0,000	0,003	0,006	0,002	0,006	0,003	0,004	-0,003	0,004	0,003	0,003	0,012	-0,004
MC30	0,006	0,048	0,004	0,008	0,024	0,006	0,010	0,012	0,017	0,006	0,032	0,017	0,015	0,071	-0,012
MC31	0,021	-0,002	-0,011	0,028	0,016	0,018	0,028	0,051	0,037	0,023	0,008	-0,004	0,031	0,054	0,003
MC32	0,070	0,109	0,068	0,059	0,103	0,033	0,026	0,049	0,055	0,040	0,082	0,019	0,054	0,131	0,021
MC33	0,009	0,058	0,012	0,010	0,016	0,004	-0,003	0,006	0,005	-0,001	0,054	-0,001	0,005	0,045	-0,009
MC34	0,036	0,088	-0,033	0,033	0,062	0,094	0,110	0,111	0,145	0,090	0,052	0,012	0,110	0,136	0,042
MC35	0,042	0,031	0,039	0,040	0,042	0,032	0,022	0,023	0,030	0,038	0,028	0,066	0,023	0,018	0,033
MC36	-0,003	0,024	-0,025	0,002	0,001	-0,002	0,010	0,025	0,023	0,009	0,021	0,005	0,006	0,062	-0,003
MC37	-0,004	0,035	-0,018	-0,004	0,012	0,020	0,029	0,024	0,014	-0,002	0,021	-0,015	0,029	0,058	-0,021
MC38	0,096	0,127	0,091	0,086	0,117	0,068	0,038	0,042	0,062	0,080	0,084	0,169	0,061	0,078	0,047
MC39	0,008	0,086	0,018	0,007	0,026	-0,013	-0,016	-0,011	-0,001	-0,009	0,060	0,017	0,001	0,139	-0,017

Source: Own work

Due to the fact that the distribution of the collected dataset is not normal, the Spearman rank-order correlation coefficients were calculated. It is a statistical measure of the strength of a link or relationship between two sets of data. The higher the value the higher the correlation between analysed variables. Results revealed that on the one hand, the highest positive correlations in the analysed dataset of movie categories and product purchase categories were identified between the *Music_Videos_and_Concerts* and *CDs_and_Vinyl* (Spearman value: 0.2217; p-value < 0.00001), *Kids_and_Family* and *Toys_and_Games* (Spearman value: 0.1692; p-value < 0.00001), *Drama* and *Books* (Spearman value: 0.1659; p-value < 0.00001), *Documentary* and *Books* (Spearman value: 0,1613; p-value < 0.00001), *Animation* and *Toys_and_Games* (Spearman value: 0,1497; p-value < 0.00001), *Documentary* and *CDs_and_Vinyl* (Spearman value: 0,1486; p-value < 0.00001), *Action* and *Electronics* (Spearman value: 0,1453; p-value < 0.00001).

Table 2. Correlations between movie preferences and purchases in different product categories on Amazon - Part 2 *(abbreviations explained in Appendix A)*

	PC16	PC17	PC18	PC19	PC20	PC21	PC22	PC23	PC24	PC25	PC26	PC27	PC28	PC29	PC30
MC1	0,026	0,027	0,025	0,024	0,033	0,001	-0,013	0,004	-0,002	0,000	0,023	0,007	-0,011	0,024	0,028
MC2	0,011	0,010	0,021	0,021	0,020	0,015	-0,001	0,008	0,005	0,005	0,018	0,005	0,002	0,024	0,019
MC3	0,014	0,006	0,029	0,029	0,033	-0,010	-0,013	0,027	0,016	0,017	0,037	0,000	0,009	0,037	0,015
MC4	-0,001	-0,004	-0,004	-0,004	0,000	0,001	-0,003	0,006	-0,001	0,002	-0,001	0,001	0,002	0,004	0,005
MC5	0,012	0,012	0,010	0,010	0,015	0,025	-0,001	0,006	0,003	0,002	0,007	0,006	0,008	0,009	0,007
MC6	0,005	0,005	0,031	0,030	0,025	-0,007	-0,017	-0,009	-0,006	-0,003	0,017	0,005	-0,026	0,024	0,023
MC7	0,013	0,014	0,037	0,036	0,038	0,002	0,003	0,004	0,006	0,008	0,032	0,012	-0,024	0,044	0,041
MC8	0,011	0,008	0,011	0,011	0,011	0,045	-0,002	0,015	0,028	0,027	0,014	-0,003	0,004	0,014	0,008
MC9	0,025	0,026	0,023	0,022	0,056	-0,008	-0,019	0,032	0,013	0,015	0,059	-0,019	0,006	0,030	0,030
MC10	0,034	0,033	0,074	0,073	0,074	-0,001	-0,018	0,030	0,006	0,008	0,067	0,010	-0,027	0,078	0,066
MC11	0,017	0,014	0,058	0,057	0,047	0,032	0,016	0,007	0,016	0,019	0,040	0,028	-0,024	0,062	0,063
MC12	0,004	0,002	0,012	0,012	0,001	-0,007	-0,017	-0,002	-0,003	-0,003	0,007	-0,003	-0,014	0,004	0,000
MC13	-0,001	-0,006	0,013	0,013	0,005	0,020	0,029	0,010	0,023	0,022	0,007	0,044	0,012	0,014	0,018
MC14	0,000	0,004	0,018	0,017	0,010	0,008	-0,002	0,001	0,001	0,001	0,011	-0,003	-0,006	0,011	0,007
MC15	0,001	-0,008	0,010	0,010	-0,001	-0,009	-0,012	-0,002	-0,003	-0,002	0,005	-0,001	-0,007	0,004	-0,002
MC16	0,007	0,000	0,001	0,001	0,007	0,005	0,000	0,015	0,017	0,015	0,018	0,000	0,010	0,009	0,004
MC17	0,062	0,071	0,052	0,052	0,077	0,092	-0,024	0,054	0,027	0,025	0,059	-0,003	0,021	0,041	0,033
MC18	0,005	0,000	0,012	0,011	0,012	0,013	0,008	0,004	0,008	0,008	0,010	0,007	-0,003	0,016	0,016
MC19	-0,015	-0,020	0,008	0,008	-0,012	-0,022	-0,011	-0,010	-0,011	-0,009	-0,008	-0,018	-0,028	-0,005	-0,009
MC20	-0,022	-0,028	0,014	0,014	0,006	-0,015	0,007	-0,014	-0,009	-0,006	0,009	0,001	-0,039	0,023	0,022
MC21	-0,007	-0,025	-0,001	0,000	-0,024	0,017	0,004	0,014	0,038	0,035	-0,008	0,034	0,013	-0,001	-0,019
MC22	0,003	-0,017	-0,036	-0,036	-0,045	0,001	-0,003	0,022	0,049	0,041	-0,016	0,015	0,046	-0,029	-0,061
MC23	-0,012	-0,010	-0,013	-0,013	-0,011	0,010	0,035	-0,024	-0,013	-0,011	-0,022	0,037	0,004	0,002	0,000
MC24	0,014	0,013	0,043	0,043	0,043	0,037	-0,001	0,012	0,013	0,011	0,032	0,013	-0,013	0,045	0,036
MC25	-0,021	-0,041	-0,038	-0,037	-0,052	0,012	0,041	-0,001	0,036	0,031	-0,036	0,061	0,041	-0,022	-0,040
MC26	0,038	0,030	0,033	0,033	0,047	0,038	0,017	0,041	0,038	0,035	0,042	0,051	0,042	0,058	0,034
MC27	0,002	0,001	0,003	0,003	-0,004	0,000	0,003	-0,006	-0,001	0,000	-0,002	-0,002	-0,005	-0,001	0,005
MC28	-0,003	-0,007	0,009	0,009	0,013	0,004	0,011	0,002	0,017	0,021	0,022	0,013	-0,013	0,028	0,027
MC29	0,001	0,001	0,002	0,002	0,005	0,000	0,003	0,000	0,005	0,006	0,007	0,001	-0,005	0,005	0,004
MC30	-0,002	-0,009	0,017	0,017	0,018	0,003	-0,002	0,014	0,014	0,014	0,024	0,003	-0,013	0,025	0,014
MC31	0,008	-0,003	0,002	0,003	0,003	0,020	0,022	0,021	0,041	0,040	0,020	0,020	0,000	0,015	0,025
MC32	0,049	0,045	0,046	0,046	0,088	0,018	0,009	0,046	0,045	0,048	0,077	0,035	0,015	0,079	0,064
MC33	0,008	0,004	0,031	0,031	0,011	0,002	-0,010	0,009	0,003	0,005	0,013	-0,001	-0,010	0,017	0,010
MC34	0,016	0,003	0,036	0,036	0,053	0,083	0,077	0,018	0,038	0,040	0,032	0,104	0,023	0,089	0,080
MC35	0,031	0,025	0,015	0,015	0,035	0,029	0,014	0,032	0,034	0,036	0,039	0,027	0,033	0,035	0,021
MC36	-0,007	-0,014	0,003	0,003	-0,003	-0,005	0,009	-0,001	0,020	0,019	0,003	0,014	-0,004	0,009	-0,002
MC37	-0,008	-0,008	0,014	0,014	0,013	0,016	0,018	-0,003	-0,007	-0,003	0,004	-0,002	-0,020	0,021	0,029
MC38	0,072	0,058	0,047	0,047	0,094	0,056	0,015	0,061	0,067	0,065	0,093	0,044	0,042	0,091	0,069
MC39	0,007	0,001	0,030	0,029	0,022	-0,015	-0,021	0,014	0,005	0,005	0,028	-0,013	-0,008	0,026	0,008

Source: Own work

On the other hand, the highest negative correlations were identified between *Science_Fiction* and *Books* (Spearman value: −0,0891; p-value < 0.00001), *Animation* and *Patio_Lawn_and_Garden* (Spearman value: −0,0613; p-value < 0.00001), *Animation* and *CDs_and_Vinyl* (Spearman value: −0,0607; p-value < 0.00001), *Animation* and *Books* (Spearman value: −0,0578; p-value < 0.00001).

5   Discussion and Implications

The above results indicate that indeed certain movie preferences are statistically significantly correlated with purchases of products in certain categories. Most of them can be reasonably explained (e.g. watching videos from *Music_Videos_and_Concerts* and purchasing *CDs_and_Vinyls*). However, there are also less obvious correlations such as positive correlation between watching *Documentaries* and purchasing *CDs_and_Vinyls* or negative correlation between watching *Science_Fiction* and purchasing *Books*. Hence, the above analysis allowed to positively answer the main research question of this paper whether movie preferences correlate with online purchases in different product categories. Information about such correlations can be used to improve existing recommender systems or to solve the cold-start problem for customers where movie preferences can be accessible. Figure 2 presents a generalized conceptual model for recommender systems that might take advantage of customer preferences. Such generalization allows to take into account the vector of customer preferences (such as music genre preferences, dining preferences, book genre preferences). Thus, this paper provides both, theoretical contributions and practical

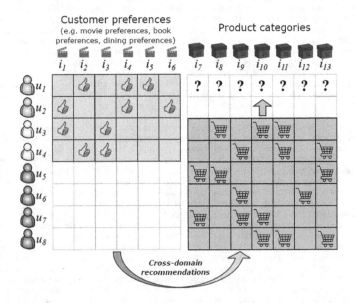

Fig. 2. Proposed conceptual model for RS based on customer preferences.   Source: Own work.

implications. Moreover, human-interpretable results are another advantage of this research, since many previous papers provide frameworks that are not interpretable by humans.

6   Limitations

Every study has limitations and this research is no exception. First of all, this experiment was based only on verified reviews. Hence, the people who watched movies without writing reviews about it are not considered in this analysis. Secondly, the number of verified reviews in particular movie categories were the only indicator of movie preferences. User ratings of the particular movies were not taken into account. Therefore, it was assumed that the more reviews in the particular category a user has written the more this user is interested in this movie category. In the situation where a user negatively rated many movies in a particular category the author assumed that this user is still interested in this category (because of many verified reviews), however, these particular movies were not good enough for him/her. Finally, the analysis was based on user accounts that might be shared with others (e.g. members of the family).

7   Future Work

First of all, future research should add the time factor to this analysis (e.g. whether movie review was created before or after product purchase). Secondly, analyzing correlations between other preferences (such as book genre, music genre, or dining preferences) and e-commerce purchases can provide interesting results regarding generalization of this model and creating a vector of customer preferences affecting purchasing behavior. Finally, the application of NLP techniques for text reviews could be a good direction to extend this model.

Appendix A

See Table 3.

Table 3. List of abbreviations used in the correlation matrix

Abbreviation	Movie category	Abbreviation	Product category
MC1	Romance	PC1	Clothing_Shoes_and_Jewelry
MC2	Faith_and_Spirituality	PC2	Books
MC3	Musicals	PC3	Women
MC4	TV_News_Programming	PC4	Clothing
MC5	Fitness_and_Yoga	PC5	Home_and_Kitchen
MC6	Art_House_and_International	PC6	Sports_and_Outdoors
MC7	Mystery_and_Thrillers	PC7	Automotive
MC8	Sports	PC8	Men
MC9	Holidays_and_Seasonal	PC9	Electronics
MC10	Drama	PC10	Accessories
MC11	Documentary	PC11	Literature_and_Fiction
MC12	Foreign_Films	PC12	Toys_and_Games
MC13	Fantasy	PC13	Tools_and_Home_Improvement
MC14	Educational	PC14	CDs_and_Vinyl
MC15	LGBT	PC15	Cell_Phones_and_Accessories
MC16	Reality_TV	PC16	Imported
MC17	Exercise_and_Fitness	PC17	Shoes
MC18	Military_and_War	PC18	Kindle_Store
MC19	Silent_Films	PC19	Kindle_eBooks
MC20	Classics	PC20	Kitchen_and_Dining
MC21	Anime	PC21	Sports_and_Fitness
MC22	Animation	PC22	Replacement_Parts
MC23	Futuristic	PC23	Jewelry
MC24	Special_Interests	PC24	Novelty_and_More
MC25	Science_Fiction	PC25	Novelty
MC26	Characters_and_Series	PC26	Home_Dcor
MC27	Performing_Arts	PC27	Computers_and_Accessories
MC28	Music_Videos_and_Concerts	PC28	Cases_Holsters_and_Sleeves
MC29	TV_Talk_Shows	PC29	Office_Products
MC30	Television	PC30	Patio_Lawn_and_Garden
MC31	Horror		
MC32	Comedy		
MC33	The_Works		
MC34	Action		
MC35	Live_Action		
MC36	Cult_Movies		
MC37	Westerns		
MC38	Kids_and_Family		
MC39	Musicals_and_Performing_Arts		

Source: Own work

References

1. Adomavicius, G., Tuzhilin, A.: Toward the next generation of recommender systems: a survey of the state-of-the-art and possible extensions. IEEE Trans. Knowl. Data Eng. **17**(6), 734–749 (2005)
2. Ayushi, S., Prasad, V.: Cross-domain recommendation model based on hybrid approach. Int. J. Mod. Educ. Comput. Sci. **10**(11), 36–42 (2018)
3. Cantador, I., Fernández-Tobías, I., Berkovsky, S., Cremonesi, P.: Cross-domain recommender systems. In: Ricci, F., Rokach, L., Shapira, B. (eds.) Recommender Systems Handbook, pp. 919–959. Springer, Boston, MA (2015). https://doi.org/10.1007/978-1-4899-7637-6_27
4. Fernández-Tobías, I., Braunhofer, M., Elahi, M., Ricci, F., Cantador, I.: Alleviating the new user problem in collaborative filtering by exploiting personality information. User Model. User-Adapt. Interact. **26**(2–3), 221–255 (2016)
5. Gorodetsky, V., Samoylov, V., Tushkanova, O.: Agent-Based Customer Profile Learning in 3G Recommender Systems: Ontology-Driven Multi-source Cross-Domain Case. In: Cao, L., et al. (eds.) ADMI 2014. LNCS (LNAI), vol. 9145, pp. 12–25. Springer, Cham (2015). https://doi.org/10.1007/978-3-319-20230-3_2
6. Hu, R., Pu, P.: A comparative user study on rating vs. personality quiz based preference elicitation methods. In: Proceedings of the 14th international conference on Intelligent user interfaces, pp. 367–372 (2009)
7. Hu, R., Pu, P.: Enhancing collaborative filtering systems with personality information. In: Proceedings of the Fifth ACM Conference on Recommender Systems, pp. 197–204 (2011)
8. Khan, M.M., Ibrahim, R.: A generic framework for cross domain recommendation. In: Huk, M., Maleszka, M., Szczerbicki, E. (eds.) ACIIDS 2019. SCI, vol. 830, pp. 323–334. Springer, Cham (2020). https://doi.org/10.1007/978-3-030-14132-5_26
9. Li, B., Yang, Q., Xue, X.: Can movies and books collaborate? Cross-domain collaborative filtering for sparsity reduction. In: Twenty-First International Joint Conference on Artificial Intelligence (2009)
10. Lian, J., Zhang, F., Xie, X., Sun, G.: CCCFNet: a content-boosted collaborative filtering neural network for cross domain recommender systems. In: Proceedings of the 26th International Conference on World Wide Web Companion, pp. 817-818 (2017)
11. Linden, G.: Two Decades of Recommender Systems at Amazon.com (2017). https://doi.org/10.1109/MIC.2003.1167344
12. Ni, J., Li, J., McAuley, J.: Justifying recommendations using distantly-labeled reviews and fine-grained aspects. In: Proceedings of the 2019 Conference on Empirical Methods in Natural Language Processing and the 9th International Joint Conference on Natural Language Processing (EMNLP-IJCNLP), pp. 188–197 (2019)
13. Nunes, M.A.S.N.: Recommender systems based on personality traits: could human psychological aspects influence the computer decision-making process? VDM-Verlag Müller (2009)
14. Ricci, F., Rokach, L., Shapira, B.: Recommender systems: introduction and challenges. In: Ricci, F., Rokach, L., Shapira, B. (eds.) Recommender Systems Handbook, pp. 1–34. Springer, Boston, MA (2015). https://doi.org/10.1007/978-1-4899-7637-6_1
15. Ristoski, P., Schuhmacher, M., Paulheim, H.: Using graph metrics for linked open data enabled recommender systems. In: Stuckenschmidt, H., Jannach, D. (eds.) E-Commerce and Web Technologies, pp. 30–41. Springer, Cham (2015). https://doi.org/10.1007/978-3-319-27729-5_3

16. Roshchina, A.: Twin: Personality-Based Recommender System. Institute of Technology Tallaght, Dublin (2012)

17. Sharma, A., Hofman, J.M., Watts, D.J.: Estimating the causal impact of recommendation systems from observational data. In: Proceedings of the Sixteenth ACM Conference on Economics and Computation, pp. 453–470 (2015)

18. Thorat, P.B., Goudar, R., Barve, S.: Survey on collaborative filtering, content-based filtering and hybrid recommendation system. Int. J. Comput. Appl. **110**(4), 31–36 (2015)

19. Tkalčič, M., Kunaver, M., Košir, A., Tasič, J.: Addressing the new user problem with a personality based user similarity measure. CEUR-WS.org (2011)

20. Webster, J., Watson, R.T.: Analyzing the past to prepare for the future: writing a literature review. MIS Q. **26**, xiii–xxiii (2002)

21. Wirth, R., Hipp, J.: CRISP-DM: towards a standard process model for data mining. In: Proceedings of the 4th International Conference on the Practical Applications of Knowledge Discovery and Data Mining, pp. 29–39. Springer, London (2000)

22. Zhang, Q., Lu, J., Wu, D., Zhang, G.: A cross-domain recommender system with kernel-induced knowledge transfer for overlapping entities. IEEE Trans. Neural Netw. Learn. Syst. **30**(7), 1998–2012 (2019)

23. Zhang, Y., Ma, X., Wan, S., Abbas, H., Guizani, M.: CrossRec: cross-domain recommendations based on social big data and cognitive computing. Mob. Netw. Appl. **23**(6), 1610–1623 (2018)

ICRM

ICRM 2020 Workshop Chairs' Message

The world has changed since the last iCRM workshop in 2019. In the face of a global pandemic, traditional business models built around direct face-to-face customer interactions are under pressure and the importance of digital channels has increased dramatically. Analysts indicate an increase of more than 8.2% of internet users within one year [1]. While large online sales platforms report a significant growth in their sales numbers, many smaller retailers without established digital sales channels are under pressure and need to adapt quickly to the new business environment. Besides the primary need for establishing digital sales channels, maintaining and enlarging a profitable customer base in the new business environment is the second challenge. Social media offer the means for reaching out to new buyers via existing relationships as well as for maintaining close personalized relationships with existing customers. In sum, the increase of 10.5% in active users since July 2019 is a substantial opportunity to pursue these goals [1].

From an organizational perspective, social customer relationship management (Social CRM) links the external world of social media with internal business processes in marketing, sales, and customer services [2]. However, social media should not be conceived solely as a new communication and marketing channel, but as a lever to transform existing CRM strategies, processes, and systems [3]. Integrated Social CRM is an approach that aims to align these aspects in integrated and digitally supported use cases. It builds on a realm of information technology that helps to turn user-generated-content (UGC) into actionable information for daily tasks as well as strategic decisions [4, 5]. Successful Social CRM also means to leverage the unique properties of social media for realizing interactive, personal, and sometimes even automated interactions [6, 7].

For the 5th time, the iCRM workshop aimed to shed light on current research and practical challenges regarding integrated Social CRM. The workshop attracted participants from different fields, such as marketing and relationship management, information systems design, and computational intelligence. Like in the previous years, all submissions underwent a double-blind peer review and the authors were offered to revise their manuscripts based on the feedback from the workshop, which was conducted virtually this year. In sum, five papers were accepted, which covered multiple perspectives.

The first three papers provide insights into the current role of social media in various industries. The first paper, "Financial Institutions and Use of Social Media: Analysis of the Largest Banks in the U.S. and Europe," examines the use of social media in customer-facing activities in major financial institutions and shows the heterogeneous exploitation of social media for CRM purposes. The second paper, "Outsourcing of Social CRM Services in German SMEs," presents findings about the outsourcing intention for Social CRM activities as well as barriers resulting from a survey among 122 small and medium sized companies. The third paper, "Customer-

Focused Churn Prevention with Social CRM at Orange Polska SA," reports research in progress on the potential use of social media for measuring and reducing the churn risk. The last two papers examine current research and IS development directions in Social CRM. The fourth paper, "Social CRM: A Literature Review based on Keywords Network Analysis," provides a network analysis of 766 research papers and shows the importance of big data technologies for Social CRM as well as a strong focus of researchers on major platforms such as Facebook and Twitter in the past. The last paper, "Social CRM Tools: A Systematic Mapping Study," examines the main features of Social CRM tools via an extensive screening of existing literature.

The papers highlight the manifold integration challenges, ranging from the availability and use of information systems for tasks in Social CRM, to the potential of speeding up the transformation of existing CRM approaches by using specialized agencies. Combining the technological and the managerial perspective has been recognized during the workshop as an important prerequisite for achieving integrated solutions in practice that help businesses to build and foster relationships in a digital environment.

Above all, the 5th iCRM workshop was a community effort in extraordinary times, and the contribution of all authors as well as Program Committee members in meeting these challenges was highly appreciated.

August 2020

Rainer Alt
Fabio Lobato
Olaf Reinhold

References

1. Kemp, S.: Digital 2020: July global statshot. https://datareportal.com/reports/digital-2020-july-global-statshot (2020). Accessed 9 August 2020
2. Alt, R., Reinhold, O.: Social Customer Relationship Management (Social CRM) - Application and Technology. Bus. Inf. Syst. Eng. 4(5), 287–291 (2012)
3. Alt, R., Reinhold, O.: Social Customer Relationship Management - Fundamentals, Applications, Technologies. Berlin, Springer (2020)
4. Choudhury, M., Harrigan, P.: CRM to social CRM: the integration of new technologies into customer relationship management. J. Strateg. Mark. 22(2), 149–176 (2014)
5. Trainor, K. J., Andzulis, J., Rapp, A., Agnihotri, R.: Social media technology usage and customer relationship performance: a capability-based examination of social CRM. J. Bus. Res. 67(6), 1201–1208 (2014)
6. Greenberg, P.: CRM at the Speed of Light: Social CRM Strategies, Tools, and Techniques for Engaging Your Customers. New York, McGraw-Hill (2009)
7. Sigala, M.: Implementing social customer relationship management: a process framework and implications in tourism and hospitality. Int. J. Contemp. Hosp. Manage. 30(7), 2698–2726 (2018)

Organization

Chairs

Olaf Reinhold   Leipzig University, Germany
Rainer Alt   Leipzig University, Germany
Fabio Lobato   Federal University of Western Pará, Brazil

Program Committee

Alan Marcel Fernandes de Souza   Amazon University, Brazil
Alireza Ansari   Leipzig University, Germany, and IORA
     Regional Center for Science and Technology
     Transfer, Iran
Antônio Jacob Jr.   State University of Maranhão, Brazil
Cristiana Fernandes De Muylder   FUMEC University, Brazil
Douglas R. Cirqueira   Dublin City University, Ireland
Emílio Arruda   FUMEC University and University of Amazon,
     Brazil
Fábio Lobato   Federal University of Western Pará, Brazil
Flavius Frasincar   Erasmus University Rotterdam,
     The Netherlands
Gültekin Cakir   Maynooth University, Ireland
José Marcos de Carvalho Mesquita   The University of Connecticut, USA
Julio Viana   Social CRM Research Center, Germany
Kobby Mensah   University of Ghana Business School, Ghana
Kwabena Obiri Yeboah   Catholic University College of Ghana, Ghana
Kwame Adom   University of Ghana Business School, Ghana
Luis Madureira   Universidade NOVA de Lisboa, Portugal
Mark Lennon   California University of Pennsylvania, USA
Mattis Hartwig   University of Lübeck, Germany
Nino Carvalho   Fundacao Getulio Vargas and European
     Institute of Management, Portugal
Omar Andres Carmona Cortes   Instituto Federal do Maranhão, Brazil
Rabi Sidi Ali   Takoradi Technical University, Ghana
Rafael Geraldeli Rossi   Universidade Federal do Mato Grosso do Sul,
     Brazil
Renato Fileto   Federal University of Santa Catarina, Brazil

Sandra Turchi   Digitalents, Brazil
Thiago Henrique Silva   Federal University of Technology of Parana,
                                   Brazil
Winnie Ng Picoto   Technical University of Lisbon, Portugal

Financial Institutions and Use of Social Media: Analysis of the Largest Banks in the U.S. and Europe

Thaís Helen Sena[1] , Cristiana Fernandes De Muylder[1]([✉]) ,
and Emilio José Monteiro Arruda Filho[2]

[1] University FUMEC, Belo Horizonte, MG, Brazil
thaishelen.sena@gmail.com, cristiana.muylder@fumec.br
[2] University of Amazon – UNAMA, Belém, PA, Brazil
emilio.arruda@unama.br

Abstract. This article aims to analyze the largest banks in the United States of America and Europe in relation to acceptance, reach, engagement and interaction with their customers on social media. A qualitative approach comparing some factors based on analysis of the banks' social media data is provided. Results show that the analyzed banks do not use the full potential of social media, incorporating these tools to their strategies. Attitudes such as lack of post, low number of followers, lack of responses to customer interactions, and pages blocking comments are some of the results that indicate possibilities for improvements in the usage of these tools for building up relationships with customers. Despite the investment in technology by banks and the strong presence of social media in contemporary life, this study shows an opportunity for new strategies related to the interaction between the banks and their consumers. Besides the analysis of social media characteristics, this paper also provides a sentiment analysis to indicate possible issues regarding brand image in the online environment.

Keywords: Social media analysis · Social CRM · Financial institutions · Consumer sentiment

1 Introduction

Social media has become an important tool in communications and customer relationships [1], as its use is one of the world's most popular online activities [2]. The number of social media users in the world went from 0.97 billion (2010) to 2.48 billion (2017) and a forecast expects this number to rise to 3.09 billion users in 2021 [3].

A concept of Social Customer Relationship Management (Social CRM) has changed the way businesses and customers build their relationships [4]. It is defined by the possibility of integrating data from social media into CRM tools to improve insights on consumers, as well as the management of these social media as instruments for interacting with customers [5]. Given this new scenario, social media research becomes relevant,

W. Abramowicz and G. Klein (Eds.): BIS 2020 Workshops, LNBIP 394, pp. 203–214, 2020.
https://doi.org/10.1007/978-3-030-61146-0_16

such as Twitter, Facebook, LinkedIn, Instagram, and YouTube, that are considered to be valuable assets for marketing strategy development [6].

Social media support the development of profitable relationships with customers, providing competitive advantage to the organization and, through Social CRM, companies are able to use social media to analyze customer engagement and understand better the online customer for stronger relationships [7]. Social networks allow companies to communicate with their customers and could be a support element considering brand [8] and a new way to establish new relationships with potential customers by promoting customer satisfaction and ensuring an increase in sales [9].

Some studies seek to understand the use of social media in different sectors [10–13], and some, focus in bank sector [14–18].

Considering the latest investments from banks in technology, the use of cell phones and the mobile Internet has brought changes to financial services, as consumers and banks have been provided with new tools to manage their money, time and company-customer relationships [19]. Consequently, in the USA, banks were expected to spend 67 billion in technology in 2019 [20].

Another study found out that social media represent great opportunities for banks to offer better individual marketing experiences [21]. However, a study with managers from different banks regarding the effect of digitization processes on customer relationship satisfaction and results point that, besides the lack of research in the field, part of the current theory of customer loyalty needs to be revised to align it with the new digitized sector [17]. Moreover, considered the first authors to investigate the consumer perspective on issues of social media adoption by financial institutions, and suggest that consumers will use social media if the industry clearly creates and articulates consumer value and develop strategies about technology security perception [22].

Based on the importance of social media in the contemporary world to improve the company-customer relationships, as well as the relevance of this novel relationships for the banking industry, this article seeks to analyze the largest banks in the United States of America (USA) and Europe in relation to acceptance, reach, engagement and interaction with their customers on social media.

To reach the proposed analysis, the paper includes the concept of Social CRM to analyze the performance of the selected banks in the five largest social media. Grounded on descriptive and qualitative approaches, this article also includes the comparison of users' sentiments based on mentions in social media regarding products and services from these banks. The originality of this paper lies on the fact that existing research did not analyze more than one social network [14, 22, 24].

The results support the alignment of bank actions to explore the value of the adoption and use of social media as a source of customer information [22, 25]. This institutional alignment with the source of customer information aims to leverage the innovation of products and services [26], understand the perceived effects on customer loyalty from the managerial aspect [17], achieve higher levels of financial performance [18, 27] and still understand the impediments related to the use of technology [21]. These analyses also highlight the importance of the client's opinion and action in social media for financial institutions [6]. The next pages will explore the methodology, results and conclusions, indicating the practical implications of the work.

2 Methodology

Through a qualitative descriptive method, looking for the information about banks' social media, this paper describes the factors based on the Social CRM concept: acceptance (adherence of the studied bank to the tool), reach (number of followers), engagement (number and recurrence of publications on the platforms) and interaction (which means dialogue, or level of responses between customers and the banks studied).

The study focuses on studying the US and European markets, since other markets, such as Asia, have different characteristics and moments, making a horizontal comparison unfeasible [20]. There were seven institutions of the fifteen largest banks in the world in assets, being: JPMorgan (6th), HSBC (7th), Bank of America (8th), BNP Paribas (9th), Credit Agricole Group (10th), Citigroup (11th) and Wells Fargo (13th) and five of the most used platforms in the world, which are: Twitter, YouTube, Facebook, LinkedIn and Instagram [3].

To collect data for acceptance analysis, reach, engagement and interaction, each page of the banks on Twitter, YouTube, Facebook, Instagram and LinkedIn was analyzed. The following software solutions were used to analyze the data from the pages: Foller.me®, Likealyzer.com®, and social-searcher® (sentiment analysis).

3 Results

3.1 Analysis of Twitter Profiles

Twitter, the seventh most used social network in the world in 2018 [3], presents itself as "the right place to know more about what's going on in the world and what people are talking about right now". With the expansion of social media and the strong need for social CRM tools, several researchers have explored this field of analysis [28, 29].

This analysis of financial institutions becomes relevant due to a new moment in relationship with their clients based on Web 2.0 platforms [22]. The five selected banks are present on Twitter with individual performance characteristics.

Regarding the time, the tool Foller.Me® shows that the banks started using the platform in different moments. The first to join was Wells Fargo in 2007 and the last, JPMorgan in 2013. Scope or location data, as well as the URL directing to the bank's website, are not disclosed on the pages from JPMorgan and Bank of America, differently from the other banks. However, all institutions provided a very attractive biography and further information regarding digital interaction through the platform. Some of them directed the user to another environment, which could be considered a customer service page. Concerning the reach, engagement and interaction factors, the scenario was different in each bank. The Credit Agricole Group had the largest number of Tweets, followed by Citigroup and Wells Fargo. The HSBC appears in the last position in this ranking. However, when it comes to followers, Citigroup appears with nearly double of the second place, Bank of America. Table 1 discloses these results, as well as the number of accounts these banks follow and the ratio between followers and following.

The strategy of following profiles is different comparing BNP Paribas and JPMorgan. While the first follows several accounts, the latter does not follow the same strategy. Listing is an important factor to determine the engagement of users with a profile on

Table 1. Reach of analyzed banks on Twitter.

Bank	Tweets	Followers	Following	Followers Ratio	Listed
JP Morgan	6.909	483.737	48	10.078	2.867
HSBC	3.210	181.036	547	331	838
Bank of America	9.398	536.232	662	810	2.051
BNP Paribas	14.785	68.481	3.617	19	1.170
Credit Agricole Group	29.398	60.249	892	68	726
Citigroup	17.387	915.986	351	2.610	3.167
Wells Fargo	17.960	311.620	329	947	2.551

Source: Research database

Twitter. It means users add a page to their list, so that they can easily access what a certain account is posting. In this sense, the most listed bank by followers is the Citigroup, followed by JPMorgan and Wells Fargo.

Despite the listing, an analysis of the last 100 Tweets from each account shows the low number of responses from banks, as Citigroup and HSBC have no registered responses, BNP Paribas and JPMorgan responded to 3% of their followers and Bank of America responded to 12%. Credit Agricole Group and Wells Fargo, however, show a moderate level of engagement as they respond to 53 and 56% followers, respectively.

This analysis indicates a high number of shared information by the banks, but a low level of interaction with users when analyzing the amount of responses. However, even with a low level of interaction, when analyzed the cloud of most common used words and hashtags linked to these accounts, results show a positive word communication (Table 2). In this case, attention is drawn to the positive and neutral words that appear in relation to the banks analyzed, as in JPMorgan: leaders, career, build, develop, entrepreneurs; HSBC: proud, global, potential; Bank of America: honor, congratulations, family, celebrate, love; BNP Paribas: groups, model, inauguration, young, celebrate; Credit Agricole Group: solution, message, layout; Citigroup: digitization, community, business, results, digital; and Wells Fargo: love, proud, communities, volunteers, opportunity and fan.

3.2 Analysis of YouTube Channels

YouTube[1] was considered the second most used social network in the world, in 2018, with the mission of "giving everyone a voice and revealing the world" [3]. The literature presents some works focused on the analysis of YouTube as a tool for customer relationships [23].

The channels from the seven selected banks are active on YouTube, demonstrating acceptance of the tool. Citigroup was the first to use the platform in 2005, while the Credit Agricole Group was the last in 2010. All other banks joined the platform in 2006.

Wells Fargo leads the number of subscribers, while Credit Agricole Group is the institution with the lowest number of followers, but also the last to join officially the platform. BNP Paribas is the leader of posts; however, the largest number of views belongs to Wells Fargo, even with the lowest number of published videos (Table 3).

---

[1] https://www.YouTube.com/about/.

Table 2. Analysis of topics and hashtags.

BANK	TOPICS	HASHTAGS
JP MORGAN	founder discusses morgan develop floor power leaders share leadership investing want advice business life home solutions look attendees WOMEN high companies people data big ceo employees help giving step les shares te ch million food annual head starting treasury future conference jpms says insights technology weekend story career joy disruptio n jpm make build leader robin businesses trading hear supporting strategies 120 entrepreneurs	#powerher #robinhoodinvestors #codeforgood #sca leupny #london #tech #cesummit #techtrends #aart #jpmcartcollection #hispanicheritagemonth
HSBC	trade campaign female join most launch lost mental potential york feel abroad quinn initiative named open financial announced helped share leaders climate sustainable weve conference team want work chief partner excited leading future launched water proud results finan ce museum colleagues talking banking group china ceo global hsbc recognised businesses hsbcs partners business executive cdle ambassadors according bank health change best world	#speakyourmind #blockchain #neweconforum #she sthebusiness #worldmentalhealthday # outrolemodels19 #hsbc #paymehk #expats #menta lhealth #hsbcgrads #startups #climatechange #inn ovation #retirement #payments #hsbcnavigator #ch ina #hk #asia
BANK OF AMERICA	helped grant bank awarded honor food retreat help matters team years executive sleeps proud make mo ment twitter india home excited 2019 family long congrats putting start better hear shopping annual season we ekends twitte planning charlotte weekend education OWNers happy america community business teamed weve inspired clients game women build working teammate world honored key celebrate global festival congratulations volunteer s helping love	#bettermoneyhabits #chicagomarathon #smallbusi ness #powerto #globalbuild #bofavolunteers #ad # artconservationproject #podcast #takeachildoutsid eweek #bofagrants #thejcast #miami #financial #cu stomerserviceweek #bettermore #texas #lableeds blue #hamptonroads #cyberaware
BNP PARIBAS	linauguration guillaume bnp engagement axis flowers ceo banks groups destals direct reaching clients cost bonnaf financial set dits deputy starting people stage work inauguration 2019 programme watch place muhammad business tonight lengagement pari bas transformation celebrate bivwaks gives jacques chapitre learn model lgbt projects bivwak supports inspiring night chance share yuniu s lot young group project gwendolyn head 2017 starts delegation revient jeanlaurent	#bivwak #bnppcsr # oyw2019 #bnpparibas #lancezvous #transformatio n #bnppresults #digitaltransformation #greenreflex #innovation #positivebanking #tech # accessart25 #agile #bnppcinema #bnpptennis #ble ndedfinance #intrapreneuring #laurastaffe #bnppgro up
CREDIT AGRICOLE GROUP	essay bonjourpourrievous carte message dsols projes pay salon bonjournous dsinstaller venez data apple aujo urdhui solution camille permis indiquer sil plus probleme crdit mpborne navons fait cotte rinstaller dire disposition grand lapp lire peut puis sign octobre agricole contraire rencontrer appert bonjourca sincrament journemcamille cest pourriezvous dun slam dcouvrez priv cela sagit navra scurise bonjour respectueuse bonne parfois confdelementamille bonjour avezvous rponse privborne	#sportecoledevie #graceausport #crditagricole #jeu #sportecoledevieamoureux #animaldecompagnie #mobilite #travail #logement #consommation #case xplique #startups #lorraine #hell #metz #photovolta que #judobrasilia #diversit #patrimoine #villagebyc a
CITIGROUP	mike disability partnership innovation information impact citis support years refugees news travel careers living performance supporting digitizati on veterans instant congratulations global wave talk named clients learn 100 head corporate discuss business com mitted data capital ding asia companies citi financial joined 2019 big leaders role ceo payments military banking digital growing C ommunity pacific annual results quarter fdns announced corbat focused proud institution	#citi #progressmakers #citis #rescuingfutures #tea mclti # fast50 #royalparkshalf #spiritday #womenempower ment #eforeducation #teamusa #loan # pathways2progress #lastree #financ #itsabouttime #progressmakersthelr #breakfastofcorporatechamp ions #veteransmonth #london
WELLS FARGO	donated serve mission plan setur proud hopes build tasting zachary voice game military official birthday wildcats happy joy love sponsors player members experience expand heres communities house grateful volunteers program support offe r excited join notice anna building brought letting noticed opportunity volunteer committed having help fan ng h gift nate sharing team wells annual dog hope announce service let community fargo going	#neighborhoodlift #wfvolunteers #montana #vetera nsday #smallbusiness #integrateds #wamerobbin s #solarthon #rolltideroll #mnwild #centraljersey #w fvolu #charlestonanimalsociety #dogtoberfest #fafs a #diversecommunitycapital #verification #californi afires #veterans #halloween

Source: Research database

Table 3. YouTube channel information for each bank.

Bank	Subscribed	Subscription	Videos	Viewing
JP Morgan	33,300	17/01/2006	483	2,953,846
HSBC	26,000	09/03/2006	431	17,733,983
Bank of America	94,800	17/03/2006	298	11,300,054
BNP Paribas	33,100	26/06/2006	1,697	34,485,362
Credit Agricole Group	21,500	06/01/2010	1,114	25,649,635
Citigroup	40,700	23/10/2005	1,047	23,227,009
Wells Fargo	284,000	02/03/2006	259	98,058,481

Source: Research Database

In terms of posting frequency, results show no specific standard. Citigroup and Wells Fargo, for example, had not posted in the last seven days when the latest videos were analyzed. On the other hand, Credit Agricole Group had posted in a period of two days and the other banks in a period inferior than 24 h.

Results show point that views tend to increase as more videos are uploaded, with exception of Wells Fargo, which manage to reach a higher number of users with less published videos.

3.3 Analysis of Facebook Pages

Facebook[2] is the most used social network in the world, with 2.38 Million monthly active users [3]. Related to this high involvement of users in this platform, there is a vast amount of work involving Facebook from different perspectives [15]. Using the tool likealyzer.com[3], results were divided in five major groups: FrontPage, page information, activity, response, and engagement.

BNP Paribas and Wells Fargo are the two banks that have provided the largest number of information in the "about" section, with over 50% of filled information, while Bank of America, HSBC and JPMorgan have about 20% completion. However, Bank of America and Credit Agricole Group have 92% of activity engagement, while HSBC demonstrate no engagement.

Nevertheless, regarding interaction, only Bank of America responds to its customers (20 responses for every 100 interactions). In 75% of cases, followers cannot post content, which leaves limited user involvement. It is worth mentioning that Bank of America is the only one that allows posting content from users. Additionally, in general, banks do not interact with other pages and do not confirm participation on Facebook event pages.

Concerning the types of posts, Credit Agricole Group had the highest percentage of posts with photos, reaching to 71%. JPMorgan had its largest interaction through notes (66%) and BNP Paribas the one with most interactions through videos (80%). JPMorgan and Bank of America have the highest recurrence of posts, reaching more than 1 per day and Bank of America interacts with over 200 other pages. In comparison with the other banks HSBC has the lowest performance as it did not post any information, neither interacted with users.

3.4 Analysis of Instagram Pages

Instagram was launched on October 6th of 2010 and was considered in 2011 the App of the Year in the Apple Store. In 2012 the platform was acquired by Facebook, reaching to over 80 Million users and adding 25 languages to its portfolio. This network achieved, in 2019, one billion monthly users and more than 500 million stories daily.

Instagram is the second most used social media by marketers, with 73%, behind only Facebook with 94%. Additionally, it is the network with the highest intention from marketers to increase advertising expenditures [3]. Due to these facts, several studies analyzed Instagram data [30, 31], making it a relevant tool to provide insights to companies regarding consumer behavior.

Table 4 shows some analysis of the Instagram accounts from the selected banks regarding the creation date of their page, as well as number of followers, latest activities and number of used hashtags.

---

[2] https://about.fb.com/.

[3] https://likealyzer.com/.

Table 4. Analysis of Instagram accounts.

Bank	Start	Posts	Followers	Following	#Hashtags
JP Morgan	20/01/2011	277	151,000	1	66,5
HSBC	-	0	27,700	26	206
Bank of America	-	650	122,000	137	319
BNP Paribas	-	706	17,500	305	51,2
Credit Agricole Group	17/06/2013	38	9,000	66	12,2
Citigroup	11/11/2013	540	46,700	82.000	232
Wells Fargo	20/03/2014	220	68,700	53	377

Source: Research Database

Despite JPMorgan being the first to join Instagram in 2011 (according to available data), Bank of America, BNP Paribas and Citigroup are the ones with the most publications. Credit Agricole Group and BNP Paribas are institutions with the lowest number of followers. Although HSBC has not posted on the platform, it has over 27 thousand followers, indicating a potential underutilization of the tool. BNP Paribas is the institution with the highest number of publications, even though it has around 17 thousand followers.

3.5 Analysis of LinkedIn Pages

LinkedIn is considered the tenth most used platform in 2018 [3], however, it is one of the largest and most popular professional networks in the world. Companies use it to attract the best talent, establishing connections and building relationships [33]. According to Anders (2012), although Facebook draws all the attention, LinkedIn defined how to generate revenue for its users [32]. The analysis of LinkedIn becomes relevant as three out of five users would like to interact with a company on LinkedIn when interested in their products [33]. Additionally, three out of four users consider LinkedIn a reliable source of industry and business information, 78% expect more companies to establish themselves on the platform and 43% consider LinkedIn a reliable source of corporate information when making a purchase [34].

The results of this paper show that Citigroup appears as a leader of followers in the tool (above 2 million) followed by HSBC, JPMorgan and Bank of America with more than 1 Million followers each. Credit Agricole Group occupies the last position with a little over 150,000 followers. Most banks are actively present on the platform, with the exception of BNP Paribas, which does not post as frequent as the other institutions. The banks use LinkedIn to disclose their job opportunities, for example, Bank of America and Citigroup had each approximately 7,000 advertised jobs.

3.6 Sentiment Analysis of the Selected Banks on Social Media

The sentiment analysis ranking technique helps companies to understand their clients and how they evaluate services using social media [35]. Several studies used social media sentiment analysis to provide them with feedback from customers [36–40]. It should be used to analyze hidden meaning of expressions or feedback in social media, that could be positive, negative, or neutral action [41].

The sentiment analysis of the selected banks were categorized in negative, positive and neutral experiences according to posts related to the banks, as well as their products and services (Table 5).

Table 5. User's perspective analysis.

Bank	Positive	%	Negative	%	Neutral	%	Ratio
JP Morgan	134	24	45	8	391	68	7:3
HSBC	117	18	56	9	479	73	7:3
Bank of America	151	29	52	9	404	66	7:3
BNP Paribas	145	25	19	3	410	72	9:1
Credit Agricole Group	129	29	30	7	293	64	4:1
Citigroup	152	23	51	8	469	69	7:3
Wells Fargo	199	35	45	8	329	57	4:1

Source: Research Database

Wells Fargo Bank has the highest number of positive posts, while HSBC ranks in the last position. Regarding negative sentiments, all banks perform similarly, except BNP Paribas with only 3% of negative sentiment. The ratio indicates the relation between positive and negative sentiments. However, the results indicate a higher tendency for neutral posts for all banks. Neutral sentiments could point to a negative or a positive sentiment when analyzed closely. Therefore, the institutions can profit from a frequent monitoring to understand what users are posting about their brands on social media.

3.7 Result Summary

To summarize the analysis of the seven selected banks according to the five biggest social media, a ranking was developed taking into consideration the number of the publications and followers (reach and engagement) of each bank in each social network, assigning a score from one to seven for each item.

Results on Table 6 indicate that:

- Banks operate more with some networks than in others.
- Citigroup was the most active institution considering number of followers and publications.
- HSBC and Credit Agricole Group performed poorly in most social media, despite the potential indicated in the previous analysis. While the HSBC uses mostly LinkedIn and Twitter to communicate with its customers, Credit Agricole Group focused its activities on Facebook.

Table 6. Analysis of reach factor on social media.

BANK	FACEBOOK	YOUTUBE	INSTAGRAM	TWITTER	LINKEDIN	TOTALLY
JP MORGAN	4,5	3,0	5,5	3,5	3,0	4,3
HSBC	1,0	3,0	2,0	5,0	6,0	3,4
BANK OF AMERICA	5,5	4,5	4,0	4,5	4,0	4,5
BNP PARIBAS	4,0	5,0	4,5	2,0	2,0	3,7
CREDIT AGRICOLE GROUP	5,5	3,5	3,5	1,0	1,0	2,9
CITIGROUP	3,5	5,0	4,5	6,0	7,0	5,2
WELLS FARGO	4,0	4,0	4,0	3,0	3,0	4

Source: Research Database

Table 7. Analysis of interaction factor on social media.

BANK	FACEBOOK	YOUTUBE	INSTAGRAM	TWITTER	LINKEDIN	TOTALLY
JP MORGAN	1,0	1,0	7,0	4,0	7,0	4,0
HSBC	1,0	1,0	7,0	1,0	7,0	3,4
BANK OF AMERICA	7,0	1,0	7,0	5,0	7,0	5,4
BNP PARIBAS	1,0	7,0	7,0	4,0	7,0	5,2
CREDIT AGRICOLE GROUP	1,0	1,0	7,0	6,0	7,0	4,4
CITIGROUP	1,0	7,0	7,0	1,0	7,0	5,0
WELLS FARGO	1,0	7,0	7,0	7,0	7,0	5,8

Source: Research Database

Considering the results of interaction factor shown in Table 7, results point that:

- The firms are not offering to listen to their customers on Facebook and YouTube as the comment feature is blocked, excluding customer-bank interaction.
- Bank of America is the only bank that allows an open channel to listen to its customers on Facebook.
- Instagram and LinkedIn were the most used platform for communication with customers.
- The best scores for Twitter response belong to Wells Fargo and Credit Agricole Group. However, results were still below 60% of response.

4 Conclusions

This paper proposed a descriptive analysis of the major banks in the USA and Europe regarding their performances on social media, indicating possible use of these tools to engage on Social CRM activities and improve the company-customer relationships. The performance was measured according to the following factors: acceptance, reach, engagement and interaction. Additionally, the use of analytical software provided a sentiment analysis, taking into consideration brand mentions on the five largest social media.

Results indicate that the analyzed banks underutilize social media. The lack of publications, low number of followers, low rate of responses to customers and blocking of comments in pages show this underutilization of social media as a communication channel with customers and as a source of insights on consumer behavior.

The acceptance factor was satisfactory for all banks analyzed, since all have registration in different social medias analyzed. However, most institutions performed low regarding reach, engagement and interaction factors in different social media. Based on these factors, specific results show that:

- Facebook could improve the communication with customers and analysis of possible insights derived from this interaction.
- YouTube could be used as interaction channel where users comment and share their videos.
- Instagram could improve brand of the company and a new way to interact with their posts.
- Twitter should be a potential tool for improvements regarding interaction with customers on the platform.
- LinkedIn posts allow the companies to interact with other users on platform.

These analyses provide insights on how the selected banks use using social media to communicate and interact with their target. Financial institutions could, therefore, profit from an increase in social media activities as these platforms are considered important channels for managing the relationship with consumers. The results highlight the opportunity to invest in interaction with customers on these social media, as well as the relevance of monitoring them in order to produce insights to stay ahead and remain competitive.

Regarding the limitations, the paper provides a temporal analysis including only publicly disclosed data and the seven largest banks in USA and Europe. Further research could address the issue crossing different data sources, as well as analyze more banks in different regions.

Acknowledgement. The Authors gratefull acknowledge to CAPES, FAPEMIG and CNPq 407907/2018-1.

References

1. Bazi, S., Filieri, R., Gorton, M.: Exploring the motivations of consumers' engagement with luxury brands in social media sites. In: 47th EMAC. European Marketing Academy, Glasgow (2018)
2. Emarketer: Fintech. Statista Digital Market (2018)
3. Statista Digital Market Outlook: Fintech Report. Statista (2018)
4. Choudhury, M.M., Harrigan, P.: CRM to social CRM: the integration of new technologies into customer relation: EBSCOhost. J. Strateg. Mark. **22**(2), 149–176 (2014)
5. Alt, R., Reinhold, O.: Social Customer Relationship Management: Fundamentals, Applications, Technologies. Springer, Heidelberg (2020)
6. Giannakis-Bompolis, C., Boutsouki, C.: Customer relationship management in the era of social web and social customer: an investigation of customer engagement in the Greek retail banking sector. Procedia – Soc. Behav. Sci. **148**, 67–78 (2014)
7. Quinton, S.: The digital era requires new knowledge to develop relevant CRM strategy: a cry for adopting social media research methods to elicit this new knowledge. J. Strateg. Mark. **21**(5), 402–412 (2013)
8. Swarts, K.M., Lehman, K., Lewis, G.K.: The use of social customer relationship management by building contractors: evidence from Tasmania. Constr. Manag. Econ. **34**(4–5), 302–316 (2016). https://doi.org/10.1080/01446193.2015.1133919

9. Koçak-Alan, A., Tumer Kabadayi, E., Eriske, T.: The new face of communication: digital marketing and social media marketing. Elektronik Sosyal Bilimler Dergisi **17**(66), 123–134 (2018). https://doi.org/10.17755/esosder.334699

10. Voorveld, H.A.M., van Noort, G., Muntinga, D.G., Bronner, F.: Engagement with social media and social media advertising: the differentiating role of platform type. J. Advert. **47**(1), 38–54 (2018). https://doi.org/10.1080/00913367.2017.1405754

11. Qiu, L., Tang, Q., Whinston, A.B.: Two formulas for success in social media: learning and network effects. J. Manag. Inf. Syst. **32**(4), 78–108 (2015)

12. Johnston, W.J., Khalil, S., Le, A.N.H., Cheng, J.M.S.: Behavioral implications of international social media advertising: an investigation of intervening and contingency factors. J. Int. Mark. **26**(2), 43–61 (2018)

13. Merry, E.: Mobile banking: a closer look at survey measures (2018). https://www.federalre serve.gov/econres/notes/feds-notes/mobile-banking-a-closer-look-at-survey-measures-201 80327.htm. Accessed 24 Jan 2020

14. Mang'unyi, E.E., Khabala, O.T., Govender, K.K.: Bank customer loyalty and satisfaction: the influence of virtual e-CRM. Afr. J. Econ. Manag. Stud. **9**(2), 250–265 (2018). https://doi.org/ 10.1108/AJEMS-08-2017-0183

15. Hodis, M.A., Sriramachandramurthy, R., Sashittal, H.C.: Interact with me on my terms: a four segment Facebook engagement framework for marketers. J. Mark. Manag. **31**(11–12), 1255–1284 (2015). https://doi.org/10.1080/0267257X.2015.1012535

16. Statista Digital Market Outlook: Fintech Report. Statista (2019)

17. Larsson, A., Viitaoja, Y.: Building customer loyalty in digital banking: a study of bank staff's perspectives on the challenges of digital CRM and loyalty. Int. J. Bank Mark. **35**(6), 858–877 (2017). https://doi.org/10.1108/IJBM-08-2016-0112

18. Diffley, S., McCole, P., Carvajal-Trujillo, E.: Examining social customer relationship management among Irish hotels. Int. J. Contemp. Hosp. Manag. **30**(2), 1072–1091 (2018). https:// doi.org/10.1108/IJCHM-08-2016-0415

19. Shevlin, R.: How much do banks spend on technology? (Hint: It Would Weight 670 Tons in $100 Bills), Forbes (2019). https://www.forbes.com/sites/ronshevlin/2019/04/01/how-much- do-banks-spend-on-technology-hint-chase-spends-more-than-all-credit-unions-combined/# 632cfe0f683a

20. Lee, S., Bowdler, C.: Banking sector globalization and monetary policy transmission: evidence from Asian countries. J. Int. Money Financ. **93**, 101–116 (2019). https://doi.org/10.1016/j. jimonfin.2018.12.011

21. Vella, J., Caruana, A., Pitt, L.: Organizational commitment and users' perception of ease of use: a study among bank managers. J. Manag. Dev. **32**(4), 351–362 (2013). https://doi.org/ 10.1108/02621711311326356

22. Dootson, P., Beatson, A., Drennan, J.: Financial institutions using social media – do consumers perceive value? Int. J. Bank Mark. **34**(1), 9–36 (2016). https://doi.org/10.1108/IJBM-06-201 4-007

23. Feng, Y., Chen, H., He, L.: Consumer responses to femvertising: a data-mining case of dove's "campaign for real beauty" on YouTube. J. Advert. **48**(3), 292–301 (2019). https://doi.org/ 10.1080/00913367.2019.1602858

24. Yoon, G., Li, C., Ji, Y., North, M., Hong, C., Liu, J.: Attracting comments: digital engagement metrics on Facebook and financial performance. J. Advert. **47**(1), 24–37 (2018). https://doi. org/10.1080/00913367.2017.1405753

25. Meire, M., Hewett, K., Ballings, M., Kumar, V., Van den Poel, D.: The role of marketer- generated content in customer engagement marketing. J. Mark. **83**(6), 21–42 (2019). https:// doi.org/10.1177/0022242919873903

26. Sophonthummapharn, K.: The adoption of techno-relationship innovations: a framework for electronic customer relationship management. Mark. Intell. Plan. **27**(3), 380–412 (2009)

27. Al-Dmour, H.H., Algharabat, R.S., Khawaja, R., Al-Dmour, R.H.: Investigating the impact of ECRM success factors on business performance: jordanian commercial banks. Asia Pac. J. Mark. Logist. **31**(1), 105–127 (2019)
28. Gunarathne, P., Rui, H., Seidmann, A.: Whose and what social media complaints have happier resolutions? Evidence from Twitter. J. Manag. Inf. Syst. **34**(2), 314–340 (2017). https://doi.org/10.1080/07421222.2017.1334465
29. Soboleva, A., Burton, S., Mallik, G., Khan, A.: 'Retweet for a chance to...': an analysis of what triggers consumers to engage in seeded eWOM on Twitter. J. Mark. Manag. **33**(13–14), 1120–1148 (2017)
30. Al-Bahrani, A., Patel, D.: Incorporating Twitter, Instagram, and Facebook in economics classrooms. J. Econ. Educ. **46**(1), 56–57 (2015)
31. Lewis, J.S., Goranson, J., Kastriba, L.: Policy point-counterpoint: the good and the bad of the social media revolution. Int. Soc. Sci. Rev. **95**(1), 18 (2019)
32. Hairston, S., Wu, D., Yu, J.: Analyzing the lindekin profiles of audit partners (March 2019)
33. Anders, G.: The other social network. Forbes **190**, 76–84 (2012)
34. Gamonar, F.: Tudo o que você ainda não sabe sobre o LinkedIn. LinkedIn (2015). https://www.Lindekin.com/pulse/tudo-o-que-voc%C3%AA-ainda-n%C3%A3o-sabe-sobre-Lindekin-incr%C3%ADvel-flavia-gamonar/
35. Peterson, R.M., Dover, H.F.: Building student networks with lindekin: the potential for connections, internships, and jobs. Mark. Educ. Rev. **24**(1), 15–20 (2014)
36. Liu, X., Burns, A.C., Hou, Y.: An investigation of brand-related user-generated content on Twitter. J. Advert. **46**(2), 236–247 (2017)
37. Chang, Y.C., Yeh, W.C., Hsing, Y.C., Wang, C.A.: Refined distributed emotion vector representation for social media sentiment analysis. PLoS ONE **14**(10), 1–23 (2019)
38. Abbasi, A., Zhou, Y., Deng, S., Zhang, P.: Text analytics to support sense-making in social media: a language-action perspective. MIS Q.: Manag. Inf. Syst. **42**(2), 427–464 (2018)
39. Ashraf, S.S., Verma, S., Tech, M.: A survey on sentiment analysis techniques on social media data. Int. J. Recent Res. Asp. **3**(3), 65–68 (2016)
40. Schweidel, D.A., Moe, W.: Listening in on social media a joint model. J. Mark. Res. **51**(4), 387–402 (2014)
41. Chen, W., Cai, Y., Lai, K., Xie, H.: A topic-based sentiment analysis model to predict stock market price movement using Weibo mood. Web Intell. **14**(4), 287–300 (2016)

Outsourcing of Social CRM Services in German SMEs

Julio Viana[1]([⊠]) [iD], Maarten van der Zandt[2], Olaf Reinhold[1,2] [iD], and Rainer Alt[1,2] [iD]

[1] Social CRM Research Center, Leipzig, Germany
julio.viana@scrc-leipzig.de
[2] Leipzig University, Leipzig, Germany

Abstract. Outsourcing is a common market practice that supports companies to focus on the development of their core business. Innovative SMEs, especially, can highly benefit from this support. Although outsourcing is commonly adopted in the field of Digital Marketing, services are changing as companies require more analytical and technological skills from suppliers. These new demands overlap with the concept of Social CRM and its integration of social media data into CRM tools to improve insights on customers. Based on Social CRM services, as well as the reasons for companies to outsource, this paper provides insights on current practices derived from a questionnaire applied to SMEs in Germany. Results show that German SMEs outsource to increase customer satisfaction, focusing on core business, service quality and cost reduction. Additionally, they invest more on services related to the interaction between brand and customers on social media. These results support companies in the field to understand the current and future demands and service suppliers to focus on specific needs when approaching German SMEs.

Keywords: Outsourcing · Social CRM · German SMEs · Digital marketing

1 Introduction

Outsourcing is a word that describes the business practice of hiring a party outside a company to preform services and create goods [30]. Companies leverage their resources through outsourcing by: (1) developing their core competencies in which they aim to be the best at; (2) focusing investment and management attention to these competencies; and (3) strategically outsourcing other activities where they either cannot or do not need to be best [26]. In recent decades, industrial companies have reduced the number of the vertical range of manufacture of their products and concentrated on those steps of the production process that generated the highest added value, leading to network formation of innovative suppliers [16].

The enormous developments in the information and communication technologies (ICT) creates a favorable scenario for an increase in outsourcing. The level of innovation of a company will influence its outsourcing decision, especially in small and medium-sized enterprises (SMEs), which tend to outsource more when they are more innovative in products and processes [1].

© Springer Nature Switzerland AG 2020
W. Abramowicz and G. Klein (Eds.): BIS 2020 Workshops, LNBIP 394, pp. 215–228, 2020.
https://doi.org/10.1007/978-3-030-61146-0_17

Besides the need to focus on their core competencies, companies outsource also to reduce costs. The cost reduction in the transfer to an external service provider results primarily from the exploitation of scale economies, as outsourcing providers specialize in the offered services and usually provide them on a much higher scale [7, 16].

Following these outsourcing strategies, companies have long outsourced creative marketing activities, such as advertising and promotion campaigns. Nevertheless, companies are now increasingly outsourcing marketing operation and analytics as well [24].

This increase emanates from the need of marketing approaches to adapt to new forms of conversation and interaction between companies and their customers in the realm of the social web [23]. Social customers are connected with their peers and they expect to interact with a company on their terms, at their time and through channels they feel comfortable with [13, 32].

Hence, the already extensive range of services provided by digital marketing agencies (DMAs) are now going through an adaptation to offer analytical services based on the integration of different data sources [31]. These new services overlap with the concept of Social Customer Relationship Management (Social CRM), which focus on the improvement of company-customer relationship based on the integration of social media data into existing Customer Relationship Management (CRM) solutions to provide better insights on customer journey, as well as to enhance the communication channels used by firms [3].

Therefore, Social CRM derives from the concept of CRM, which combines people, processes and technology throughout an organization to understand its customers [6, 17] and provide an infrastructure to support companies defining and increasing value for customers and motivate them to stay loyal [10]. Social CRM benefits from emerging Information Systems (IS) solutions [17, 22], as well as the increase of company-customer interactions derived from advances, which started in the Web 2.0 era [12, 13, 23]. For that, it is necessary to integrate large amount of data into CRM systems [2] and use information technology (IT) techniques to analyze this data [27].

This scenario influences how DMAs define their services in an increasing scenario in favor of analytical services [18, 32]. On the other hand, Social CRM can provide an integrated approach that supports DMA and companies to establish new processes [31]. This new integration and service supply might also influence outsourcing strategies from companies, especially innovative SMEs, that tend to focus on their core business, outsourcing specialized knowledge.

The importance of SMEs has been increasing across the globe because of their specific potentials, such as high specialization, the ability to react faster than larger companies [15], and capacity to create jobs [21]. In Germany, the contribution from SMEs to the economy increased from 745 billion Euros in 2011 to 907.6 billion Euros in 2017 [9] and they account for two-thirds of employment and two-fifths of all investments [21], making it relevant to analyze the German market.

Based on the aforementioned facts, this paper analyzes the factors influencing German SMES in the adoption of outsourcing strategy towards Social CRM services. For that, the following research questions were established: (1) what are the reasons for German SMEs to outsource Social CRM services?; and (2) what Social CRM services are mostly outsourced by German SMEs? This study presents the results and analysis of a survey conducted with SMEs in Germany, as well as the practical implications of the concluding facts.

2 Related Concepts

2.1 Outsourcing Strategies and Influencing Factors

Different strategies are pointed out by the literature as belonging to the outsourcing decision-making process. The three most mentioned are (1) Single Vendor - when companies prefer one provider for most services; (2) Competition - when few providers provide most services; and (3) Best-of-breed - when companies choose the best provider for each service [16, 20, 25].

Besides the strategy, it is important to define the factors that influence companies when deciding to outsource services. Besides the two main factors or reasons pointed out by scholars – cost and focus on core competencies [19, 26] – other aspects might also represent chances and risks regarding outsourcing [16]. Therefore, the following reasons/factors were selected and adapted to be part of the survey:

- Focus on Core Business
- Promotion of New Business Models
- Increase in Service Quality
- Cost Reduction
- Use of Innovative [CRM] Methods
- Improve Customer Satisfaction

These factors, together with the outsourcing strategies, supported the design of questions related to answering the first research question about the reasons for German SMEs to outsource.

2.2 Social CRM Services

The Social CRM framework depicted in Fig. 1 supports the categorization of services by allocating and grouping the offerings from DMAs into Integration, Analysis and Management/Interaction. Results from [31], who analyzed the supply of Social CRM services through the analysis of the offerings from DMAs, pointed to a higher offering of interaction services, followed by services in the analysis category and, lastly, the Management/Integration category.

The same framework guides the analysis to find whether the current supply from DMAs matches the demands from German SMEs. Table 1 presents the Social CRM services according to the categories shown in Fig. 1.

The remaining 24 services and their categories supported the design of survey questions that support answering the second research question about the types of outsourced Social CRM services.

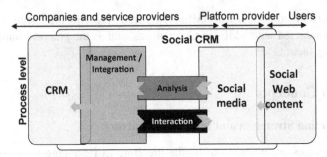

Fig. 1. Categorization and Ranking of Social CRM Services based on [2] and [31]

3 Methodology

In 2019, German SMEs participated in a survey developed and applied to answer the proposed research questions. The survey was based on the literature from the topics of outsourcing and Social CRM, as described on the previous chapter. This chapter introduces aspects from the survey design and insights on the participating sample.

The study took place using a developed questionnaire applied through an online survey tool called Typeform[1]. The questionnaire was divided into three sections: (1) Outsourcing and Reasons; (2) Outsourcing of Social CRM Services; and (3) Demographic Data. The 20 questions were built according to the aforementioned Social CRM services and categories, as well as to outsourcing strategies and reasons, divided into open, semi-open and closed options.

As recommended, before its application, the questionnaire was tested by selected experts to uncover methodological gaps and inconsistencies before the research design was carried out on a large number of observation units [28]. This test group consisted of six experts and was composed of three employees from the marketing department of an SME and three experts in the field of Social CRM, who provided feedback on the structure and comprehensibility.

The target group of this empirical research consists of German micro, small and medium-sized enterprises. Thus, an SME definition based on Recommendation 2003/361 of the EU Commission was used to ensure the selection of participants. Hence, this study used the definition of SME, according to a company's number of employee and turnover, as follows: (1) Micro – up to ten employees and two million Euros; (2) Small – up to 50 employees and 10 million Euros; and (3) Medium – up to 250 employees and 50 million Euros [11]. The study considered the size of the companies in the analysis to identify possible differences when companies choose a reason for outsourcing and the category of services they invest more.

Since innovative SMEs tend to outsource more [1], companies participating in large technology-oriented trade fairs in Germany were contacted. For that, participation lists from the Forum Deutscher Mittelstand 2019[2] and DMEXCO 2019[3] were used. Based

---

[1] https://www.typeform.com/.

[2] https://forumdigitalermittelstand.de/.

[3] https://dmexco.com/de/.

Table 1. Social CRM services and categories

Interaction
01. Content Production/ Copywriting
02. Community Management
03. Social Media General Interaction
04. Crisis Management
05. Setup of Services within Social Platforms
06. Omnichannel Interaction
07. Product/Service Offerings
08. Traffic Increase to Landing Pages[*]

Analysis
09. Brand Mention Search[*]
10. Social Web General Search[*]
11. Competition Analysis
12. Keyword Analysis
13. Trend Monitoring and Detection
14. Analysis of Brand and Campaign Impact
15. Lead Opportunity, Detection, and Classification
16. Data Analysis Reports

Management/Integration
17. Development of a Communications Strategy
18. Integration of CRM and Social Media Data
19. Feedback and Complaint Management
20. Social Media Accounts Management
21. Track and Monitor Leads/Prospects through Social Media
22. Conversion Rate Comparison
23. Establishment of Process Roadmap for Customer Interaction[*]
24. Educational Trainings on Data Integration
25. Software Development and/or Setup
26. Information Architecture and Design
27. Digital Influencer Marketing

Source: Adapted from [5, 31]

[*]To provide a faster questionnaire, these services were excluded as they are strongly connected to other services and do not influence the results and conclusions.

on these lists, a total of 700 potential participants were selected and contacted via email. Out of this total, 122 companies completed the questionnaire.

Regarding the size, 33 (28%) were microenterprises, 63 (52%) were small and 23 (20%) were medium size. When comparing with the total population of SMEs in Germany (approximately 2.45 million), the vast majority consists of micro-sized enterprises which employed up to nine people, while small enterprises are around 357 thousand and 58.8 thousand are medium-sized enterprises. However, microenterprises do not outsource as much as larger ones [29], indicating the possible difference in the study sample.

Most of the companies belonged to the Information Technology (IT) sector (24%), followed by Telecommunications and Media (12%), Consumer and Industrial Goods (11%) and Construction (10%). The other sectors corresponded to less than 10% of respondents.

The next chapters explore the results of the survey descriptively and highlight the practical implications of outsourcing practices concerning Social CRM.

4 Results

Out of the 122 companies participating in the study, only 55 are currently outsourcing any of the Social CRM Services. Nevertheless, the 67 remaining companies were asked to choose their reason for not outsourcing. Results displayed in Fig. 2 indicate that the high efforts to adapt their internal work, together with a small cost advantage, are the main reasons for companies to not outsource Social CRM activities (Fig. 3).

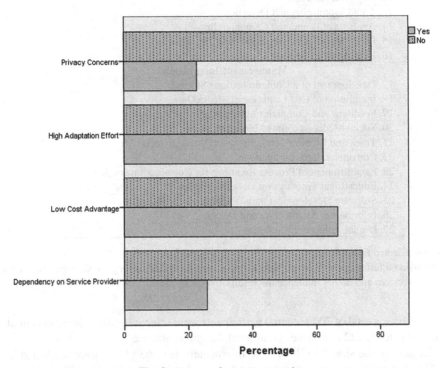

Fig. 2. Reasons for not outsourcing

The next results explore the answers from participants who claimed to outsource Social CRM services. Respondents were asked regarding their outsourcing strategies and reasons. The results seek to answer the first research question concerning factors influencing the decision to seek for services outside the company.

Regarding outsourcing strategies, most respondents (45.5%) claimed to follow the Best-of-breed strategy, followed by Competition (27.3%) and Single Vendor (12.7%).

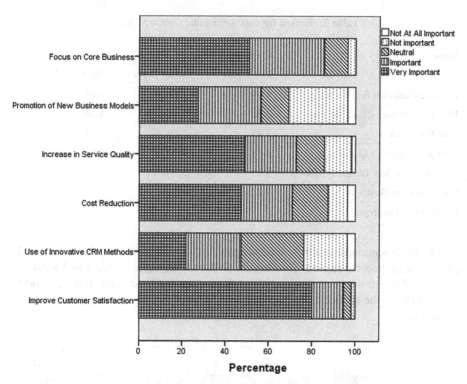

Fig. 3. Reasons for outsourcing and their importance

14.5% of the respondents claimed to not have a specific strategy. Results show that the majority of participants adopt the strategy to outsource activities to more than one company, choosing the best provider for each service. These results are aligned with the offer analysis of Social CRM services from DMAs. Many agencies choose to specialize themselves in specific services [31] and this specialization reverberates in demands from companies seeking the best provider according to their outsourcing strategies. It creates a favorable scenario for multi-vendor strategies, envisioning different service providers within a company's value chain [4].

Moreover, the participants were asked to choose the reasons for them to outsource according to the importance of each reason.

Overall, participants seem to be concerned with improving customer satisfaction by outsourcing their Social CRM activities. Focusing on core business, increasing service quality and reducing costs follow customer satisfaction also as important reasons. However, it is important to understand whether the size of the company plays a role in reasoning outsourcing. Table 2 shows this analysis, indicating that medium-sized enterprises consider also the use of innovative methods and service quality as important reasons for outsourcing activities.

Table 2. Reason for outsourcing and company size

Reason	Company size		
	Micro (Mean)	Small (Mean)	Medium (Mean)
Improve Customer Satisfaction	4.69	4.71	4.88
Use of Innovative CRM Methods	2.94	3.42	**4.38**
Cost Reduction	3.75	4.19	3.88
Increase in Service Quality	3.88	4.00	**4.63**
Promotion of New Business Models	3.25	3.68	3.25
Focus on Core Business	4.25	4.39	4.25

Source: Own Database

The aforementioned results provide insights that support answering the first research question regarding the reasons for German SMEs to outsource Social CRM services. The second question concerns the services these companies are outsourcing the most. Figure 4 shows the service categories and proportion of digital marketing budget spent on each one.

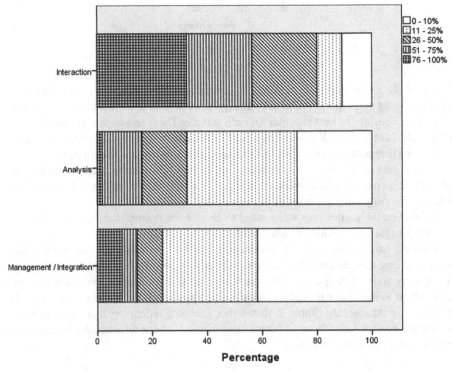

Fig. 4. Proportion of digital marketing budget spent on each service category

Participants invest more of their outsourcing budget in the Interaction category. However, participants indicated the category Analysis as the greatest potential for an increase in outsourcing.

Nevertheless, it is important to assess which specific services these companies are outsourcing or planning to outsource in the near future. The latter can support service providers to detect possible demands in the short term. Table 3 provides some insights regarding the current outsourcing scenario and imminent demands.

Table 3. Outsourcing of social CRM services

	Services	Yes	No	Not yet
Interaction	Content Production/Copywriting	**56.4%**	32.7%	10.9%
	Community Management	43.6%	34.5%	21.8%
	Social Media General Interaction	**70.9%**	20.0%	9.1%
	Crisis Management	21.8%	**38.2%**	**40.0%**
	Setup of Services within Social Platforms	14.5%	**65.5%**	20.0%
	Omnichannel Interaction	**65.5%**	23.6%	10.9%
	Product/Service Offerings	**66.7%**	20.4%	13.0%
Analysis	Competition Analysis	29.1%	41.8%	29.1%
	Keyword Analysis	32.7%	38.2%	29.1%
	Trend Monitoring and Detection	21.8%	**47.3%**	**30.9%**
	Analysis of Brand and Campaign Impact	**52.7%**	27.3%	20.0%
	Lead Opportunity, Detection, and Classification	21.8%	**49.1%**	29.1%
	Data Analysis Reports	34.5%	38.2%	27.3%
Management/Integration	Development of a Communications Strategy	22.2%	**64.8%**	13.0%
	Integration of CRM and Social Media Data	**52.7%**	32.7%	14.5%
	Feedback and Complaint Management	38.2%	40.0%	21.8%
	Social Media Accounts Management	47.3%	40.0%	12.7%
	Track and Monitor Leads through Social Media	25.5%	**54.5%**	20.0%
	Conversion Rate Comparison	29.1%	**52.7%**	18.2%

(*continued*)

Table 3. (*continued*)

	Services	Yes	No	Not yet
	Educational Trainings on Data Integration	18.2%	**74.5%**	7.3%
	Software Development and/or Setup	20.0%	**70.9%**	9.1%
	Information Architecture and Design	25.5%	**54.5%**	20.0%
	Digital Influencer Marketing	10.9%	**76.4%**	12.7%

Source: Own Database

Participants were asked if they are currently outsourcing a Social CRM service (Yes), if they are currently not outsourcing the indicated service (No) and if they plan to outsource that service in the short term (Not Yet). When answering 'No', companies can infer that they do this activity internally, as well as that they do not perform this activity in the company.

In general, companies outsource mostly services within the Interaction category. However, two services in this category are not highly outsourced as the others. The first one regards the setup of company accounts on social media. Besides indicating that this activity is mostly done inside the company, the result shows that it will remain so, as the number of companies willing to outsource this service in the near future is also low. On the contrary, German SMEs divide themselves mostly between not outsourcing or planning to outsource crisis management. It reflects an increasing concern from companies related to this service, aligned with an increase in social media activities and brand relationship in this environment [8]. In the same line, participants showed interest in outsourcing activities related to trend analysis and monitoring, which also points out concerns regarding the social media environment.

Many other services, not highly outsourced, relate to activities sensitive to the disclosure of company data, such as conversion rate comparison, lead management and monitoring. This points out to a barrier that DMAs might face due to the lack of internal data from clients. Additionally, training, software development and design of information architecture related to Social CRM seem to be done internally or not done in the company.

Another result points to the lack of outsourcing related to the management of digital influencers in social media. Over 75% of the companies claim to not outsource such service. It indicates that they are either managing these activities inside the company or that German SMEs do not use influencers for supporting their marketing activities. Moreover, SMEs may experience barriers towards some forms of influencer marketing that require substantial investments, especially when seeking access to influencers who have access to the target audience continuously [14].

Besides the indication of outsourcing activities, the mean from the investments declared by participants on a five-point scale (0–10, 11–25, 26–50, 51–75 and 76–100) was analyzed according to the company size. These points in the scale indicate the percentage spent on digital marketing activities and Table 4 presents the mean of these results, taking into consideration the service categories and the size of the participating SMEs.

Table 4. Mean of spent budget according to company size and service category

Service category	Company size		
	Micro (Mean)	Small (Mean)	Medium (Mean)
Interaction	3.56	3.55	3.75
Analysis	2.00	**2.45**	1.88
Management/Integration	1.62	2.19	**2.38**

Source: Own Database

As shown before, the results point to higher investments in services related to the Interaction category. However, services related to the Analysis category receive secondary attention from small companies, while medium-sized enterprises allocate budget to the category of Management/Integration as a second option.

It is also relevant to look closely in the data to find which services within these categories are currently outsourced according to the company sizes. In this sense, small companies claim to outsource analytical activities, which are mostly related to data analysis reports and analysis of brand and campaign impact as show on Table 3. Moreover, medium companies claim to outsource Integration/Management activities related to the comparison of conversion rate and integration, feedback and complain management and integration of social media and CRM data. The latter, however, is considered an important challenge for companies, as well as providers of information system solutions in the field of Social CRM [22].

5   Conclusions and Further Work

The aforementioned results provide insights concerning the outsourcing of Social CRM services among participants. Based on a questionnaire with German SMEs, results answered the two proposed research questions by (1) providing insights on the reasons for these companies to outsource and (2) highlighting the main services outsourced by them. Moreover, the analyses ground the following conclusions:

- SMEs, which do not outsource Social CRM services, are mostly concerned with the effort to adapt their internal processes and the low-cost advantage of outsourcing these services.
- German SMEs indicate concerns towards improving customer satisfaction, focusing on core business, increasing service quality and reducing costs as the main reasons for outsourcing Social CRM services.
- When looking at the differences between company sizes, medium-sized enterprises consider also the use of innovative methods and service quality as important reasons for outsourcing.

- Participating SMEs invest more in activities within the Interaction category, which correspond to efforts in communicating campaigns with customers via social media managing their feedback to solve problems and improving satisfaction.
- Services like Crisis Management and Trend Monitoring and Detection are indicated as possible demands in the short term, as respondents showed interest in outsourcing them in the near future.
- Considering the differences in company size, small companies claimed to outsource analytical activities in a secondary step, mostly related to data analysis reports and analysis of brand and campaign impact.
- However, medium companies claimed to outsource Integration/Management activities in a secondary step, mostly related to the comparison of conversion rate and integration, feedback and complain management and integration of social media and CRM data.

The Social CRM framework and the results of services supplied by DMAs, shown in Fig. 1, indicate a higher offer of services related to interaction, followed by the categories Analysis and Management/Interaction, respectively. These results match with the current demands from German SMEs, as services related to interaction receive more investments, followed by services related to analysis and later to management and integration. However, results also support the identification of possible future demands from SMEs in Germany and indicate a future increase in demand for services related to analytics.

Some practical implications derive from these results, as the conclusions can support companies taking decisions towards outsourcing. German SMEs can understand the differences in outsourcing according to company size, as well as the most outsourced services from each category. This can impact their strategies when seeking to outsource some digital marketing and Social CRM activities. Service suppliers, especially DMAs, can benefit from these results when approaching German SMEs by better understanding their current and short-term demands. The indication of current needs, as well as the potential to increase investments in other services in the short term could support agencies to revise their services, adapting them to the demands of SMEs.

The study also has some limitations and factors that can ground further work on this topic. The number of participants does not represent satisfactory the population of SMEs in Germany and, therefore, could be improved. A higher number of participants could also provide better statistical analysis by crossing different variables. Future work could adapt the survey to focus on different markets as well as provide qualitative analysis of the reasons for outsourcing or insourcing.

Acknowledgement. The authors gratefully acknowledge the financial support of this research by the German Academic Exchange Service (DAAD) within the project 'Social CRM as Specialization Subject in Brazilian Universities' (57449332) and the Sächsische Aufbaubank and European Union within the ESF project SPE4CRM (100362354).

References

1. Alderete, M.V.: Do information and communication technology access and innovation increase outsourcing in small and medium enterprises? JISTEM **10**(2), 303–322 (2013). https://doi.org/10.4301/S1807-17752013000200007
2. Alt, R., Reinhold, O.: Social-customer-relationship-management (social-CRM). Wirtschaftsinformatik **54**(5), 281–286 (2012). https://doi.org/10.1007/s11576-012-0330-6
3. Alt, R., Reinhold, O.: Social Customer Relationship Management. Fundamentals, Applications, Technologies. Management for Professionals. Springer, Cham (2020). https://doi.org/10.1007/978-3-030-23343-3
4. Alt, R., Reitbauer, S.: Towards an integrated architecture and assessment model for financial sourcing. In: Rabhi, F.A., Veit, D.J., Weinhardt, C. (eds.) Second International Workshop on Enterprise, Applications and Services in the Finance Industry, pp. 67–74. IEEE, Regensburg (2005)
5. Barata, G.M., Viana, J.A., Reinhold, O., Lobato, F., Alt, R.: Social CRM in digital marketing agencies: an extensive classification of services. In: 2018 IEEE/WIC/ACM International Conference on Web Intelligence (WI), pp. 750–753. IEEE (2018)
6. Brown, S.A. (ed.): Customer Relationship Management. A Strategic Imperative in the World of e-Business. Wiley, Toronto (2000)
7. Cachon, G.P., Harker, P.T.: Competition and outsourcing with scale economies. Manag. Sci. **48**(10), 1314–1333 (2002). https://doi.org/10.1287/mnsc.48.10.1314.271
8. Civelek, M.E., Cemberci, M., Eralp, N.E.: The role of social media in crisis communication and crisis management. IJRBS **5**(3), 111–120 (2016). https://doi.org/10.20525/ijrbs.v5i3.279
9. Clark, D.: Number of SMEs in Germany from 2011 to 2017 (2019)
10. Dyché, J.: The CRM Handbook. Addison-Wesley, Harlow (2001)
11. European Commission: Commission Recommendation of 6 May 2003 concerning the definition of micro, small and medium-sized enterprises (2003)
12. Faase, R., Helms, R., Spruit, M.: Web 2.0 in the CRM domain: defining social CRM. IJECRM **5**(1), 1 (2011). https://doi.org/10.1504/ijecrm.2011.039797
13. Greenberg, P.: CRM at the Speed of Light. Social CRM Strategies, Tools, and Techniques for Engaging Your Customers, 4th edn. McGraw-Hill, New York (2010)
14. Gustavsson, A.-S., Suleman Nasir, A., Ishonova, S.: Towards a world of influencers: exploring the relationship building dimensions of influencer marketing (2018)
15. Hamburg, I., Brien, E.O., Engert, S.: Engaging SMEs in cooperation and new forms of learning. CIS **7**(1) (2013). https://doi.org/10.5539/cis.v7n1p1
16. Hermes, H.-J., Schwarz, G.: Outsourcing, 1st edn. Chancen und Risiken, Erfolgsfaktoren, rechtssichere Umsetzung. Rudolf Haufe Verlag, München (2005)
17. Chen, I.J., Popovich, K.: Understanding customer relationship management (CRM). Bus. Process Manag. J. **9**(5), 672–688 (2003). https://doi.org/10.1108/14637150310496758
18. Jobs, C.G., Aukers, S.M., Gilfoil, D.M.: The impact of big data on your firms marketing communications: a framework for understanding the emerging marketing analytics industry. Acad. Mark. Stud. J. **19**(2), 81–92 (2015)
19. Kakabadse, A., Kakabadse, N.: Outsourcing: current and future trends. Thunderbird Int. Bus. Rev. **47**(2), 183–204 (2005). https://doi.org/10.1002/tie.20048
20. Kotlarsky, J., Oshri, I., Willcocks, L.P.: Governing Sourcing Relationships. A Collection of Studies at the Country, Sector and Firm Level. LNBIP, vol. 195. Springer, Cham (2014). https://doi.org/10.1007/978-3-319-11367-8
21. Lauder, D., Boocock, G., Presley, J.: The system of support for SMEs in the UK and Germany. Eur. Bus. Rev. **94**(1), 9–16 (1994). https://doi.org/10.1108/09555349410050712

22. Lobato, F., Pinheiro, M., Jacob, A., Reinhold, O., Santana, Á.: Social CRM: biggest challenges to make it work in the real world. In: Abramowicz, W., Alt, R., Franczyk, B. (eds.) BIS 2016. LNBIP, vol. 263, pp. 221–232. Springer, Cham (2017). https://doi.org/10.1007/978-3-319-52464-1_20

23. Malthouse, E.C., Haenlein, M., Skiera, B., Wege, E., Zhang, M.: Managing customer relationships in the social media era: introducing the social CRM house. J. Interact. Mark. 27(4), 270–280 (2013). https://doi.org/10.1016/j.intmar.2013.09.008

24. McGovern, G., Quelch, J.: Outsourcing marketing. Harv. Bus. Rev. 83, 22–26 (2005)

25. Oshri, I., Kotlarsky, J.: Global Sourcing of Information Technology and Business Processes. LNBIP, vol. 55. Springer, Heidelberg (2010). https://doi.org/10.1007/978-3-642-15417-1

26. Quinn, J.B., Hilmer, F.G.: Strategic outsourcing. Sloan Manag. Rev. 35(4), 43 (1994)

27. Rodrigues Chagas, B.N., Nogueira Viana, J.A., Reinhold, O., Lobato, F., Jacob, A.F.L., Alt, R.: Current applications of machine learning techniques in CRM: a literature review and practical implications. In: 2018 IEEE/WIC/ACM International Conference on Web Intelligence (WI), pp. 452–458. IEEE (2018)

28. Schilderman, H.: Quantitative method. In: Miller-McLemore, B.J. (ed.) The Wiley-Blackwell Companion to Practical Theology, pp. 123–132. Wiley-Blackwell, Malden (2012)

29. Thyagarajan, S., Nambirajan, T., Chandirasekaran, G.: Outsourcing decision of micro-small-medium enterprises (MSME). Pragyaan. J. Manag. 15(1), 23–32 (2017)

30. Twin, A.: Outsourcing. Investopedia Business Essentials (2019)

31. Viana, J., van der Zandt, M., Reinhold, O., Alt, R.: Social CRM services in digital marketing agencies: a preliminary study on service offerings in Germany. In: Abramowicz, W., Corchuelo, R. (eds.) BIS 2019. LNBIP, vol. 373, pp. 383–395. Springer, Cham (2019). https://doi.org/10.1007/978-3-030-36691-9_32

32. Woodcock, N., Green, A., Starkey, M., The Customer Framework™: Social CRM as a business strategy. J. Database Mark. Custom. Strategy Manag. 18(1), 50–64 (2011). https://doi.org/10.1057/dbm.2011.7

Customer-Focused Churn Prevention with Social CRM at Orange Polska SA (Research in Progress)

Ewelina Szczekocka[1,2](✉) (iD)

[1] Faculty of Informatics and Electronic Economy,
Poznan University of Economics, Poznań, Poland
[2] R&D Labs, Orange Polska, Warsaw, Poland
ewelina.szczekocka@orange.com

Abstract. This article presents research in progress on customer-focused churn prevention. The research aims at providing a customer-centric methodology for churn prevention based on customer social data (social CRM), as a significant complement to an approach based on customer data derived from their calling activities (e.g. discovering calling patterns) from Orange systems (like CRM). In scope of the presented research a customer churn tendency is intended to be inferred as knowledge from customers' messages posted on a company portal (such as Orange one) thanks to discovering customer's emotions, opinions and sentiment of customers' messages. As a result of the present stage of research works a literature review was provided, as well as a first conceptual model derived from the review in order to underpin a role of social CRM for churn prediction. A research agenda was also elaborated. This article provides an attempt of presenting a concept of a measure of customer churn tendency based on a customer experience, in particular customer satisfaction. This particular work considers a business case for a telecom industry, as an example of an industry for which reducing customer churn is one of the fundamental requirements. It is a work in progress concerned on analysing data from a social media channel related to one of the telecom companies (Orange) aiming at discovering signals hidden in textual messages, which can be signs of a potential churn.

Keywords: Social CRM · Customer churn prevention · Customer experience

1 Introduction

The purpose of this article is to present a research work in progress conducted by the author. The work is focused on providing methods of early customer churn prediction based on information from social channels of communication with a customer (social CRM). It can be treated as an element of a more complex approach, which is deriving knowledge on a customer from customer's data available

© Springer Nature Switzerland AG 2020
W. Abramowicz and G. Klein (Eds.): BIS 2020 Workshops, LNBIP 394, pp. 229–236, 2020.
https://doi.org/10.1007/978-3-030-61146-0_18

in a company (e.g. CRM data). Orange in scope of its project concentrates on discovering activities which may result in churn (statistical approach) and it investigates to some extent CRM data to discover these kinds of activities. It also uses NPS (Net Promoter Score) based on questionnaires of customers to discover a level of customer's satisfaction (NPS presents a level of satisfaction of a customer expressed by a willingness to recommend a company to other people). In scope of the present research the author of this article is going to enrich Orange approach by proposing and elaborating methods of discovering knowledge on customer satisfaction from Orange portal data, given that knowledge on customer's satisfaction is crucial for the company (and presently investigated via NPS). A motivation behind this research is concerned with the role of social CRM and a significance of churn prevention presented below. In general, investigating customer relationships with a company, retention and customer engagement is a strategical issue. Decreasing customer churn impacts on preserving enterprise's revenue and requires ensuring an appropriate customer relationship management. CRM together with social CRM provide a key tool set which can help to reduce churn and it is important to depict a significant role of social CRM in this process. In practical terms, the motivation is concerned with the fact that social CRM can be a good source of knowledge on a customer, this data is available in the company and it is underused. It will allow acquiring a new kind of knowledge on a customer. It should be also noticed that CRM data are less accessible due to legal and privacy constraints. A research question of this research has been formulated as follows: what methods are appropriate for predicting churn risks. Results of first works carried out in scope of this research are described in the section related to the research approach. A first conceptual model is proposed. A set of first methods was discovered during the literature review and these methods will be examined in details in order to assess their potential relevance to the approach as well new methods will be proposed based on results of the examinations and experiments. It corresponds to the research question. A research gap has been identified thanks to those first works. It is related to deriving churn risks from customer messages available in social channel of communication with a company. Moreover, there is a research gap for churn prediction from data in Polish identified (churn prediction based on customer messages expressed in Polish) which gives an opportunity for potential contributions.

2   Related Works

A literature review was provided by the author at the first phase of the research and it was aimed at finding some trends as well as inspirations for creating a solution. Three major areas of the review can be distinguished. The first part of the review was focused on understanding important elements of current major research approaches on CRM and Social CRM information systems, as well as their role for a company and its customers. A fundamental research in this area is presented in [1]. This research article provides a comprehensive view on CRM system and describes capabilities-based perspective of a social CRM system, allowing us to understand role of each of them as well as their mutual relations. It explains that a social CRM system will not replace traditional one, but it is a significant enrichment of the first one, enabling new quality of knowledge. In addition, [17] provides analysis of business value of CRM systems. Articles [18,20] and [25] – explore a role of CRM systems (in particular [20] - in developing countries' context, [25] - in a context of value drivers). Articles [19,23] - describe a role of CRM in banking system for a customer perception and quality of relationships between a company and a customer. Works presented in [21] and [22] provide a process-oriented perspective on CRM and describe an impact of some factors on the processes. Finally, the following articles consider CRM performance and effectiveness: [24] - proposes and describes CRM performance and measurement framework (this framework was also experimentally implemented), [26] – provides model of CRM effectiveness. All the above mentioned articles were helpful in understanding a role of CRM and moreover [1] provided knowledge on an important enrichment. The second part of the review was concerned with existing approaches of a churn and a churn prediction (e.g. what churn factors are taken into consideration, which data is used for the modelling, what are the methods of prediction and assumed variables derived from data). In summary, this review shows some trends, like a trend of considering a churn prediction through an assessment of a likelihood of a customer to leave a company. A prediction based on different methods (ML and statistical methods), churn factors and different sets of data are proposed in the set of articles. For instance [2] studies aspects of customer choices and choice models vs customer lifetime value (CLV) models and different prediction methods considering their pros and cons, [3] - provides model based on CLV: it describes CLV and CLV-based churn model developed on account transactions from Belgian financial service company considering different classifiers: logistic regression, decision trees, neural networks. Almost every article in the considered set contains description of experiments on methods which are performed using concrete data. For instance few articles describe research related to telecommunication churn approach proposing concrete methods of analysis based on telecommunication data derived from calls ([4] - rough set approach, [6] - neural networks in Indian telecom market, [14] - churn prediction, segmentation and fraud detection using supervised ML, the issue of data and appropriate selection of target variables were considered, [16] - describes experiments with a use of CNN method performed on phone call transcriptions). Other exemplary prediction methods are presented in the following articles. The article [5] - proposes improved balanced random forests, [12] - derives a method based on order and social networks (mixed), [13] - provides an anti-churn model using pca method. Some research works considers predicting churn based on analysis of social groups ([8,9] and [11] - in mobile networks, [10,15]). In one of the examples some interesting methods enabling personalization of services for a customer are proposed ([7] - mining customer ratings). All the proposed herein analytical methods are based on numerical data (deriving calling patterns), except [16]. The final part of the literature review was dedicated to discovering existing methods of knowledge prediction from customer textual data, (like data from social communication channels). It allows inferring properly churn factors from customer messages. Very little research is available on churn factors derived from textual data (only [16] discovered so far). As it is concerned with inferring knowledge from textual data, there is some interesting and emerging research based on deep neural networks (methods like e.g. Bert). Moreover, there is very little research dedicated to Polish, which is also a good opportunity to contribute to churn analysis based on texts in Polish. Below there is a synthetic view on this part of the research. There are few works on discovering emotions from texts identified for further research, like [27] – describing user profiles that include explanatory features, like emotions and personality for service personalization, [28] - presenting experiments with a large dataset of fine-grained emotions, [29] - emotion-enriched word representations, [30] – presenting system submitted to IEST-2018 task, focused on learning implicit emotion from Twitter data, [31] - describing the task of emotion detection in textual conversations in SemEval-2019.

3   Business Case: Telecom Churn Prevention with Social CRM

According to McKinsey report [32] the telecom industry continues to face growing pricing pressure worldwide and it is forced to respond through more competitive offers, bundles, and price cuts. Based on above mentioned report, analysing telecom companies around the world reveals that implementing a comprehensive, analytics-based approach can reduce their churn by as much in average as 15%. The figure Fig. 1 shows that customer retention is the most significant retail revenues' driver. In the same time, the success rate of selling to an existing customer is 60–70%, while the success rate of selling to a new customer is only 5–20%. The author conducts research on novel churn prediction methods taking a telecom company as a business case, concretely Orange Polska (a part of Orange Group). This local operator has implemented among others a CRM system, a Business Intelligence platform for automation of operational and analytical support of key processes in the area of accounting, planning, reporting and sales management as well as a portal called "My Orange" dedicated to contacts with its customers

Fig. 1. Retail revenue drivers

Anonymous data from these systems can be used in order to derive knowledge and propose key churn measures as well as churn prediction methods.

4   Research Approach

Design Science together with guidelines for Information Systems was decided to be a methodology used in the present research. A work is provided iteratively in a four steps' approach (research, analysis, synthesis and realization). These first findings are a result of author's research activities based on the literature review and examination of Orange case. Outcomes of this stage of the research (artefacts) consist of a literature review, as well as a conceptual model and a research agenda updated thanks to the works performed. In the next stage, a major artefact will be delivered as a set of methods for a customer churn prediction based on a measurement of customer experience and satisfaction level. A conceptual model for the approach was proposed as an outcome from the literature review, see Fig. 2. It is an attempt of depicting the role and place of social CRM in companies' environment. It was among other inspired by [1]. The proposed research agenda assumes the following tasks: providing state-of-the-art analysis - on churn (general perspective), on anti-churn strategies ("as it is") for a business case company (telecom), on a churn prediction from a customer data, on methods for knowledge discovery from texts (emotions, sentiment, opinions), building a model of a churn prediction for a business case company (telecom) ("to be") - as data-based anti-churn solution (delivering relevant methods), evaluation of the solution empirically. In this research scope the author will deliver among others: a theoretical model on churn prediction, results of experiments with selected state-of-the-art methods, evaluation of the experiments and proposal of the enrichment of methods to be finally tested and evaluated, final set of methods based on data sets from a company social media communication

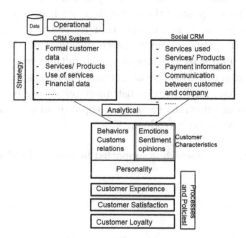

Fig. 2. Conceptual model of churn from CRM and social CRM

channel. The present literature review helps to answer earlier research questions arisen from Orange needs: how to deal with a customer churn involving a customer experience perspective, how to predict a customer churn tendency with a use of customer's information from a social CRM. This review shows that social enhancement of CRM can lead to better customer understanding. In CRM system predominantly structured data is collected which allows deriving a certain kind of knowledge (e.g. communication patterns). Involvement of social media opens up for potentially new kind of knowledge, which is mostly based on unstructured information, e.g. information in form of texts in a natural language. It provides opportunity to derive customers' feelings, moods, opinions (e.g. on services, brand), questions toward a company. It is a potential rich source of knowledge on a customer, relatively difficult to extract from data. However, some interesting works were found, based on which further experiments on discovering specific knowledge on customer's feelings can be provided.

5 Conclusions and Outlook

The author of this article addresses a challenge of providing novel methods of extracting knowledge based on social data of customers in Orange Polska. Knowledge on customers' emotions, moods, opinions can fuel a churn model. It could be later envisioned as an integral part of an enterprise solution. A literature review on existing research shows that there is a research gap in above described approach concerning extracting knowledge from Social CRM in order to fuel churn predictions, in particular a lack of practical cases. The author presents a work in progress on methods that can fuel a model of a customer churn. At this stage the author works on some algorithms of neural networks which can retrieve knowledge on emotions and sentiment from data (e.g. Bert for Polish). This knowledge can help in defining successful measures of the level of customer experience and satisfaction. At this stage some data from published Polish corpora are used, like[1,2] adjusted to the purpose of the research. Data from the Orange portal will be used in the next phase. It is assumed that its characteristics will be similar to the data used during the present research provided by the author: it will be short messages (2–3 sentences in a message). An initial verification of some texts from "My Orange" was provided by the author. There is a variety of information contained in those texts, for instance customers references to a particular topic, expressions of different emotions (around a topic or in general), expressions of opinions e.g. on a service, on a trademark.

Acknowledgement. This research is supported by Orange Polska SA.

---

[1] http://clarin-pl.eu/en/.

[2] http://zil.ipipan.waw.pl/.

References

1. Trainor, K.J.: Relating social media technologies to performance: a capabilities-based perspective. J. Pers. Sell. Sales Manag. **32**(3), 317–331 (2012). https://doi.org/10.2753/PSS0885-3134320303
2. Kamakura, W., et al.: Choice models and customer relationship management. Mark. Lett. **16**(3–4), 279–291 (2005)
3. Glady, N., Baesens, B., Croux, C.: Modeling churn using customer lifetime value. Eur. J. Oper. Res. **197**(2009), 402–411 (2009)
4. Amin, A., Shehzad, S., Khan, C., Ali, I., Anwar, S.: Churn prediction in telecommunication industry using rough set approach. In: Camacho, D., Kim, S.-W., Trawiński, B. (eds.) New Trends in Computational Collective Intelligence. SCI, vol. 572, pp. 83–95. Springer, Cham (2015). https://doi.org/10.1007/978-3-319-10774-5_8
5. Xie, Y.Y., Li, X., Ngai, E.W.T., Ying, W.Y.: Customer Churn prediction using improved balanced random forests. Expert Syst. Appl. **36**(3), 5445–5449 (2009)
6. Sharma, A., Panigrahi, P.K.: A neural network based approach for predicting customer churn in cellular network services. Int. J. Comput. Appl. **27**, 26–31 (2011). https://doi.org/10.5120/3344-4605
7. Cheung, K.W., Kwok, J.T., Law, M.H., Tsui, K.C.: Mining customer product ratings for personalized marketing. Decis. Support Syst. **35**, 231–243 (2003)
8. Richter, Y., Yom-Tov, E., Slonim, N.: Predicting customer churn in mobile networks through analysis of social groups. In: Proceedings of the SIAM International Conference on Data Mining, pp. 732–741 (2010)
9. Motahari, S., Mengshoel, O.J., Reuther, P., Appala, S., Zoia, L., Shah, J.: The impact of social affinity on phone calling patterns: categorizing social ties from call data records. In: Proceedings of the Sixth Workshop on Social Network Mining and Analysis (2012)
10. Ngonmang, B., Viennet, E., Tchuente, M.: Churn prediction in a real online social network using local community analysis. In: 2012 IEEE/ACM International Conference on Advances in Social Networks Analysis and Mining (ASONAM), pp. 282–288. IEEE (2012)
11. Polepally, A., Mohan, S.: Behavior analysis of telecom data using social networks analysis. In: Cao, L., Yu, P. (eds.) Behavior Computing, pp. 291–303. Springer, London (2012). https://doi.org/10.1007/978-1-4471-2969-1_18
12. Birtolo, C., Diessa, V., De Chiara, D., Ritrovato, P.: Customer churn detection system: identifying customers who wish to leave a merchant. In: Ali, M., Bosse, T., Hindriks, K.V., Hoogendoorn, M., Jonker, C.M., Treur, J. (eds.) IEA/AIE 2013. LNCS (LNAI), vol. 7906, pp. 411–420. Springer, Heidelberg (2013). https://doi.org/10.1007/978-3-642-38577-3_42
13. Xin, Z., Yi, W., Hong-Wang, C.: A mathematics model of customer churn based on PCA analysis. In: International Conference on Computational Intelligence and Software Engineering, CiSE 2009, pp. 1–5 (2009)
14. Rehman, A., Raza Ali, A.: Customer churn prediction, segmentation and fraud detection in telecommunication industry. In: 2014 ASE BigData/SocialInformatics/PASSAT/BioMedCom 2014 Conference (2014)
15. Verbeke, W., Martens, D., Baesens, B.: Social network analysis for customer churn prediction. Appl. Soft Comput. **14**, 431–446 (2014). https://doi.org/10.1016/j.asoc.2013.09.017

16. Zhong, J., Li, W.: Predicting customer churn in the telecommunication industry by analyzing phone call transcripts with convolutional neural networks, pp. 55–59 (2019). https://doi.org/10.1145/3319921.3319937

17. Dong, S., Zhu, K.: The business value of CRM systems: a resource-based perspective (2008). 1530-1605/08 © 2008 IEEE

18. Khodakarami, F., Chan, Y.E.: Exploring the role of customer relationship management (CRM) systems in customer knowledge creation. Elsevier (2014)

19. Dubey, N.K., Sangle, P.: Customer perception of CRM implementation in banking context scale development and validation. J. Adv. Manag. Res. **16**(1), 38–63 (2019)

20. Akroush, M.N., Dahiyat, J.S.E., Gharaibeh, J.H.S., Abu-LailAL-Hilal, B.N.: Customer relationship management implementation: an investigation of a scale's generalizability and its relationship with business performance in a developing country context. Int. J. Commer. Manag. **21**(2), 158–191 (2011)

21. Santouridis, I., Tsachtani, E.: Investigating the impact of CRM resources on CRM processes: a customer life-cycle based approach in the case of a Greek bank. Proc. Econ. Finan. **19**, 304–313 (2015)

22. Keramati, A., Mehrabi, H., Mojir, N.: A process-oriented perspective on customer relationship management and organizational performance: an empirical investigation. Ind. Mark. Manag. **39**(7), 1170–1185 (2010)

23. Sivaraks, P., Krairit, D., Tang, J.C.S.: Effects of e-CRM on customer-bank relationship quality and outcomes: the case of Thailand. J. High Technol. Manag. Res. **22**, 141–157 (2011)

24. Kim, H.S., Kim, Y.G.: A CRM performance measurement framework: its development process and application. Ind. Mark. Manag. **38**(4), 477–489 (2009)

25. Richards, K.A., Jones, E.: Customer relationship management: finding value drivers. Ind. Mark. Manag. **37**(2), 120–130 (2008)

26. Kim, J., Suh, E., Hwang, H.: A model for evaluating the effectiveness of CRM using the balanced scorecard. J. Interact. Mark. **17**(2), 5–19 (2003)

27. Tkalčič, M., De De Carolis, B., de de Gemmis, M., Odić, A., Košir, A. (eds.): Emotions and Personality in Personalized Services, Models, Evaluation and Applications. HIS. Springer, Cham (2016). https://doi.org/10.1007/978-3-319-31413-6. ISBN 978-3-319-31411-2

28. Abdul-Mageed, M., Ungar, L.: EmoNet: fine-grained emotion detection with gated recurrent neural networks. In: Proceedings of the 55th Annual Meeting of the Association for Computational Linguistics. Association for Computational Linguistics (2017)

29. Agrawal, A., An, A., Papagelis, M.: Learning emotion-enriched word representations. In: Proceedings of the 27th International Conference on Computational Linguistics. Association for Computational Linguistics (2018)

30. Alhuzali, H., Elaraby, M., Abdul-Mageed, M.: UBC-NLP at IEST 2018: learning implicit emotion with an ensemble of language models (2018)

31. Zhong, P., Miao, Ch.: Ntuer at SemEval-2019 Task 3: emotion classification with word and sentence representations in RCNN (2019)

32. McKinsey: Reducing churn in telecom through advanced analytics (2017)

Social CRM: A Literature Review Based on Keywords Network Analysis

Fábio M. F. Lobato[1,3]($\boxtimes$) (iD), Jorge L. F. Silva Junior[1] (iD), Antônio Jacob Jr.[3] (iD), and Diego Lisboa Cardoso[2] (iD)

[1] Engineering and Geoscience Institute, Federal University of Western Pará (UFOPA), Santarém, Brazil
fabio.lobato@ufopa.edu.br
[2] Federal University of Pará (UFPA), Belém, Brazil
[3] State University of Maranhão, São Luís, Brazil
antoniojunior@professor.uema.br

Abstract. The ways of relationship between companies and customers have changed dramatically due to web-users engagement with Social Media. This phenomenon brought up a new concept, Social Customer Relationship Management (Social CRM), a multidisciplinary and promising research topic. This multidisciplinarity requires the combination of different perspectives from the many research areas involved. In this sense, having in mind that undertaking a literature review is a cornerstone step to conduct reliable scientific research, this paper contributes to existing research by performing a structured keywords literature review based on network analysis. The results obtained revealed the most investigated topics in this field and their relations. Additionally, it was possible to identify the main areas investigating Social CRM related subjects.

Keywords: Social CRM · Customer Relationship Management · Literature review · Knowledge networks · Network analysis

1 Introduction

Nowadays, it is unquestionable the importance of Online Social Networks (OSN) for the companies to establish a relationship with their customers [1,18]. The Social Customer Relationship Management (Social CRM) concept arouses some years ago as "a new paradigm for integrating social networking in more traditional CRM systems" [4], or yet "a philosophy and a business strategy, supported by a technology platform, business rules, processes and social characteristics, designed to engage the customer in a collaborative conversation in order to provide mutually beneficial value in a trusted and transparent business environment. It's the company's response to the customer's ownership of the conversation" [15].

More pragmatically, Social CRM aims to use and integrate information from social media and the traditional CRM systems, enhancing the results reliability

© Springer Nature Switzerland AG 2020
W. Abramowicz and G. Klein (Eds.): BIS 2020 Workshops, LNBIP 394, pp. 237–249, 2020.
https://doi.org/10.1007/978-3-030-61146-0_19

as well as to provide new sort of insights [6,11,19]. In this context, *Electronic Word-of-Mouth* plays a significant role since customers are more exigent and having easy access to steadily more information about products, services and brands reputation [3,24]. The big social media data seen in platforms such as forums, blogs, OSN and news reports contain a sea of thoughts and opinions [10,21,24]. Analyzing such data may reveal new insights for companies aiming to improve their offers [3,14,22,24], and to become more competitive in the market [23,26].

In face of these facts, academia and industry have been proposing techniques to deal with that scenario, including analytic tools for Social CRM [11,19,22]. That field focuses on best practices for relationships management with customers via OSN [6,12]. Besides that, the behavioral economics boom, altogether with OSN pervasiveness that brought to the table more information about consumers, sheds light on Social CRM.

It is already observed that the scientific literature is growing steadily all over the years [9]. Recent bibliometrics shows that the number of published scientific papers has climbed by 8–9% each year over the past several decades [17]. This phenomenon is clearly observed in multidisciplinary, emerging and dynamic research topics. Knowing that undertaking a literature review is a cornerstone step to conduct reliable scientific research [13]. There are some review types, such as critical review, meta-analysis, overview, mapping approaches, systematic review, comprehensive *etc.* [8]. The review soundness depends on many factors, usually looking for a compromise between the analysis coverage and depth. The whole process can be considered time-consuming, especially in the scenario described previously regarding multidisciplinary, emerging and dynamic research topics [16].

Moreover, in multidisciplinary studies, the researcher might be not familiarized with all subjects, making a need for automatic/semi-automatic tools to guide the efforts in the exploration of a new field, evaluating its relationship between concepts and research areas, and guiding further in-depth qualitative analysis. There are several strategies for mapping knowledge structure, wherein the Keyword Co-occurrence Network, or, simply, Keyword Network (KN) is noteworthy due to its comprehensibility, clarity, and easiness of implementation [25]. KN are networks constructed from articles keywords correlations; therefore, two words which belong to the same paper are related to each other, forming a network edge. The analysis of these networks can be done by social network analysis tools as well as Ego Network analyzers, which enable the interpretation of the keywords correlation [27].

Retaking the issue mentioned before, Social CRM deserves attention, since it fulfills all three characteristics mentioned above. First, Social CRM embraces very diverse areas, like Marketing, Computer Science, Business, Communication, and Information Systems, for instance; Then, it is directly related to Online Social Networks (OSN), which is a new and intrinsic dynamic investigation field [19].

In the light of such facts, this ongoing research paper addresses the issue mentioned above by conducting a key-word literature review on Social CRM subject. As there is relatively little research in the multidisciplinary relation of Social CRM efforts, this paper identifies the works gateways, most frequent topics, and its relationships. Hence, the following research questions were raised:

1. What are the main topics that have been addressed in Social CRM literature?
2. How these topics are related to each other?

The main method applied to answer these research questions was the key-word co-occurrence network, in which the keywords are modeled as a network graph and then it is analyzed. For this paper, first, social network analysis is conducted to identify the main areas of topics discussed and their relations. A more specific analysis in Social CRM concept was made by performing an Ego Network investigation, analyzing the Social CRM as the main node.

The remainder of this paper is organized as follows. In the Sect. 2 the methodology adopted is described. The results are discussed in the Sect. 3. Finally, the paper concludes with a summary, threats to the research validity and further research questions to coped in future work.

2 Methodology

In this paper the steps were structured accordingly to systematic literature review scheme presented by [13], which includes: (i) Identification of a topic of interest, (ii) searching and retrieving the appropriate literature, (iii) analyzing and synthesizing the findings and, (iv) reporting the methodology and conclusions. The tasks conducted are described below.

1. **Identification of a topic of interest:** As mentioned, the topic of interest here is Social CRM. The motivation to adopt the keyword review is because this topic is multidisciplinary, emerging and dynamic, with a considerable number of publication in recent years.
2. **Search and retrieving:** In this step, the search query, inclusion criteria and data collection was conducted as follows:
   (a) *Query definition:* The query was built using Boolean operators commonly supported by scientific databases systems:
   (b) *Inclusion criteria:* Considering the number of publications, the following criteria were established:
      - Only journal articles were considered;
      - Only papers written in English were evaluated;
      - Papers without keywords were not included.
   (c) *Data organization:* The meta-data was defined aiming to enable the envisioned analysis. The features considered were: Title, Digital Object Identifier (DOI) code, Publication Year, Journal, Abstract, and Keywords. These data were organized in a Comma-Separated Values (CSV) file.

(d) *Data acquisition:* Aiming to reduce the human effort and the time required to conduct the review, this step was automated through a Web Crawler written in Python using the Scrapy Framework. The input of the data acquisition tool developed consists of web page links for the searched query. The output was the CSV with the data collected.

3. **Analysis and synthesis:** This step comprises of three main tasks, to know: pre-processing, analysis and interpretation, which are described below:

(a) *Pre-processing (filtering):* In this first pre-processing stage, *spurious* keywords that represents noise are removed. Examples of *spurious* are journal predefined keywords codes (*e.g.:* M100, M212), or keywords in other languages than English. Both, keyword codes and foreign languages keywords, founded in the case study were duplicated versions of the keywords of interest, for this reason, they were removed.

(b) *Pre-processing (Normalization):* Basically, this task aims to unify synonyms and create the network structure for further analysis. This is the most time-consuming task. The detailed sub-processes are described below, then, an example is given.

   i. Preparation: The list obtained from the previous task were converted in a single column list containing all keywords (KW). Then, the items were sorted lexicographically to facilitate the comparisons.

   ii. Synonym identification: The KW were compared to each other and the synonyms were identified and a reference term was chosen/created. An example for the Synonym identification process is the occurrence of these keywords across the papers: { "artificial neural network", "ann", "artificial neural network (ann)", "artificial neural network analysis (ann)", "artificial neural networks", "neural network", "artificial neural networks (anns)"}. The reference term chosen was the first, 'artificial neural network".

   iii. Synonym substitution: The synonym list is then combined with the keyword list of each paper, where synonyms were substituted by the respective reference term. For the previous example, all occurrences of the given list were substituted by "artificial neural network".

   iv. Binary combination: For each keyword list corresponding to a paper, a binary KW combination was constructed. Taking into account a paper with the keywords { "agility", "altruism", "animation"}, the Binary combination result is { "agility - altruism", "agility - animation", "altruism - animation"}. The pair of keywords produced in the previous step represents the existence of a connection (edge) between them.

(c) *Analysis:* The data was analyzed using Gephi, an open source software for graph and network analysis [7]. The CSV file containing the KW pairs was imported as adjacency list of a non-directed graph, with the sum as edge merging strategy. Self-loops that might occur due to the Synonym substitution process were ignored. The distribution algorithm used to plot the network was the Fruchterman Reingold, with default parameters, since this algorithm produced easy-to-read graphs. The filtering strategies adopted were based on the graph topology. The first was the degree range, aiming to improve the model comprehensibility by reducing the number of nodes with few connections. The other filter was applied to create the Ego Network to be analyzed.

(d) *Interpretation:* With the network statistics and the graphic visualization it was possible to identify *hubs, outliers* and most important *research topics* and its relationships.

4. **Reporting:** In this point all previously tasks are described and the results are discussed. This paper represents an outcome of this step.

3   Results

Following the methodology described in the previous section, the first step was to define the topic of interest and the *Search Query*. Considering the "Social Customer Relationship Management" subject, some related topics were added to the query aiming to improve the search coverage. Therefore, the following *search query* was adopted:

(''Social networks'' or ''Social media'') AND (''CRM'' OR ''Customer relationship management'') OR (''Social CRM'')

The ScienceDirect database was adopted in this study since its reuniting prominent journals related to Social CRM field. Additionally, its portal allowed the usage of a Web Crawler to automate the data acquisition process.

3.1   General Results

The ScienceDirect search engine returned 791 papers for the query above, including Research and Review articles. The period considered was from January 2017 to June 2020. From these, 25 papers were not meeting the inclusion criteria, remaining 766 papers that covered close to 3,000 unique keywords extracted from papers, which where distributed in around 162 scholarly journals. In the Fig. 1 it is shown the distribution of publications in the last four years.

Analyzing the Fig. 1 it is possible to perceive a crescent interest in this area. It is important to highlight that the data collection covers till mid 2020. The data collected also allowed the identification of journals that are publishing more papers in this areas. For this, journals with less than 20 papers published during this period were filtered, resulting in 10 frequent scholarly journals as shown in Table 1.

The most relevant journals for Social CRM field in ScienceDirect database represents five main areas: (i) **Marketing**, with the journal *Industrial Marketing Management*; (ii) **Management and Business**, represented by *Information & Management* and *Journal of Business Research* journals; (iii) **Computer Science**, with the *Procedia Computer Science*; (iv) **Hospitality/Tourism** with *International Journal of Hospitality Management* and, finally, (v) **Psychology**,

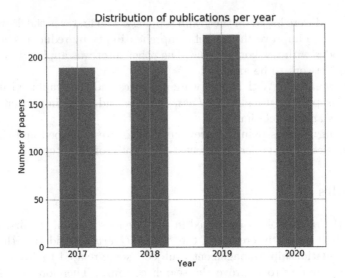

Fig. 1. Distribution of publication per year.

Table 1. Distribution of journals publications.

Journal	Number of papers
Industrial Marketing Management	61
Journal of Business Research	59
International Journal of Information Management	45
Journal of Retailing and Consumer Services	40
Technological Forecasting and Social Change	34
Information & Management	33
Procedia Computer Science	32
Computers in Human Behavior	24
International Journal of Hospitality Management	23
Decision Support Systems	22

represented by the *Computers in Human Behavior* - just to mention the four most relevant. The later is gaining even more attention with the behavioral economics and the seminal work of Richard Thaler. It is important to point out that all of these journals have more than 20 papers mentioning Social CRM topics in the last four years. These facts demonstrates that Social CRM is a multidisciplinary and promising research topic.

3.2   Keywords Pre-processing

From the 766 papers that met the inclusion criteria, 2,951 unique keywords were extracted. Following up the methodology, the first step applied was the filtering process (3.a), in which non-related keywords were removed. Around 90 keywords were removed, most of then were the translation of the original keywords for other language than the English and some predefined codes.

The remaining were passed to the Normalization step (3.b). This was, by far, the most time consuming step. For the main keywords, the synonyms were identified in order to reduce the dimensionality. For instance, the *"big data"* was chosen as the node which represents the following keywords:

> *big data* • *big data analytics* • *big data analytics capability* • *big data analytics system* • *big data commerce* • *big data framework* • *big data visualization* • *big data warehouse* • *big data-enhanced database marketing* • *big-data*

Thus, for this example, these 9 keywords were substituted (step 3.b.iii) by one, *"big data"*. Consequently, the number of unique keywords was reduced considerably, with 397 keywords being substituted by a representative ones. It means a reduction of around 15%, aiding the next step, the binary combination (3.b.iv), which resulted in 9,598 peers from the 2,554 keywords.

3.3   Keyword Network Analysis

The peers of keywords works as an adjacency list, where the keywords are the nodes and the peer represents a connection (edge) between then. This data structure allowed to build the correspondent network. In the Table 2 is given a network summary and basic statistics.

Table 2. Keywords network summary statistics.

Feature	Quantity
Number of nodes	2,554
Number of edges	9,598
Average degree	7.516
Avg. weighted degree	8.096
Network diameter	8
Graph density	0.003
Connected components	66
Avg. clustering coefficient	0.895
Avg. path length	3.73

The statistics presented in Table 2 were obtained by Gephi V0.9.2 and following the methodology described previously. Through its analyses it is possible to assert that the network follows the free-scale one [5], considering that, despite the majority of the nodes have a stable number of connections, however, few keywords have a very high connection degree. This fact is also related to the diameter (8), which is the shortest distance between the two most distant nodes in the network; and to the proportion of direct ties in a network relative to the total number possible - density (0.003) [20]. Figure 2 shows the degree distribution among the nodes.

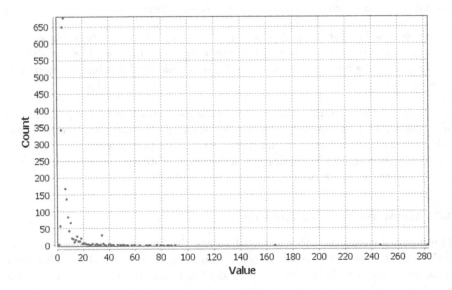

Fig. 2. Network degrees distribution.

Another feature shown in Table 2 that deserves attention is the number of Connected Components (66), it means that the networks has 6 components connected, demonstrating that the network has clusters formed by subject. It is also corroborated by the Avg. Clustering Coefficient. Despite the present of clusters, it is difficulty to visualize due to the scale-free property, in which a lot number of nodes have just few connections.

Aiming to allow a better visualization, a filtering strategy based on the Topology was applied. In the Figure a network topology is shown, using a filter by degree with the lower bound equal to 70. Analyzing the Fig. 3, it is possible to perceive that *"big data"*, *"social media"*, *"decision support systems"*, *"cloud computing"*, *"business-to-business"*, and *"Customer Relationship Management"* are the ones most frequent in this field.

It is also possible to verify some relations, for instance, *"big data"*, a well-known concept in this area, is heavily connected to *"customer relationship management"* and *"social media"*, where *"social media"* is connected with *"small and medium enterprises"*, *"business-to-business"*, *"machine learning"* and *"tourism"*. An interesting point is that *"social media"* is more used than *"social networks"*, which forms a disconnected component - connected only with *"trust"*.

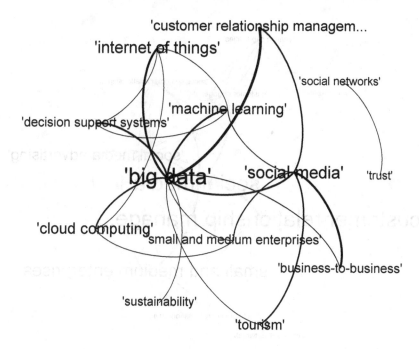

Fig. 3. Filtering nodes with degree below 70.

Finally, the Ego Network for the node "*Customer Relationship Management*" was also built, but using the Force Atlas distribution algorithm instead the Fruchterman Reingold adopted in the previous topologies. Basically, this plot getting only the nodes directly connected to a chosen node. Again, due to the scale-free property and aiming to improve its readability, a filtering strategy based on the node degree was applied. The obtained Ego Network is given in the Fig. 4.

Analyzing Fig. 4 it is possible to perceive the similarities with non-ego networks previously given. It happens for two reasons, (i) the network follows a power-law and (ii) the research queries prioritized the "*Social Media*" and "*Customer Relationship Management*".

The most relevant social platform for this application field is "*Facebook*" with degree equals to 66. This results is also reaffirming the conclusions presented in [2]. It is important to highlight that Twitter is the overall most mentioned platforms, especially in researches related to social sciences, for instance. However, Facebook fan-pages, Reviews tool (with score and comments) and chats are relevant features to perform Social CRM [19]. The Facebook relationships can be viewed in the Fig. 5.

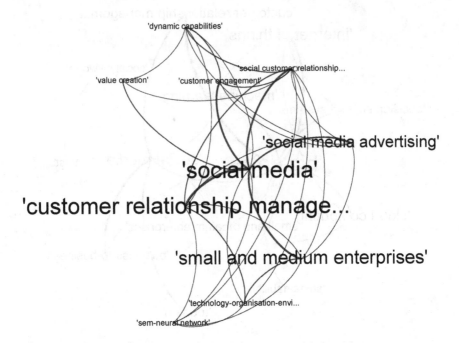

Fig. 4. Ego Network for the node "Social Customer Relationship Management".

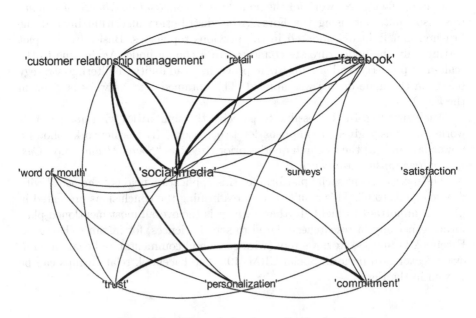

Fig. 5. Ego Network for the node "facebook".

4   Final Remarks

Social Customer Relationship Management is a relevant and promising research topic considering that social media users are getting even more engaged over the years. Being a prolific research field, it is difficult to follow up and cover all published material. In this sense, automatic or semi-automatic strategies arise as an elegant solution, reducing the efforts to figure out a broad area panorama and to identify important relations between the topics addressed a research area. On that account, in this paper, the conduction of a keywords literature review on Social Customer Relationship Management was reported. The review was planned to answer two research questions: (RQ1) *"What are the main topics that have been addressed in Social CRM literature?"*; and (RQ2) *"How these topics are related to each other?"*.

Altogether, 766 papers with almost three thousands unique keywords were obtained from the ScienceDirect database. The resulted keywords network follows a power law, permitting its classification as a free-scale network. That is, just few keywords are presented in the majority of the papers. These information allows to answer the RQ1, trough network analysis regarding the node size, it was possible to perceive the five most relevant areas in Social CRM studies, to know: Marketing, Management, Computer Science, Hospitality/Tourism and Psychology, being represented by the following keywords: *"Social Media"*, *"Marketing"*, *"Tourism"*, *"Big Data"*, *"Behavioral Analysis* and *"customer relationship management"*, for instance. The RQ2 was answered by the network analysis based on the edges. Stronger edges means a higher connection between the keywords. The results reinforces the above mentioned areas and the interdependence between them.

4.1   Treats to the Research Validity

The adoption strict methodology can reduce the research bias, even though some treats to the research validity exists and should be mentioned. For this work, a semi-automatic approach was adopted, in which a few steps are, in some measure, subjective. The most relevant one in this context is the step 3.a.ii (Synonym identification), conducted in the pre-processing phase. To reduce an undesired bias, an independent judge was asked to evaluate and validate the obtained list. In addition, a supplementary material with the papers list and the output of each step is given in a GitHub Repository[1] to facilitates the research replicability.

4.2   Future Works

Regarding the potential extensions of this work, there exist several aspects which can be addressed in future developments. For instance, a next step might be the

---

[1] https://github.com/fabiolobato/keywords_socialcrm.

inclusion of more databases and the abstract analysis for a systematic categorization. Other relevant extension would be the expansion of the search queries for inclusion of the terms discovered in this papers results.

Acknowledgment. The authors gratefully acknowledge the financial support of this research by the German Academic Exchange Service (DAAD) within the project Social Customer Relationship Management as Specialization Subject in Brazilian Universities, the Maranhão Foundation for the Protection of Research and Scientific and Technological Development (FAPEMA), and the National Scientific and Technological Development Council (CNPq).

References

1. Agnihotri, R., Dingus, R., Hu, M.Y., Krush, M.T.: Social media: influencing customer satisfaction in B2B sales. Ind. Mark. Manag. **53**, 172–180 (2016)
2. Alalwan, A.A., Rana, N.P., Dwivedi, Y.K., Algharabat, R.: Social media in marketing: a review and analysis of the existing literature. Telematics Inform. **34**(7), 1177–1190 (2017)
3. de Almeida, G.R.T., Lobato, F., Cirqueira, D.: Improving Social CRM through electronic word-of-mouth: a case study of ReclameAqui. In: XIVWorkshop de Trabalhos de Iniciação Científic (2017)
4. Askool, S., Nakata, K.: A conceptual model for acceptance of social CRM systems based on a scoping study. AI SOCIETY **26**(3), 205–220 (2011)
5. Barabási, A.L., Albert, R.: Emergence of scaling in random networks. Science **286**(5439), 509–512 (1999). https://doi.org/10.1126/science.286.5439.509
6. Barata, G.M., Viana, J.A., Reinhold, O., Lobato, F., Alt, R.: Social CRM in digital marketing agencies: an extensive classification of services. In: 2018 IEEE/WIC/ACM International Conference on Web Intelligence (WI), pp. 750–753. IEEE (2018). https://doi.org/10.1109/WI.2018.00009
7. Bastian, M., Heymann, S., Jacomy, M.: Gephi: an open source software for exploring and manipulating networks. In: Third International AAAI Conference on Weblogs and Social Media (2009)
8. Booth, A., Sutton, A., Papaioannou, D.: Systematic Approaches to a Successful Literature Review, 2nd edn. SAGE Publications, London (2016)
9. Bornmann, L., Mutz, R.: Growth rates of modern science: a bibliometric analysis based on the number of publications and cited references. J. Assoc. Inf. Sci. Technol. **66**(11), 2215–2222 (2015)
10. Chae, J.: Reexamining the relationship between social media and happiness: the effects of various social media platforms on reconceptualized happiness. Telematics Inform. **35**(6), 1656–1664 (2018)
11. Chagas, B.N., Viana, J., Reinhold, O., Lobato, F.M., Jacob, A.F., Alt, R.: A literature review of the current applications of machine learning and their practical implications. Web Intell. **18**(1), 69–83 (2020). https://doi.org/10.3233/web-200429
12. Cirqueira, D., et al.: Improving relationship management in universities with sentiment analysis and topic modeling of social media channels: learnings from UFPA. In: Proceedings of the International Conference on Web Intelligence, pp. 998–1005. ACM (2017)
13. Cronin, P., Ryan, F., Coughlan, M.: Undertaking a literature review: a step-by-step approach. Brit. J. Nurs. **17**(1), 38–43 (2008)

14. Feldman, R.: Techniques and applications for sentiment analysis. Commun. ACM **56**(4), 82–89 (2013)
15. Greenberg, P.: Social CRM comes of age (2009)
16. Hart, C.: Doing a Literature Review: Releasing the Social Science Research Imagination. SAGE Publications, London (1998). Open university set book
17. Landhuis, E.: Scientific literature: information overload. Nature **535**(7612), 457–458 (2016)
18. Lindsey-Mullikin, J., Borin, N.: Why strategy is key for successful social media sales. Bus. Horiz. **60**(4), 473–482 (2017)
19. Lobato, F., Pinheiro, M., Jacob, A., Reinhold, O., Santana, Á.: Social CRM: biggest challenges to make it work in the real world. In: Abramowicz, W., Alt, R., Franczyk, B. (eds.) BIS 2016. LNBIP, vol. 263, pp. 221–232. Springer, Cham (2017). https://doi.org/10.1007/978-3-319-52464-1_20
20. McGlohon, M., Akoglu, L., Faloutsos, C.: Statistical properties of social networks. In: Aggarwal, C. (ed.) Social Network Data Analytics, pp. 17–42. Springer, Boston (2011). https://doi.org/10.1007/978-1-4419-8462-3_2
21. Ravi, K., Ravi, V.: A survey on opinion mining and sentiment analysis: tasks, approaches and applications. Knowl.-Based Syst. **89**, 14–46 (2015)
22. Rodrigues Chagas, B.N., et al.: Current applications of machine learning techniques in CRM: a literature review and practical implications. In: 2018 IEEE/WIC/ACM International Conference on Web Intelligence (WI), pp. 452–458. IEEE (2018). https://doi.org/10.1109/WI.2018.00-53, https://ieeexplore.ieee.org/document/8609630/
23. Silva, W., Santana, Á., Lobato, F., Pinheiro, M.: A methodology for community detection in Twitter. In: Proceedings of the International Conference on Web Intelligence, pp. 1006–1009 (2017). https://doi.org/10.1145/3106426.3117760
24. de Sousa, G.N., Almeida, G.R., Lobato, F.: Social network advertising classification based on content categories. In: Abramowicz, W., Corchuelo, R. (eds.) BIS 2019. LNBIP, vol. 373, pp. 396–404. Springer, Cham (2019). https://doi.org/10.1007/978-3-030-36691-9_33
25. Su, H.N., Lee, P.C.: Mapping knowledge structure by keyword co-occurrence: a first look at journal papers in technology foresight. Scientometrics **85**(1), 65–79 (2010)
26. Yu, Y., Duan, W., Cao, Q.: The impact of social and conventional media on firm equity value: a sentiment analysis approach. Decis. Support Syst. **55**(4), 919–926 (2013)
27. Zhu, W., Guan, J.: A bibliometric study of service innovation research: based on complex network analysis. Scientometrics **94**(3), 1195–1216 (2013)

Social CRM Tools: A Systematic Mapping Study

Jorge L. F. Silva Junior[1,2]([✉]) [ID], Julio Viana[2,3] [ID], Olaf Reinhold[2,3] [ID],
Antônio F. L. Jacob Jr.[4] [ID], Rainer Alt[2,3] [ID], and Fábio M. F. Lobato[1,4] [ID]

[1] Engineering and Geoscience Institute, Federal University of Western Pará
(UFOPA), Santarém, Brazil
jorgeluizfigueira@gmail.com, fabio.lobato@ufopa.edu.br
[2] Social CRM Research Center (SCRC), Leipzig, Germany
[3] Leipzig University, Leipzig, Germany
[4] State University of Maranhão, São Luís, Brazil

Abstract. The evolution of Web 2.0 has allowed the development and
success of social media, which has brought about drastic changes in how
people interact in the online environment. One of the phenomena related
to this fact is the massive production of user-generated content that is
seen as a powerful data source on customer behavior. These data allowed
innovation in existing approaches such as Customer Relationship Management (CRM), which led to a concept known as Social CRM. This
concept aims to use and integrate information from social media and
traditional CRM systems through IT tools, to improve the reliability of
results, and to provide new types of analysis. Social CRM tools, however, has different challenges and limitations concerning their analytical
potential. Given this gap, this paper explores the current tools used for
Social CRM through a systematic mapping study.

Keywords: Social media · Customer relationship management ·
Social CRM · Systematic mapping study

1 Introduction

In the last ten years, there has been a significant change in the way people interact online [1]. The rise of Web 2.0 allowed the development of several systems,
which resulted in the success of social media. These platforms are understood
as a set of web-based applications that allow the creation and exchange of user-
generated content (UGC) [22]. Sharing photos, videos, thoughts, opinions, and
participating in review websites are examples of UGC [28].

When analyzing social media from a consumer perspective, these platforms
represent a channel of rich information about brands and services, as well as
opinions, which support the decision-making process [19]. From the perspective
of companies, social media are powerful sources of data on customer behavior,
in addition to being a dynamic and low-cost communication channel [11]. Given

© Springer Nature Switzerland AG 2020
W. Abramowicz and G. Klein (Eds.): BIS 2020 Workshops, LNBIP 394, pp. 250–261, 2020.
https://doi.org/10.1007/978-3-030-61146-0_20

the influence provided by social media, companies of all sizes are being forced to adopt new strategies in order to survive the competition in the hyper-connected world [33].

The interaction between companies and their customers on social media, influenced by Web 2.0, promotes the innovation of the existing approach to Customer Relationship Management (CRM) [11]. The integration between CRM and social media is a growing research topic in Information Systems, which led to the concept of Social Customer Relationship Management (Social CRM) [19].

Social CRM aims to use and integrate information from social media and traditional CRM systems, in addition to improving the reliability of results and providing new types of analysis. Among the most significant benefits that could be achieved with the adoption of Social CRM are increased sales and revenue from associations, exhibitions, and sponsorships, as well as a direct reduction of service and marketing costs [29]. As a consequence, these systems are causing considerable changes in the market [11,19,38].

To achieve the mentioned goals and benefits, companies can make use of Social CRM tools to process and analyze the information generated by customers on social media [12,23,37]. However, as it is a new and promising research topic, there are still many open challenges, including the limitation of these tools [27]. c. As the main contribution, this article provides the current technologies used in Social CRM.

The remainder of this paper is structured as follows. Section 2 briefly describes Social CRM tools. Section 3 describes the research methodology and systematic mapping process, supported by the analysis' results in Sect. 4. Finally, in Sect. 5 the conclusions are presented.

2   Social CRM Tools

The practice of Social CRM aims to obtain improved support of the marketing, sales, and services processes through social media. Companies can enter these processes with the adoption of simple IT tools [4,10]. These tools are known as Social CRM application systems, or simply Social CRM tools, and allow them to collect social media data in order to offer organizations the acquisition and continuous generation of customer knowledge [2,15].

Social media monitoring, social marketing campaigns, and event management in the social domain are some of the activities that organizations can use to generate valuable customer information with the help of Social CRM tools [18, 35]. Due to a great diversity and applicability of features, there is no single application that covers the whole concept of Social CRM, but several types of application systems are relevant [3]. According to Alt and Reinhold [4], Social CRM tools can be classified into seven types, and these are presented in Table 1.

Table 1. Types of Social CRM Tools (following Alt and Reinhold 2012)

Type of Social CRM Tool	Functional description
Business Intelligence (BI)	Consolidation and visualization of operational data for decision-making. Due to the focus on structured data, analyses of keywords or product mentions in postings may be analyzed with BI tools.
Community Management	Creation and administration of communities or forums, e.g. to plan and track activities on multiple social media platforms.
CRM System	Support of core operative CRM processes to obtain an overall view on customer activities. Social media are considered as additional interaction channel that needs to be aligned with other channels.
Social Media Management	Administration of profiles and distribution of content on multiple platforms (e.g. sharing of postings on various platforms).
Social Media Monitoring	Evaluation of pre-defined public and restricted social content (UGC) by topics, opinions/trends, sentiments or user activity (e.g. opinion leader or "influencers").
Social Network Analysis	Tracking of authors on several social media platforms to provide evaluations on relationships among authors, authors and topics as well as effect of single contents on discussions.
Social Search	Search of blogs via keywords or topics and navigation through social web postings and identification of relevant content.

3   Research Methodology

The Systematic Literature Review (SLR) is a formal review of the literature adopted to identify, evaluate, and summarize evidence of empirical results to answer one or more research questions [24]. However, when the field of interest is broad, and the purpose is to get an overview of what is being developed in the research field, a Systematic Mapping Study (SMS) is highly recommended [32].

This paper reports an SMS carried out to obtain the current tools used in Social CRM. The review process followed the guidelines proposed by Kitchenham et al. [25,32], and consists of three main phases illustrated in Fig. 1. In the first phase, keywords, research questions, and the process protocol are developed; in the second phase, the search for related studies is conducted and then summarized in order to answer the research questions. Finally, the result of this mapping is presented.

3.1   Research Questions

The main purpose of this paper is to glimpse the use of current Social CRM tools. More specifically, to list these tools by conducting SMS in the current works that address the adoption of Social CRM with software support; then analyze them. These objectives allow us to develop the following research questions(RQ):

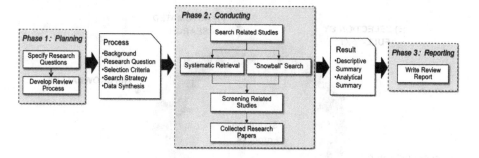

Fig. 1. Systematic mapping process [13]

- **RQ-1 What are the current tools used for Social CRM?**
  The goal is to provide an updated list of the main tools used in Social CRM.
- **RQ-2 What are the main types of these tools?**
  The goal is to get an overview of how these tools are targeted.
- **RQ-3 What are the main features of these tools?**
  The goal is to understand which tasks or analyzes are most offered.

3.2   Search Process

As digital databases have both large quantities of publications and variability in their types, it was decided to automatically retrieve from each database, all publications that match the keywords adopted. Subsequently, in order to enrich the results of the SMS, the "snowball" search method was used to provide relevant literature manually [40]. In summary, this paper uses two search strategies to collect existing literature. Figure 1 illustrates this process.

(1) Systematic retrieval
Since Social CRM's topic belongs to the field of Information Systems, the following digital libraries [26] are used to identify studies: ACM Digital Library, IEEE Xplore, Science Direct, Web of Science, and Springer Link.

Define a Search Key String. The research focused on its main objective as a determining factor for choosing the keywords for systematic search. Therefore, "social crm" and "tools" were chosen as keywords initially. Then, combinations of these keywords were made, based on the use of their synonyms and their variants (singular and plural), which resulted in the following search string: *("social crm" OR "social customer relationship management" OR "social information systems" OR "crm 2.0") AND ("tools" OR "tool" OR "application" OR "applications" OR "software" OR "system" OR "systems")*

(2) "Snowball" search

In order to obtain the most significant number of tools found in the literature search, it was decided to complement this with the addition of relevant literature through a "snowball" search [40] in the papers retained after screening, a process described in Subsect. 3.3.

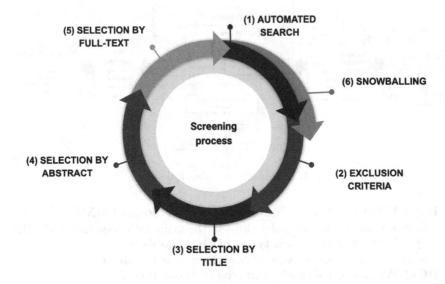

Fig. 2. Screening process

3.3   Screening Process

Based on the definitions of the research questions, inclusion, and exclusion criteria were developed in the search process in order to refine the results found. Both criteria are shown in Table 2. Automatic searches sometimes tend to incur failures, such as duplication of results. To correct such failures, after collection, the papers were checked automatically and manually (EC1). Looking forward to obtaining high-quality papers, research published in other languages than English were excluded (EC2).

A period of publication limit was also stipulated, ignoring papers published before 2015 in order to obtain the current solutions for Social CRM (EC3). Besides, analyzing the papers of the last five years is considered to be good practice in review studies in the information systems field. Among the publications found in the automatic search, some of them are not published in their entirety or still do not have relevant content. Therefore, master's and doctoral theses, textbooks, websites, and unfinished works were ruled out (EC4). Publications whose web access is unavailable were ignored (EC5). Hence, this study selected Social CRM papers that included, reported, or mentioned the use of software tools in their implementation process (IC1).

Table 2. The inclusion and exclusion criteria

Inclusion criteria
IC1:Papers that mention the use of tools in some Social CRM area
Exclusion criteria
EC1: Exclusion of repeated papers
EC2: The paper was not written in English
EC3: The paper was published before 2015
EC4: Exclusion of master and doctoral thesis, textbooks, websites, and unfinished working papers
EC5: The paper is not available on the web.

In Fig. 2, we have the screening process, which consists of 6 steps. The first refers to the identification of studies in the digital databases previously cited, using the search string defined in Subsect. 3.2. The second step consists of the elimination of studies according to the exclusion criteria mentioned above. The selection of SMS papers is carried out along steps three to five. In the third step, the selection is made by evaluating the title of the work. When the lack of relevance in the title is verified, the item is removed directly.

When a relevant contribution is detected, the paper moves forward to the fourth step, which deals with the selection, evaluating the abstract of the work. Finally, in the fifth step, the remaining works are carefully read through. In addition to the screening process, there is the sixth stage that deals with the "snowball" search, this is done based on the references of the articles that passed the screening stages (third to the fifth step). It is worth noting that articles manually selected in the search for "snowball" also go from third to fifth steps.

4   Results and Analysis

In this section, the results of conducting the SMS are summarized, considering the research questions defined in Subsect. 3.1. First, an overview of the selected studies is presented, and then the answers to each of the research questions are discussed.

Figure 3 details the screening process and results. Initially, 922 papers were selected in the systematic search phase, and eleven papers were retained. Then, in the screening using the "snowball" search under the references of the selected papers, five papers were retained. Finally, a total of sixteen papers were obtained.

After studies were selected on SMS, they were carefully analyzed. Regarding **RQ-1**, a total of 49 tools applied to Social CRM were identified. It is important to mention that in this process, deprecated tools were omitted in this survey. Deprecated tools are understood to be tools that had their support or activities currently terminated on the market. In addition, tools that were acquired by

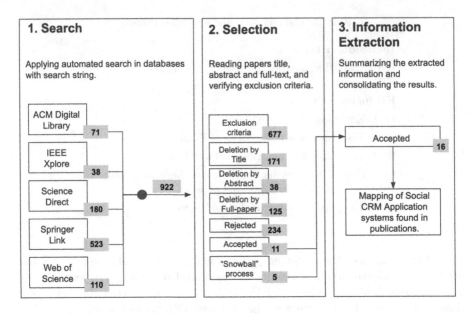

Fig. 3. Papers screening results

third parties, merged with another company, or had their names changed, were updated accordingly to their current status. The list with the tools found is presented below.

Identified tools: Social Baker [30], Fanpage Karma [30], SimilarWeb [30], Voziq [21], Social Studio (Salesforce) [14,20,38], Tlab [14], Neo4j [14], Tweets Stats [5], Mention [4,5,20], SentiStrength [7], Discover Text [7], Digimind [16], Synthesio [4,16,31], Sindup [16], HootSuite [20,36,38], BrandWatch [20], SproutSocial [20,31,36], Twitonomy [20], Simplify360 [20], PipelineDeals [37], Desk (Salesforce) [37], Piksel [37], Sysomos [37], Tweet Deck [37], SAS [4], Microstrategy [4], Qlikview [4], Get Satisfaction [4,36,38], Jive [4,38], Salesforce CRM [4,8,36,39], Microsoft Dynamics CRM [4,39], Falcon.IO [4], RD Station [4], Adobe Analytics [4], Novomind [8], Obi4wan [8], Talkwalker[9], Conversocial [9,36], Boomsonar [17], Nimble [38], Netbase [38], Khoros [4,38], Hubspot [36,38,39], Zoho [31,39], Act! [39], SAP [39], Sage [39], Social Cloud (Oracle) [39], Keap [39].

Since most of the papers do not describe details about the mentioned tools, it was necessary to consult the website of each tool to obtain a better description, especially of its features. The information obtained made it possible to conduct a classification of the tools according to the categorization of their types [6] described in Sect. 2. The results of this classification are shown in Fig. 4.

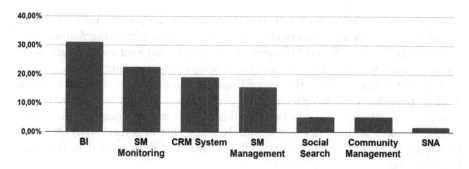

Fig. 4. Classification of Social CRM tools

According to the classification in the tools found, the results demonstrate that Social CRM tools have a strong foundation in Business Intelligence, Social Media Monitoring, and CRM Sytems technologies, thus answering **RQ-2**.

Finally, an analysis of the features tools was conducted. During the analysis, it was noticed that the tools had similar features but with different names. In order to obtain a consistent analysis, these features had their names standardized and are presented with their description in Table 3.

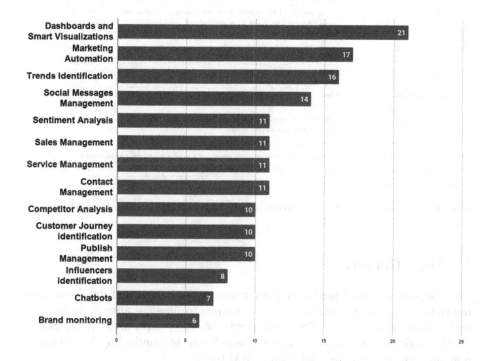

Fig. 5. Features of Social CRM tools

Analyzing the features of each tool described in the papers, we built Fig. 5. The obtained result allows us to state that Dashboards, Marketing Automation, Trends identification, Social Messages Management, and Sentiment Analysis are the main features offered by the analyzed tools, thus answering **RQ-3**.

One factor that justifies the use of Business Intelligence technologies is that it often provides various analyzes and performance metrics using dashboards to report the summary of results. The dashboards are excellent for managers' decisions makes as they offer a clear and dynamic view of several analyzes simultaneously. Besides, it facilitates the interpretation of data related to costumer behavior [34].

Table 3. Main features of Social CRM tools

Features	Description
Dashboard and Smart Visualizations	Panels that show important metrics and indicators to achieve objectives and goals traced visually, facilitating the understanding of the information generated.
Sentiment Analysis	Contextual mining of a text that identifies and extracts subjective information. It helps companies to understand the social sentiment of their brand, product or service.
Competitor Analysis	Analysis of competition to find out what their strengths and weaknesses are, and how those strengths and weaknesses compared to your own from a social media perspective.
Marketing Automation	It allows to identify, monitor and act in a personalized way with the public, improving the relationship in general automatically.
Chatbots	Automation of repetitive and bureaucratic tasks, such as frequent questions, in the form of a predefined dialogue between the user and a "robot"
Brand Monitoring	It is a process to strategically and proactively monitor the reputation, growth, and topics associated with a brand or set of brands.
Trends Identification	Identification of the main topics related to the brand, product or a specific publication.
Influencers Identification	Identification of digital influencers that best match the brand or product identity.
Customer Journey Identification	Customer identification and segmentation from the purchase intention to the current consumer stage.
Publish Management	Management of publication content.
Social Messages Management	Management of messages received on one or more social media.
Sales Management	Definition of objectives, planning and control of the entire sales process.
Service Management	It offers product or service management, in its various aspects, such as inventory control, supplier and others.
Contact Management	Process of recording contact details and tracking their interactions with a company

5 Final Remarks

The adoption of social media as a communication channel and data source to understand customer needs is crucial for keeping companies alive in an increasingly competitive market. There are plenty of tools, depending on the Social CRM activities. The paper tackles a research gap by conducting a systematic mapping study on commercial Social CRM tools.

The results obtained in this study contribute to better visualization of current Social CRM solutions available in the market. Besides, the analysis conducted is capable of providing insights for businesses in the decision-making process. Our conclusions also help practitioners, since it was presented a comprehensive and accessible list of tools categorization and their features, helping the selection of the appropriate tools according to the companies' needs. The integration of industry and academic information resulted in an analysis aligned with a market-oriented vocabulary.

The study reported has some limitations, mainly regarding the search queries and the validation of the results. The addition of more synonyms for Social CRM should be considered in future works. Moreover, some social media management tools, despite not belonging to the Social CRM scope, could be included in the analysis.

Due to an ongoing process and the low amount of expert feedback, the results were not analyzed by practitioners using a proper/replicable methodology. Certainly, it should be tackled in future works. Finally, we would like to include some pricing and freemium/premium analyses, correlating with features and tools categories.

Acknowledgments:. The authors gratefully acknowledge the financial support of this research by the German Academic Exchange Service (DAAD) within the project Social Customer Relationship Management as Specialization Subject in Brazilian Universities, the Maranhão Foundation for the Protection of Research and Scientific and Technological Development (FAPEMA), and the National Scientific and Technological Development Council (CNPq).

References

1. Afify, E.A., Eldin, A.S., Khedr, A.E., Alsheref, F.K.: User-generated content (UGC) credibility on social media using sentiment classification. FCI-H Inform. Bull. **1**(1), 1–19 (2019)
2. Al-Azzam, A., Khasawneh, R.T.: Social customer relationship management (SCRM): a strategy for customer engagement. In: Strategic Uses of Social Media for Improved Customer Retention, pp. 45–58. IGI Global (2017)
3. Alt, R., Reinhold, O.: Social-customer-relationship-management (Social-CRM). Wirtschaftsinformatik **54**(5), 281–286 (2012)
4. Alt, R., Reinhold, O.: Social CRM: tools and functionalities. Social Customer Relationship Management. MP, pp. 57–80. Springer, Cham (2020). https://doi.org/10.1007/978-3-030-23343-3_3
5. Arora, D., Malik, P.: Analytics: key to go from generating big data to deriving business value. In: 2015 IEEE First International Conference on Big Data Computing Service and Applications, pp. 446–452. IEEE (2015)
6. Barata, G.M., Viana, J.A., Reinhold, O., Lobato, F., Alt, R.: Social CRM in digital marketing agencies: an extensive classification of services. In: 2018 IEEE/WIC/ACM International Conference on Web Intelligence (WI), pp. 750–753. IEEE (2018)
7. Calderon, N.A., et al.: Mixed-initiative social media analytics at the world bank: observations of citizen sentiment in Twitter data to explore "trust" of political actors and state institutions and its relationship to social protest. In: 2015 IEEE International Conference on Big Data (Big Data), pp. 1678–1687. IEEE (2015)

8. Carnein, M., et al.: Towards efficient and informative omni-channel customer relationship management. In: de Cesare, S., Frank, U. (eds.) ER 2017. LNCS, vol. 10651, pp. 69–78. Springer, Cham (2017). https://doi.org/10.1007/978-3-319-70625-2_7

9. Carnein, M., Homann, L., Trautmann, H., Vossen, G., Kraume, K.: Customer service in social media: an empirical study of the airline industry. Datenbanksysteme für Business, Technologie und Web (BTW 2017)-Workshopband (2017)

10. Chagas, B.N., Viana, J., Reinhold, O., Lobato, F.M., Jacob Jr, A.F., Alt, R.: A literature review of the current applications of machine learning and their practical implications. In: Web Intelligence, pp. 1–15, No. Preprint. IOS Press (2020)

11. Choudhury, M.M., Harrigan, P.: CRM to social CRM: the integration of new technologies into customer relationship management. J. Strateg. Mark. **22**(2), 149–176 (2014)

12. Cirqueira, D., Pinheiro, M.F., Jacob, A., Lobato, F., Santana, A.: A literature review in preprocessing for sentiment analysis for Brazilian Portuguese social media. In: 2018 IEEE/WIC/ACM International Conference on Web Intelligence (WI), pp. 746–749. IEEE (2018)

13. Dong, X., Li, T., Li, X., Song, R., Ding, Z.: Review-based user profiling: a systematic mapping study. In: Reinhartz-Berger, I., Zdravkovic, J., Gulden, J., Schmidt, R. (eds.) BPMDS/EMMSAD -2019. LNBIP, vol. 352, pp. 229–244. Springer, Cham (2019). https://doi.org/10.1007/978-3-030-20618-5_16

14. Ducange, P., Pecori, R., Mezzina, P.: A glimpse on big data analytics in the framework of marketing strategies. Soft Comput. **22**(1), 325–342 (2018). https://doi.org/10.1007/s00500-017-2536-4

15. Foltean, F.S., Trif, S.M., Tuleu, D.L.: Customer relationship management capabilities and social media technology use: consequences on firm performance. J. Bus. Res. **104**, 563–575 (2019)

16. Fourati-Jamoussi, F.: E-reputation: a case study of organic cosmetics in social media. In: 2015 6th International Conference on Information Systems and Economic Intelligence (SIIE), pp. 125–132. IEEE (2015)

17. Furman, I.: Algorithms, dashboards and datafication: a critical evaluation of social media monitoring. In: Bilić, P., Primorac, J., Valtýsson, B. (eds.) Technologies of Labour and the Politics of Contradiction. DVW, pp. 77–95. Springer, Cham (2018). https://doi.org/10.1007/978-3-319-76279-1_5

18. Gonçalves, A.: Dedicated vs. hybrid tools. In: Social Media Analytics Strategy, pp. 67–77. Springer, Heidelberg (2017). https://doi.org/10.1007/978-1-4842-3102-9_5

19. Greenberg, P.: CRM at the Speed of Light: Social CRM Strategies, Tools, and Techniques. McGraw-Hill, New York (2010)

20. Harrigan, P., Soutar, G., Choudhury, M.M., Lowe, M.: Modelling CRM in a social media age. Australas. Mark. J. (AMJ) **23**(1), 27–37 (2015)

21. He, W., Zhang, W., Tian, X., Tao, R., Akula, V.: Identifying customer knowledge on social media through data analytics. J. Enterp. Inf. Manage. 32 (2019)

22. Kaplan, A.M., Haenlein, M.: Users of the world, unite! the challenges and opportunities of social media. Bus. Horiz. **53**(1), 59–68 (2010)

23. Kar, A.: Applications of analytics in social media. In: Foundation Innovation Tech. Transfer Forum Newsletters (IIT Delhi), vol. 21, pp. 6–8 (2015)

24. Kitchenham, B., Brereton, O.P., Budgen, D., Turner, M., Bailey, J., Linkman, S.: Systematic literature reviews in software engineering-a systematic literature review. Inf. Softw. Technol. **51**(1), 7–15 (2009)

25. Kitchenham, B., Charters, S.: Guidelines for performing systematic literature reviews in software engineering (2007)

26. Kitchenham, B., et al.: Systematic literature reviews in software engineering-a tertiary study. Inf. Softw. Technol. **52**(8), 792–805 (2010)
27. Lobato, F., Pinheiro, M., Jacob, A., Reinhold, O., Santana, Á.: Social CRM: biggest challenges to make it work in the real world. In: Abramowicz, W., Alt, R., Franczyk, B. (eds.) BIS 2016. LNBIP, vol. 263, pp. 221–232. Springer, Cham (2017). https://doi.org/10.1007/978-3-319-52464-1_20
28. Malthouse, E.C., Calder, B.J., Kim, S.J., Vandenbosch, M.: Evidence that user-generated content that produces engagement increases purchase behaviours. J. Mark. Manag. **32**(5–6), 427–444 (2016)
29. Orenga-Roglá, S., Chalmeta, R.: Social customer relationship management: taking advantage of web and big data technologies. SpringerPlus **5**(1), 1462 (2016). https://doi.org/10.1186/s40064-016-3128-y
30. Păvăloaia, V.D., Teodor, E.M., Fotache, D., Danileţ, M.: Opinion mining on social media data: sentiment analysis of user preferences. Sustainability **11**(16), 4459 (2019)
31. Pereira, V.C., Fileto, R., de Souza, W.S., Wittwer, M., Reinhold, O., Alt, R.: A semantic BI process for detecting and analyzing mentions of interest for a domain in tweets. In: Proceedings of the 24th Brazilian Symposium on Multimedia and the Web, pp. 197–204 (2018)
32. Petersen, K., Feldt, R., Mujtaba, S., Mattsson, M.: Systematic mapping studies in software engineering. In: 12th International Conference on Evaluation and Assessment in Software Engineering (EASE) 12, pp. 1–10 (2008)
33. Shuen, A.: Web 2.0 : A Strategy Guide: Business Thinking and Strategies Behind Successful Web 2.0 Implementations. O'Reilly Media, Newton (2018)
34. Sönmez, F., et al.: Technology acceptance of business intelligence and customer relationship management systems within institutions operating in capital markets. Int. J. Acad. Res. Bus. Soc. Sci. **8**(2) (2018)
35. Tiwari, A., Misra, M.: Analysis of operative factors and practices in social CRM. Int. J. Digit. Enterp. Technol. **1**(1–2), 135–176 (2018)
36. Turban, E., King, D., Lee, J.K., Liang, T.-P., Turban, D.C.: Social commerce: foundations, social marketing, and advertising. Electronic Commerce. STBE, pp. 285–324. Springer, Cham (2015). https://doi.org/10.1007/978-3-319-10091-3_7
37. Turban, E., Strauss, J., Lai, L.: Social customer service and CRM. Social Commerce. STBE, pp. 155–178. Springer, Cham (2016). https://doi.org/10.1007/978-3-319-17028-2_7
38. Van Looy, A.: Social customer relationship management. Social Media Management. STBE, pp. 87–111. Springer, Cham (2016). https://doi.org/10.1007/978-3-319-21990-5
39. Wachtler, V.M.: From information transaction towards interaction: social media for efficient services in CRM. In: Kryvinska, N., Greguš, M. (eds.) Data-Centric Business and Applications. LNDECT, vol. 30, pp. 371–407. Springer, Cham (2020). https://doi.org/10.1007/978-3-030-19069-9_15
40. Wohlin, C.: Second-generation systematic literature studies using snowballing. In: Proceedings of the 20th International Conference on Evaluation and Assessment in Software Engineering, pp. 1–6 (2016)

QOD

QOD 2020 Workshop Chairs' Message

Preface

The Third Workshop on Quality of Open Data (QOD 2020), organized in conjunction with the 23rd International Conference on Business Information Systems (BIS 2020), was affiliated at the University of Colorado, USA. The specific focus was on bringing together different communities working on quality in Wikipedia, DBpedia, Wikidata, OpenStreetMap, Wikimapia, and other open knowledge bases and data sources.

There were 13 papers submitted for the workshop and the Program Committee decided to accept 6 papers (an acceptance rate of 46%). There were 17 members in the Program Committee, representing 16 institutions from 9 countries.

The first paper "Open Data Quality Dimensions and Metrics: State of the Art and Applied Use Cases," focused on questions related to open data quality indicators and illustrated it with some case studies from the industry. In this work, the authors described a brief state of the art of the quality dimensions that can help in assessing web and open data quality.

The second paper "Analyzing OpenStreetMap Contributions at Scale: Introducing OSM-interactions tilesets," showed how tilesets can be generated for specific objects such as highways or buildings to support OSM researchers, especially when conducting intrinsic data quality assessment. OSM interactions were reconstructed from the full history of OpenStreetMap and converted to vector tilesets with open source tools, including the OpenStreetMap Historical Database (OSHDB) and "tippeanoe."

The third paper, "Synthesizing Quality Open Data Assets from Private Health Research Studies," focused on HealthGAN approach, which generated high-quality synthetic data and produced similar results while preserving patient privacy. By creating synthetic versions of the private datasets, the authors created valuable open-health data assets for future research and education efforts.

The fourth paper, "Models for Quality Assessment of Arabic Web Documents," proposed a general model that can assess the quality of Arabic documents that lack Wikipedia metadata with acceptable accuracy. The model was trained and built using features from documents collected from Arabic online news sites and blogs, and annotated in collaboration with university students.

The fifth paper, "Materia: A Data Quality Control Embedded Domain Specific Language in Python," focused on evaluation of Materia using two metrics: productivity and a quantitative performance analysis. Authors showed how Materia can simplify complex descriptions of tests in Pandas and mirror natural language descriptions of common QC tests.

The last paper, "Enhancing the interactive visualisation of a data preparation tool from in-memory fitting to Big Data sets," presented an approach which was used to overcome web-browsers' client-side data handling limitations and to avoid information overload when using granular information charts from existing in-memory data preparation tools with big data sets. The developed solution provided the user with an acceptable GUI interaction time.

August 2020

Mouzhi Ge
Włodzimierz Lewoniewski
Krzysztof Węcel

Organization

Chairs

Mouzhi Ge (Chair) — Masaryk University, Czech Republic
Włodzimierz Lewoniewski (Co-chair) — Poznań University of Economics and Business, Poland
Krzysztof Węcel (Co-chair) — Poznań University of Economics and Business, Poland

Program Committee

Maribel Acosta — Karlsruhe Institute of Technology, Germany
Riccardo Albertoni — CNR-IMATI, Italy
Ioannis Chrysakis — FORTH-ICS, Greece, and Ghent University, Belgium
Vittoria Cozza — ENEA, Italy
Gianluca Demartini — The University of Queensland, Australia
Anastasia Dimou — Ghent University, Belgium
Suzanne Embury — The University of Manchester, UK
Ralf Härting — Hochschule Aalen, Germany
Antoine Isaac — Europeana and VU University Amsterdam, The Netherlands
Dimitris Kontokostas — Leipzig University, Germany
Jose Emilio Labra Gayo — Universidad de Oviedo, Spain
Maristella Matera — Politecnico di Milano, Italy
Finn Årup Nielsen — Technical University of Denmark, Denmark
Matteo Palmonari — University of Milano-Bicocca, Italy
Simon Razniewski — Max Planck Institute for Informatics, Germany
Anisa Rula — University of Milano-Bicocca, Italy
Blerina Spahiu — University of Milano-Bicocca, Italy

Analyzing OpenStreetMap Contributions at Scale: Introducing OSM-Interactions Tilesets

Jennings Anderson$^{(\boxtimes)}$ ⓘ

University of Colorado Boulder, Boulder, USA
jennings.anderson@colorado.edu

Abstract. OSM-interaction tilesets are vector tiles containing GeoJ-SON features that represent interactions between mappers (contributors) and objects in OpenStreetMap (OSM). Interactions are abstractions of edits to OSM elements called nodes, ways, and relations. Example interactions are contributors "adjusting the corners of a building" or "re-aligning a road" while the edit to the database is recorded as a "modification to the coordinates of a node." This abstraction to an OSM-object-level description makes these changes easier to understand without having to explain the OSM data structure. Each interaction is represented as an individual GeoJSON feature with properties containing information about the editor and the changeset in which the interaction between the data and the editor occurred. OSM Interactions are reconstructed from the full history of OpenStreetMap and converted to vector tilesets with open source tools including the OpenStreetMap Historical Database (OSHDB) and `tippecanoe`. This demo shows how osm-interaction tilesets can be generated for specific objects such as highways or buildings to support OSM researchers, especially when conducting intrinsic data quality assessment.

Keywords: OpenStreetMap · Intrinsic data quality · Volunteered geographic information

1 OSM-Interactions

1.1 Introduction and Background

More than a map, OpenStreetMap (OSM) has grown into the world's largest free and open geospatial data repository. Comprised of billions of geographic points that describe hundreds of millions of individual geographic features around the globe, this database has been edited by more than 1.4M contributors in the last 15 years. Today, this community of global contributors is made up of hobbyists, humanitarians, and professionals alike, who work together to improve the map [2].

This short demo-paper discusses the development and potential uses of an analysis-specific dataset built for OpenStreetMap research called *osm-interaction*

© Springer Nature Switzerland AG 2020
W. Abramowicz and G. Klein (Eds.): BIS 2020 Workshops, LNBIP 394, pp. 267–271, 2020.
https://doi.org/10.1007/978-3-030-61146-0_21

tilesets. Created in the Mapbox vector tile format, these tilesets are comprised of both OSM objects and the metadata about the historical edits to the objects that describe the interaction between the contributors and the map. The availability of these tilesets is a significant step forward in the creation of an *OSM Geo-Information Science* toolkit to support a wide-variety of OSM-related research.

1.2   Contributor-Centric OpenStreetMap Data Analysis

OSM-Interactions tilesets enable *contributor-centric* OSM data analysis that prioritizes information about an edit to an OSM object over the object itself. For example, knowing which user–and when–added the name to a highway or an address to a building or how specifically they altered the geometry of an object is more pertinent than knowing the actual coordinates of the highway or building itself [1]. These attributes allow for intrinsic data quality assessment of OSM when external datasets are not available [3], as is the case for much of the world where OSM data has become the authoritative geospatial data source [5].

Scalable analysis of the entire history of OSM is possible when used in conjunction with open-source vector-tile processing tools such as `tile-reduce`[1].

1.3   OSM-Interactions Tilesets

Built from the *Contribution View* provided by the OpenStreetMap Historical Database (OSHDB) [6], an osm-contribution-tile is an enhanced record of edits to individual OSM objects organized into vector tiles. Each object in an osm-contribution tile is a GeoJSON representation of a specific interaction that a mapper had with an OSM element such as *creation*, *modification*, or *deletion*; these objects are one-step abstracted from the original *OSM elements* as found in the OSM database because they resolve embedded geometries and slight modifications to those geometries, as returned from OSHDB [6].

Accounting for Minor Versions. Previous work in this area refers to these slight modifications as *minor versions*; these occur when child nodes of an object are moved but the metadata of the parent object remains unchanged [1]. Examples of minor versions include moving the individual points within a road to better align to satellite imagery or squaring the corners of a building to more accurately represent the built environment. These are a common edit type in OSM, with over 120M OSM objects containing at least 1 minor version as of mid-2019 [1].

Tracking minor versions of an object is imperative for intrinsic data quality assessment in user generated content. If, for example, a mapper re-aligns the nodes in a road to improve the positional accuracy of a road's geometry when access to new imagery is made available, then only the minor version of this object is aware of this update. The timestamp associated with the latest edit

---

[1] github.com/mapbox/tile-reduce.

to the actual OSM way element is ignorant to this update, making the data appear more stale than it is. Only a calculated minor version of this OSM object contains the metadata (timestamp and user) of this latest update to the object's geometry.

Highway Interactions Tileset. Known as a *way element* in OSM, a LineString geometry that represents any type of road object contains the `highway` tag with a value such as `primary`, `secondary`, `footway`, etc. Together, all of the highway objects in OSM represent a global road network. There are over 65M km of OSM objects with the `highway` tag in OSM. The *highway-interactions* tileset consists of individual GeoJSON features per edit to highway objects around the globe. Each feature includes metadata for the edit such as the timestamp and username along with specifics of the edit: a record of any tag changes and whether or not a geometry change occurred. The resulting geometry from the edit defines the GeoJSON feature. These tiles are built only at zoom level 14 to reduce the number of features in each tile.

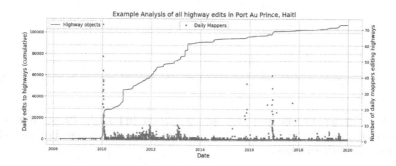

Fig. 1. Example analysis with osm-interactions in Port-Au-Prince, Haiti

Figure 1 shows an example analysis of the history of road editing in Port-Au-Prince, Haiti afforded by the osm-interaction tileset containing the history of all highway objects. It shows the first major spike in road editing activity in response

Fig. 2. Sample rendering of the osm-interactions `highway` tileset. Saturation denotes age of edit. The multitudes of lines shows the numerous geometry changes.

to the 2010 earthquake followed by a period of consistent maintenance mapping as the local mapping community evolved [7]. The spike in mapper activity in 2017 is likely in response to Hurricane Matthew, however, the volume of edits does not increase proportionally to 2010, suggesting that mappers edited fewer highways overall in 2017 (Fig. 2).

Building Interactions Tileset. Buildings are the most common *way elements* in OSM with over 386M objects in OSM containing the `building` tag. Similar to the *highways tileset*, the *osm-interactions buildings tileset* contains a GeoJSON feature per edit to a building object in OSM with the metadata describing the edit. In addition to the new geometry and a record of any changes, the difference in *squareness* of a building is also included (Figs. 3 and 4).

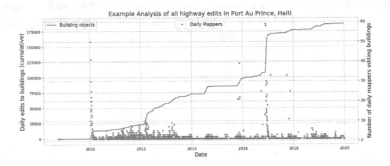

Fig. 3. Example building analysis with osm-interactions in Port-Au-Prince, Haiti shows that disaster mappers edited more buildings in response to Hurricane Matthew in 2017 than after the 2010 earthquake.

Fig. 4. History of buildings in Tamale, Ghana. Orange represent new minor versions of buildings while purple represents the previous version of an object. Objects in red were deleted. Each object is encoded with a timestamp and can be combined with a time-slider to see the complete history of the map at second-resolution. (Color figure online)

1.4   Summary

Highways and buildings are just examples of the capabilities of analysis-specific OSM metadata tilesets. Both the code and tilesets are available on github[2].

Acknowledgements. This work is supported by NSF Grant IIS-1524806 and uses the ChameleonCloud Computing Testbed for both generating tilesets and analysis [4].

References

1. Anderson, J.: Contributor-centric Analytics for OpenStreetMap: Approaches to Full Stack, Metadata-driven Analysis Infrastructure for an Open Geospatial Data Platform. Ph.D. thesis, University of Colorado at Boulder (2019)
2. Anderson, J., Sarkar, D., Palen, L.: Corporate editors in the evolving landscape of OpenStreetMap. ISPRS Int. J. Geo-Inf. **8**(5), 232 (2019). https://doi.org/10.3390/ijgi8050232, https://www.mdpi.com/2220-9964/8/5/232
3. Barron, C., Neis, P., Zipf, A.: A comprehensive framework for intrinsic OpenStreetMap quality analysis. Trans. GIS **18**(6), 877–895 (2014). https://doi.org/10.1111/tgis.12073, http://doi.wiley.com/10.1111/tgis.12073
4. Keahey, K., et al.: Chameleon: a scalable production testbed for computer science research, 1 edn. In: Vetter, J. (ed.) Contemporary High Performance Computing: From Petascale Toward Exascale. Chapman & Hall/CRC Computational Science, vol. 3, chap. 5, pp. 123–148. CRC Press, Boca Raton, FL (May 2019)
5. Palen, L., Soden, R., Anderson, J., Barrenechea, M.: Success & scale in a data-producing organization: the socio-tecnical evolution of OpenStreetMap in response to humanitarian events. In: Proceedings of the 33rd Annual ACM Conference on Human Factors in Computing Systems, CHI 2015, pp. 4113–4122. ACM Press, New York (April 2015). https://doi.org/10.1145/2702123.2702294
6. Raifer, M., et al.: OSHDB: a framework for spatio-temporal analysis of OpenStreetMap history data. Open Geospatial Data Softw. Stand. **4**(1), 3 (2019). https://doi.org/10.1186/s40965-019-0061-3, https://opengeospatialdata.springeropen.com/articles/10.1186/s40965-019-0061-3
7. Soden, R., Palen, L.: From crowdsourced mapping to community mapping: the post-earthquake work of OpenStreetMap Haiti. In: Rossitto, C., Ciolfi, L., Martin, D., Conein, B. (eds.) COOP 2014 - Proceedings of the 11th International Conference on the Design of Cooperative Systems, 27–30 May 2014, Nice, France, pp. 311–326. Springer, Cham (2014). https://doi.org/10.1007/978-3-319-06498-7_19

---

[2] github.com/jenningsanderson/osm-interactions.

Enhancing the Interactive Visualisation of a Data Preparation Tool from in-Memory Fitting to Big Data Sets

Gorka Epelde[1,2(✉)], Roberto Álvarez[1,2], Andoni Beristain[1,2], Mónica Arrúe[1,2], Itsasne Arangoa[1,2], and Debbie Rankin[3]

[1] Vicomtech Foundation, Basque Research and Technology Alliance (BRTA), Mikeletegi 57, 20009 Donostia-San Sebastián, Spain
`{gepelde,ralvarez,aberistain,marrue,iarangoa}@vicomtech.org`
[2] Biodonostia Health Research Institute, eHealth Group, 20014 San Sebastián, Spain
[3] School of Computing, Engineering and Intelligent Systems, Ulster University, Derry~Londonderry, Northern Ireland, UK
`d.rankin1@ulster.ac.uk`

Abstract. In order to derive reliable insights or make evidence-based decisions, the starting point is to assess and meet a minimum quality of data, either by those that publish the data (preferably) or alternatively by those that prepare data for analysis and develop specific analytics. Much of the (open) data shared by governments and different institutions, or crowdsourced, is in tabular format, and the amount and size of it is increasing rapidly. This paper presents the challenges faced and the solutions adopted while evolving the web-based graphical user interface (GUI) of a tabular data preparation tool from in-memory fitting to Big Data sets. Traditional standalone processing and rendering solutions are no longer usable in a Big Data context. We report on the approach adopted to asynchronously pre-compute the visualisations required for the tool, in addition to the applied visualisation aggregation strategies. The implementation of this approach has allowed us to overcome web-browsers' client-side data handling limitations and to avoid information overload when using granular information charts from our existing in-memory data preparation tool with Big Data sets. The developed solution provides the user with an acceptable GUI interaction time.

Keywords: Big data visualisation · Data preparation · Data quality · Exploratory data analysis · Visual information cluttering · Data reduction · Asynchronous pre-processing

1 Introduction

With the advent of mobile technology and the Internet of Things, together with the trend to share, either publicly or under request, different datasets for research and analysis, has led to data sets that are too large and complex for traditional data processing and data management applications.

W. Abramowicz and G. Klein (Eds.): BIS 2020 Workshops, LNBIP 394, pp. 272–284, 2020.
https://doi.org/10.1007/978-3-030-61146-0_22

The Big Data era has brought massive datasets that are noisy and heterogeneous, requiring new processing and visualisation approaches, given that traditional databases and architectures are not able to efficiently store and process them. The heterogeneous data sources have to be accessed using different protocols, transmission rates, with different data quality levels and schema representations.

In the field of Big Data information visualisation and data management, research has classified the wishful characteristics of a Big Data Visualisation tool [1]. The first desirable characteristic is defined as scalable data management to handle and enable real-time interaction over datasets with a huge number of objects. Coupled with data management, scalable and efficient visualisation of large and dynamic sets of volatile raw data is an advisable feature to have. Regarding the consumer of such tools, the other two recommended attributes are: visual scalability to avoid problems related to visual information cluttering, and customisation capabilities of visualisations to meet the expectations and analysis needs of different user types.

Moreover, increasing data democratization, i.e. societal and technological evolution making data accessible to everyone, is leading to the availability of very diverse and large datasets for analysis, to people that might lack data analysis expertise (e.g. as research scientists, policy makers, or individuals).

Visualisation techniques, used in data visualisation tools, provide users with intuitive means to interactively explore the content of the data, identify interesting patterns, infer correlations, and support sense-making activities.

Currently, the challenge is to implement the best combination of underlying data-management technologies and visualisation techniques to enable end-users to gain value and insights out of the data quickly, minimizing the role of IT-experts in the loop. This is especially critical in the Big Data context and for data preparation and Quality of Data (QoD) improvement tools, where users are not limited to exploration and analyse of data and therefore need to be able to transform the dataset to meet their goals.

The contribution of this paper is the description of detected problems and implemented solutions for the visualisations of a Data Preparation Tool for Big Data sets.

In this paper, we first present related work in the visualisation and data preparation domains (Sect. 2). Then, we introduce our original in-memory data preparation tool in Sect. 3. In Sect. 4, we describe the challenges faced and solutions adopted when evolving its visualisations from in-memory fitting to Big Data sets. Finally, we present our conclusions and discuss potential directions of future work in Sect. 5.

2 Related Work

Traditional data visualisation tools are usually restricted to small datasets, processed offline and limited to accessing and visualising pre-processed sets of static data.

In an attempt to handle the characteristics of the Big Data era, the research community has proposed different visualisation approaches [1].

The most common techniques are those of data reduction, which aim to summarise the dataset by using different approximations. The approaches followed for data summarisation include sampling (i.e. visualising a representative subset or filtering non-contributing sets of data) [2, 3] and aggregation (i.e. visualising an aggregated or abstracted version of the dataset by using binning or clustering techniques) [4, 5].

The next set of proposed techniques target the hierarchical exploration of a dataset, allowing the visual exploration of large datasets at different levels of detail [6, 7]. These are computed by a hierarchical aggregation of the dataset, which allows the user to get a synopsis of the dataset and retrieve details of the data at different levels.

These two types of strategies (i.e. data reduction and hierarchical techniques) aim to contribute to the visual scalability characteristic discussed in the introduction. Other research has targeted the real-time interaction (with the dataset) feature by working on the progressive result delivery and different caching and prefetching strategies. Regarding progressive techniques, these tend to combine both user interaction-based dynamic result calculations [8] and incremental computation and delivery of results [9].

Moreover, visualisation approaches have been developed to tackle the dynamic nature of datasets by implementing incremental and adaptive strategies that allow for an on-the-fly exploration of large and dynamic datasets [6, 10].

Finally, regarding visualisation techniques, another area of research has focussed on assisting the user by recommending visualisations that are more appropriate for the specific characteristics of the data (or identified trends) [11], or the user behaviour and preferences [12].

Regarding commercial tools, a body of research work has analysed some popular visualisation tools (i.e. Tableau, PowerBI, Plotly, Gephi and Excel) and techniques, and how well they fit into the size, heterogeneity and dynamism properties of Big Data [13]. These tools are more focussed on visualising data prepared for analysis.

A recent market analysis report has analysed data preparation tools [14], studying the integration and exploration features, data manipulation features and user experience and user interface features among others, as part of the technical assessment of these tools. As the studied features prove, Big Data management and visualisation techniques are key to these data preparation and QoD improvement tools, whereas commercial tools are focussed on data preparation following the extract, transform, load (ETL) procedure, and not in the data exploration task for QoD assessment and improvement.

In this state-of-the-art context, we report on our initially developed in-memory data preparation and QoD improvement TAQIH tool, and on the experience of enhancing this tool from in-memory dataset visualisation and preparation to Big Data sets.

3 TAQIH – in-Memory Data Preparation Tool

TAQIH [15] is a data preparation tool developed to support non-technical users on 1) the exploratory data analysis (EDA) process of tabular health data, and 2) the assessment and improvement of its quality. A web-based tool was implemented with a simple yet powerful visual interface.

First, it provides interfaces to understand the dataset, to gain an understanding of the content, structure and distribution. Then, it provides data visualisation and data quality improvement utilities for the dimensions of completeness, accuracy, redundancy and readability [16].

TAQIH was designed and developed with in-memory data preparation technologies, so visualisation, data management and data transformations are limited to a single computer's memory and web-browser capabilities. Experimentally we have been able to manage datasets under 200 MB using a desktop machine with 8 GB of RAM, but for providing the end-user with an acceptable interaction time (10 s for keeping the user's attention [17]), this is reduced to few tens of MBs. Data transformations are synchronously applied as they are requested, and visualisations updated accordingly.

TAQIH contains a main navigation bar at the top of the GUI, where items are placed from left to right following the usual iterative pipeline in EDA. First, 'General Stats' and 'Features' menu items provide global and detailed views of the data to gain insights about content, distribution and quality. Then, the 'Missing Values' section deals with the completeness dimension of data quality. After that, the 'Correlations' section presents the statistical relationship among variables, to help the identification of possible redundancies among variables or incoherent data, related to the redundancy and accuracy dimensions of data quality. Next, the 'Outliers' section identifies observations that differ significantly from others in the features and instances axes which is also related to accuracy, redundancy, readability and trust dimensions in data quality. All views include a small sample of the dataset (following data reduction by sampling) to help interpreting the dataset and identifying the transformation actions needed.

TAQIH is composed of both pre-processed property visualisation and summary visualisation (histograms or density plots), but also includes binary heatmaps (for missing values) or boxplots (for outliers) representing instance level data, which can be cumbersome when moving to very large datasets.

4 Enhancing Data Preparation Tool Visualisations for Big Data

Volumes considered in the Big Data context (i.e. large datasets not fitting in a computer's memory and expected to be increasing) impose new challenges over traditional datasets which could be totally managed in a computer's memory. When it comes to data preparation and QoD assessment, traditional Python-based or R-based methods do not directly handle datasets that do not fit into a computer's memory.

Additionally, many traditional general statistics or quality assessment algorithms need to keep global variables for their computation, for example, cardinality calculations might require expansion as large as the data source size. This makes existing data quality algorithms unsuitable for distributed parallel computing.

We have also identified two more issues when moving QoD assessment to large datasets: the visualisations used to allow the users to explore the data to evaluate its quality and that data preparation tasks cannot be run synchronously anymore.

Traditional visualisations (e.g. missing values or outliers) mainly work by plotting all the instances of the dataset, which requires pulling all instances from the dataset, and having the user's client applications manage all the data to visualise and respond to users' interactions. This is no longer feasible and it is unrealistic to expect the user to wait until a data cleansing task, over a large dataset that might require hours, is complete.

4.1 Migration to an Asynchronous Distributed Architecture

To overcome the data volume challenges identified, we have opted to use algorithms that provide approximations to evolve the TAQIH tool into an asynchronous processing framework. Big Data computing infrastructures have been used, for those algorithms which have distributable or parallelized versions, whilst for those requiring adaptations, state-of-the-art proposals have been implemented following Big Data computing approaches where possible, and per-chunk processing where more fine-grain control of shared global variables is required.

Figure 1 illustrates the previous TAQIH architecture contrasted to the architecture redesign for the GYDRA architecture.

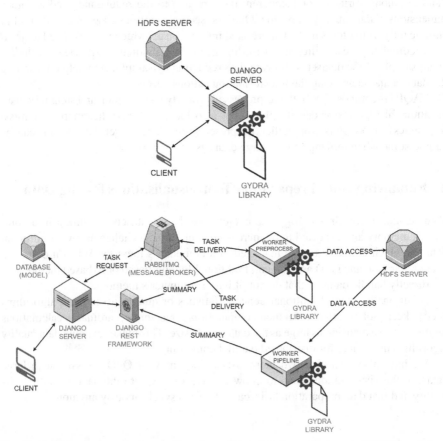

Fig. 1. TAQIH solution single machine focussed architecture (top) and GYDRA solution distributed architecture (bottom)

The GYDRA tool's user interface has been developed using the Django framework, HTML5, Asynchronous JavaScript (AJAX) and Bootstrap responsive web library. Distributed asynchronous tasking is managed through the Celery Distributed Task Queue tool with RabbitMQ as a message broker to implement the real task queue. The Celery worker (depicted as Worker Preprocess and Worker Pipeline in Fig. 1) can either run data pre-processing or transformation tasks by using the GYDRA Python library reading and handling HDFS stored datasets per chunks or by submitting applications to an Apache Spark cluster.

4.2 QoD Assessment Visualisation Updates

For the Big Data QoD indicators visualisation issues, approximations requiring a limited and controlled but representative amount of data to be displayed have been implemented. The computation and generation of the visualisation is performed by the asynchronous workers to reduce processing load and smooth the user experience on the client side. Subsequently, the different visualisation components of the GYDRA preparation tool are analysed individually.

Tab-Based Navigation Approach and General Stats
GYDRA has retained the main tab-based navigation approach and sample of the dataset described for TAQIH, while a new data transformation pipeline section has been added across the different views, to better understand the dataset and dynamically add transformations during EDA (since transformations have to be processed offline now).

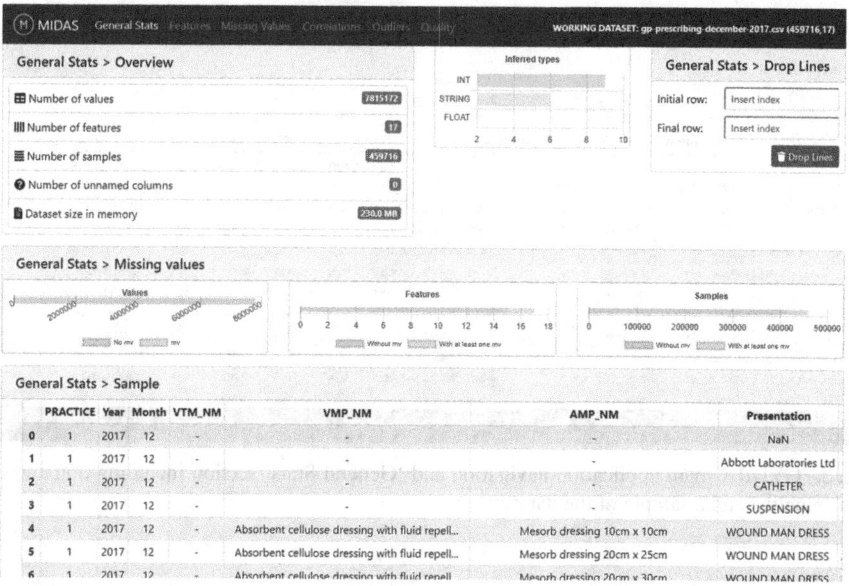

Fig. 2. TAQIH navigation and the 'General Stats' section. This figure is reproduced from [16]

Information summarisation is provided through a navigation approach by providing a hierarchical view, starting from 'General Stats' and moving to the 'Features' section, and by including a raw data sample across all sections. Next, the 'Missing Values', 'Correlations' and 'Outliers' sections are placed to assist the user through common EDA tasks. Figure 2 and Fig. 3 depict the navigation approach and 'General Stats' sections of TAQIH and GYDRA tools respectively.

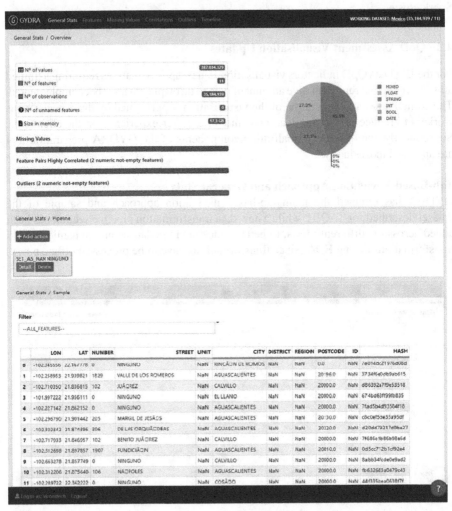

Fig. 3. GYDRA main application navigation and 'General Stats' section including transformations pipeline and a sample of the dataset

Features

Concerning the 'Features' section, visualisation remains quite similar, moving feature-related transformations from a feature specific section to the common transformation specification pipeline section. Additionally, cardinality and the number of appearances per each feature variable has been dropped, considering the scalability limitations when moving to big data sets. We have replaced this feature with the visualisation of the Top 10 values. Figure 4 and Fig. 5 depict the feature analysis section of the TAQIH and GYDRA tools respectively.

Fig. 4. TAQIH per feature analysis section. This figure is reproduced from [16]

Missing Values

For the Missing Values section, previously a binary heatmap with individual data was displayed, while in the GYDRA tool, percentages of missing values have been computed and visualised. TAQIH had a specific Missing Value imputation section, while in GYDRA, this has been moved to the common transformation pipeline. Figure 6 depicts the TAQIH tool's Missing values section, while Fig. 7 shows the GYDRA tool's implementation.

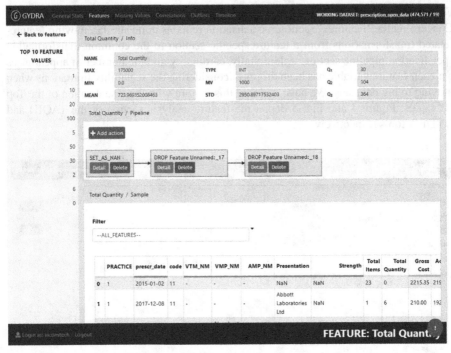

Fig. 5. GYDRA per feature analysis section.

Fig. 6. TAQIH tool's missing values section. This figure is reproduced from [16]

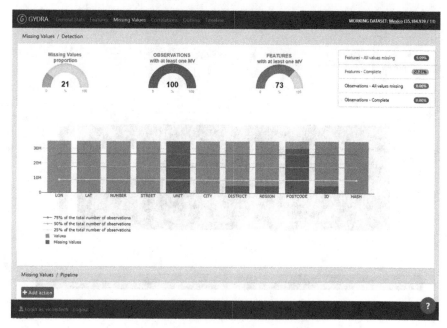

Fig. 7. GYDRA tool's missing values section

Correlations

The Correlations section remains similar considering its most important visualisation is a correlation matrix, displaying feature association values (see Fig. 8). The density plot used in TAQIH to compare two features among their value set was dropped.

Outliers

Regarding the Outliers section, in the TAQIH tool both a traditional box plot and a histogram with each different value occurrence (classified as in or out layer) were used. For GYDRA, a novel box plot visualisation has been proposed (see Fig. 9). The outlier distribution is plotted as two histograms, one for low values and the other one for high values (suspected outliers and outliers according to Tukey algorithm). Multivariate outlier detection was dropped from TAQIH while an alternative for Big Data was implemented.

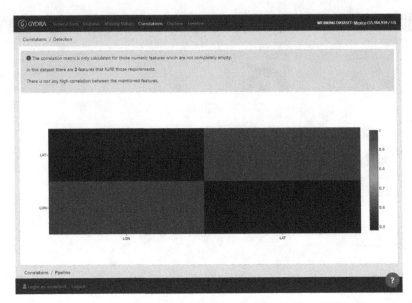

Fig. 8. GYDRA tool's correlations section

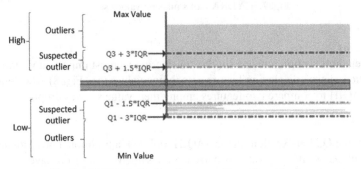

Fig. 9. GYDRA's Outlier diagram with outliers binning

5   Conclusions and Future Work

In this paper, we report on the enhancements implemented to a tabular data preparation and QoD improvement tool, focussed on its visualisation features, when moving from in-memory datasets to Big Data sets. Architectural and visualisation solutions adopted have been described accordingly. It was necessary to move from an in-memory synchronous architecture to an asynchronous distributed processing architecture to be able to operate the datasets, as well as the remote creation of visualisations. Asynchronous pre-processing of visualisations was adopted instead of on-the-fly processing, given the need to compute general statistics and per feature indicators (e.g. top n values or correlations). For Big Data sets it is not possible to calculate these indicators and provide a timely response to the user otherwise. Next, we have adopted different data reduction approaches to ensure visual scalability and meet local memory limitations for visualisation, introducing a novel box plot for Big Data.

Future work is planned on researching and implementing features lost in the transition from TAQIH to GYDRA. Next, the focus will be on the integration of streaming data and researching exploration techniques to help the user in understanding the dynamic of the sources to better define their ingestion pipelines. In line with this, enabling users to navigate through the dataset to find missing values should be enhanced to help users identify and understand underlying trends. Related to user assistance, machine learning techniques are being researched to support the user by suggesting appropriate preparation tasks.

Acknowledgments. This work was supported by the European Union's Horizon 2020 research and innovation programme under grant agreement No 727721 (MIDAS).

This work was supported by the Gipuzkoan Science, Technology and Innovation Network Programme funding of the HIDRA project.

References

1. Bikakis, N.: Big data visualization tools. In: Sakr, S., Zomaya, A.Y., (eds.) Encyclopedia of Big Data Technologies. pp. 336–340. Springer, Cham (2019). https://doi.org/10.1007/978-3-319-77525-8_109
2. Battle, L., Stonebraker, M., Chang, R.: Dynamic reduction of query result sets for interactive visualizaton. In: 2013 IEEE International Conference on Big Data, pp. 1–8 (2013). https://doi.org/10.1109/BigData.2013.6691708
3. Park, Y., Cafarella, M., Mozafari, B.: Visualization-aware sampling for very large databases. In: 2016 IEEE 32nd International Conference on Data Engineering (ICDE), pp. 755–766 (2016). https://doi.org/10.1109/ICDE.2016.7498287
4. Jugel, U., Jerzak, Z., Hackenbroich, G., Markl, V.: VDDA: automatic visualization-driven data aggregation in relational databases. VLDB J. **25**(1), 53–77 (2015). https://doi.org/10.1007/s00778-015-0396-z
5. Lins, L., Klosowski, J.T., Scheidegger, C.: Nanocubes for real-time exploration of spatiotemporal datasets. IEEE Trans. Vis. Comput. Graph. **19**, 2456–2465 (2013). https://doi.org/10.1109/TVCG.2013.179
6. Bikakis, N., Papastefanatos, G., Skourla, M., Sellis, T.: A hierarchical aggregation framework for efficient multilevel visual exploration and analysis. Semantic Web. **8**, 139–179 (2017). https://doi.org/10.3233/SW-160226
7. Elmqvist, N., Fekete, J.-D.: Hierarchical aggregation for information visualization: overview, techniques, and design guidelines. IEEE Trans. Vis. Comput. Graph. **16**, 439–454 (2010). https://doi.org/10.1109/TVCG.2009.84
8. Stolper, C.D., Perer, A., Gotz, D.: Progressive visual analytics: user-driven visual exploration of in-progress analytics. IEEE Trans. Vis. Comput. Graph. **20**, 1653–1662 (2014). https://doi.org/10.1109/TVCG.2014.2346574
9. Im, J.-F., Villegas, F.G., McGuffin, M.J.: VisReduce: Fast and responsive incremental information visualization of large datasets. In: 2013 IEEE International Conference on Big Data, pp. 25–32 (2013). https://doi.org/10.1109/BigData.2013.6691710
10. Zoumpatianos, K., Idreos, S., Palpanas, T.: Indexing for interactive exploration of big data series. In: Proceedings of the 2014 ACM SIGMOD International Conference on Management of Data, pp. 1555–1566. Association for Computing Machinery, Snowbird, Utah (2014). https://doi.org/10.1145/2588555.2610498

11. Mackinlay, J., Hanrahan, P., Stolte, C.: Show me: automatic presentation for visual analysis. IEEE Trans. Vis. Comput. Graph. **13**, 1137–1144 (2007). https://doi.org/10.1109/TVCG. 2007.70594

12. Gotz, D., Wen, Z.: Behavior-driven visualization recommendation. In: Proceedings of the 14th international conference on Intelligent user interfaces. pp. 315–324. Association for Computing Machinery, Sanibel Island, Florida, USA (2009). https://doi.org/10.1145/1502650.150 2695

13. Ali, S.M., Gupta, N., Nayak, G.K., Lenka, R.K.: Big data visualization: tools and challenges. In: 2016 2nd International Conference on Contemporary Computing and Informatics (IC3I), pp. 656–660 (2016). https://doi.org/10.1109/IC3I.2016.7918044

14. Ovum: Ovum Decision Matrix: Selecting a Self-Service Data Prep Solution, 2018–19. (2018)

15. Álvarez Sánchez, R., Beristain Iraola, A., Epelde Unanue, G., Carlin, P.: TAQIH, a tool for tabular data quality assessment and improvement in the context of health data. Comput. Methods Programs Biomed. SI Data Qual. Assess. **181**, 104824 (2019). https://doi.org/10. 1016/j.cmpb.2018.12.029

16. The Dama UK Working Group: The Six Primary Dimensions For Data Quality assessment, https://www.dqglobal.com/wp-content/uploads/2013/11/DAMA-UK-DQ-Dim ensions-White-Paper-R37.pdf. Accessed 08 Mar 2018

17. Nielsen, J.: Usability Engineering. Morgan Kaufmann, Amsterdam (1993)

Materia: A Data Quality Control Embedded Domain Specific Language in Python

Connor Scully-Allison[✉]

University of Arizona, Tucson, AZ, USA
cscullyallison@email.arizona.edu

Abstract. Current solutions for data quality control (QC) in the environmental sciences are locked within propriety platforms or reliant on specialized software. This can pose a problem for data users when attempting to integrate QC into their existing workflows. To address this limitation, we developed an embedded domain specific language (EDSL), Materia, that provides functions, data structures, and a fluent syntax for defining and executing quality control tests on data. Materia enables developers to more easily integrate QC into complex data pipelines and makes QC more accessible for students and citizen scientists. We evaluate Materia via two metrics: productivity and a quantitative performance analysis. Our productivity examples show how Materia can simplify complex descriptions of tests in Pandas and mirror natural language descriptions of common QC tests. We also demonstrate that Materia achieves satisfactory performance with over 200,000 floating-point values processed in under three seconds.

1 Introduction

Quality Control (QC) on scientific data is a critical step in the data preprocessing pipeline and is essential to performing good science. This "QC process" describes any systematic approach undertaken by data experts to check for errors in datasets. It can be done manually or automatically. In the manual case an expert scientist scans through columns of records and notes any problems they find. For example, they may discount a temperature value for being too cold for the season or when visualizing data they will see that a value spikes up dramatically compared to its preceding measurements, or drifts over time.

If these anomalies are not recorded and properly associated with their corresponding dataset, the utility of collected scientific data is significantly limited. Errenous values in a dataset do not accurately reflect the ground truth they are supposed to measure and, accordingly, cannot be used in scientific models to enhance our understanding of natural phenomena. In recent years, automation has transformed both the data collection and QC process but has not fully solved the QC problem [11,12].

© Springer Nature Switzerland AG 2020
W. Abramowicz and G. Klein (Eds.): BIS 2020 Workshops, LNBIP 394, pp. 285–296, 2020.
https://doi.org/10.1007/978-3-030-61146-0_23

Increased automation in the data collection process has introduced novel vectors for the propagation of data errors. Old sensors, dying batteries, poor wiring, network failures, and more, can contribute to a missed data point or an incorrect value being logged. In the environmental sciences, these opportunities for failure are compounded by the relative remoteness and wild characteristics that pervade data collection sites [2]. Everything from weather events to wild animals can cause sensors to log incorrect data or stop data feeds altogether.

To combat this problem of quality issues introduced by autonomous data collection, a handful of technologies have been devised and deployed into modern workflows to provide quality control. Among them are the GCE Toolbox [7], Loggernet [9], ArcGIS [4], and Pandas [8]. The GCE Toolbox was produced to provide a dedicated toolkit for managing QC processes in the earth and environmental sciences. Loggernet's support software and ArcGIS provide tools and modules which enable users to perform limited Quality Control on data. Pandas, by contrast, is a general purpose statistics and data analytics library embedded in Python which was not built for data Quality Control but provides functionality which supports QC tasks very well, like binding metadata to data frames and allowing users to define time-series indices.

Together these libraries, software packages and frameworks give data managers a variety of options to choose from for their QC needs. Unfortunately, they have drawbacks that hinder their utility to specific individuals or prevent easy integration into existing data collection workflows. For the GCE Toolbox, Loggernet and ArcGIS, each of these solutions are reliant on proprietary or expensive software to run. For the GCE Toolbox, a subscription to the Matlab language is required. The ArcGIS software requires a licence costing a minimum of 200 dollars per year. Loggernet QC solutions only work with Campbell Scientific sensors: which are expensive, enterprise-level instruments used for collecting data, and are often eschewed for cheaper "iButton" devices, which are used in projects as prominent as the NSF funded McMurdo Dry Valleys Long Term Ecological Research Network (LTER) in Antarctica [1].

```
1  dfcopy['repeat-ids'] = (dfcopy[self.column] !=
2                          dfcopy[self.column].shift(1))
3                          .cumsum()
4  counts = dfcopy[['repeat-ids',self.column]]
5          .groupby(['repeat-ids']).agg('count')
6  counts_less = counts.loc[counts[self.column] > 1]
7  dfcopy = dfcopy
8          .join(counts_less,
9              on='repeat-ids',
10             lsuffix='_caller',
11             rsuffix='_other')
12 dfcopy[self.column] = dfcopy[self.column + '_other']
13                       .map(lambda x: x >= 3)
```

Listing 1.1. Code which checks for repeat values in Pandas and returns a Boolean array indicating if a value is a repeat or not if it exceeds the repeat threshold of "3".

Pandas does not suffer from these limitations being a free, open-source library implemented in a free programming language: Python. Instead, Pandas' problem is one of domain expressiveness. The abstractions provided in that library do not always map well to defining tests in the quality-control space. An example of such a mapping can be seen in Listing 1.1, which depicts a test that checks for values that do not vary across multiple time steps. This implementation, while efficient, can be difficult to parse even as an experienced coder, making it that much more difficult for a domain scientist.

To fill these gaps which are unmet by existing solutions we developed Materia: a domain specific language embedded in Python. This language provides several data structures which simplifies the development effort required to read in tabular time-series data, manage quality control flags and perform the logical comparisons required for QC workflows. Furthermore, Materia leverages Python's built in method chaining to support a fluent syntax for managing tests. Finally, Materia provides a functional syntax which abstracts away iterative operations and enables tests to be defined over an "arbitrary" value at a single point in time. And, although the default assumption is that a function will be testing a single value, these tests also keep track of the context around a single value enabling more complex operations on one series or multiple.

The remainder of this paper is organized as follows: related and prior work are presented in Sect. 2; design and implementation details on Materia are presented in Sect. 3; evaluations of Materia and discussions of results are presented in Sect. 4. Finally, conclusions and future work are presented in Sect. 5.

2 Related Work

Relevant to the motivating problem of this research, there are several papers that discuss the importance of data quality control in the environmental sciences. Recent scholarship on this topic has emphasized an increased need for programmatic solutions to this problem. In the paper "Quantity is Nothing Without Quality: Automated QA/QC Streaming for Environmental Sensor Data," the motivation and execution of automateable quality control processes are discussed at length [2]. This paper informs domain expectations of general quality control processes and helps circumscribe the domain problem space.

In addition to this, several other works in the environmental science domain describe what standards are expected from quality controlled data [6,14]. Specifically, the handbook by Gouldman et al. provides examples of flag codes in use by the National Oceanic and Atmospheric Administration (NOAA). This resource exemplifies what a QC-focused programming language should support. It also details many types of automated tests which are standard for their organization: "Rate of change in time", "Spike Test", "Regional Range Test", "Stuck Value Test". These tests map directly to several of the tests we will evaluate for performance and brevity, and reinforces the validity of our chosen tests.

The development of the GCE toolkit, which represents the closest prior implementation of an EDSL for QC is detailed in Sheldon et al. [13]. In this work, Sheldon provides implementation and design details for the GCE toolkit which were leveraged to build up Materia. Specifically, the Dataset data structure used in Materia can be thought of as a simplified version of the GCE toolkit's "Dataset" object with some implementation features drawn from Pandas. This work also provides context about the use and impact of the GCE toolkit in the earth science domain and provides insight into the specific communities which could benefit from Materia. Sheldon indicates that it is used by various Long Term Ecological Research (LTER) sites, the United States Geologic Survey (USGS), and by GCE itself in addition to many others.

In addition to the above paper, the paper associated with the Pandas library was also mined for insight on how to construct an efficient data management/-analytics library in Python [8]. This work helped us understand how to organize the columnar data in our datasets and use numpy arrays most efficiently. Specifically, we organized our arrays into numpy matrices with shared types mirroring their "block manager" and provide support for "label based data access." Using Pandas as a reference in this way enables us to build upon state-of-the-art work and provide interfaces people are familiar with.

The concept of an embedded domain specific language (EDSL) has been steadily growing in attention and scholarship over the past two decades. For an overview of embedded domain specific languages, see Gill [5]. This classification describes programming languages built within and using an existing programming language. This construct allows for ease of development on the part of the language developer and allows language users to use syntax and supporting libraries which are familiar to them when working on domain problems.

3   Materia

Materia is a domain specific language (DSL) shallowly embedded in Python. The term, "shallowly-embedded," means that it extends the host programming language with additional functionality and makes-use of language provided constructs. It provides data structures, an abstraction which removes loop-definition clutter from function definitions, and a fluent syntax. Together this functionality enables users to quickly define and execute QC tests on tabular, time-series data. In addition to these features, Materia was designed to work with a wide variety of datasets and arbitrary flag codes to enhance its utility to various organizations. The following subsections will describe in detail these specific aspects of Materia. For more details on the usage and implementation of Materia, source code and documentation can be found at [10].

3.1   Data Structures

Materia is comprised of 2 key data structures: a DataSet and a TimeSeries. Both of these structures were designed to mitigate overhead of array creation and management, flag storage and data alignment on the part of the user.

The first data structure, a DataSet, can be thought of as a lightweight Pandas dataframe. Compared to it's bulkier cousin, it provides some domain-specific functionality which the dataframe lacks. First, it stores header metadata found in many tabular data files in a dedicated row-major matrix. This header metadata can be used for reference at any point in the development of a quality control script and can also be used to produce unique and semantically meaningful header names for label-based array access of specific data columns. An abridged example of a tabular data file which would work with Materia can be seen in Table 1.

Second, a DataSet operates under the assumption that data will be indexed by a series of datetime values. Accordingly, it performs automatic detection and conversion of any datetime columns found in the provided data file. It also automatically sets one of these columns as the primary index for the DataSet. This timeseries index is used in Quality Control tests to find missing values and align time series for comparison when they have the same temporal range but may not have the same number of array elements. All vectors in the DataSet object are numpy arrays and are stored in column major format.

To provide a simple and familiar interface for extracting an individual Time-Series object from a DataSet, the DataSet class overloads the "[]" operators to enable label-based access of individual columns. The returned TimeSeries object manages the execution of tests and storage of resultant flags. A time series object is comprised of three numpy arrays: the global time series index which was defined at the DataSet level, the values array for this series and an array which keeps track of quality flags resulting from tests. The TimeSeries object tracks what tests were run on it and other related metadata like its label in the DataSet and flag codes which were originally defined over the DataSet object.

Table 1. An example of a tabular dataset, downloaded from the Nevada Research Data Center. Materia was designed for datasets like this one.

Site name:	Rockland summit	Rockland summit
Deployment:	Air temperature	Air temperature
Monitored system:	Climate	Climate
Measured property:	Temperature	Temperature
Vertical offset from surface:		
Units:	degC	degC
Measurement type:	Maximum	Minimum
Measurement interval:	00:01:00	00:01:00
Time stamp (UTC-08:00)		
2019-04-01T16:41:00.0000000-08:00	−9999	5.235
2019-04-01T16:42:00.0000000-08:00	5.124	4.97
2019-04-01T16:44:00.0000000-08:00	5.165	5.046

3.2   Managing Flag Codes

Contrary to some popular conceptions of the purpose of Quality Control, the main output of the initial phases of a quality control process is not repaired data. Instead, it is only metadata. This metadata comes in the form of "flags" – codes that indicate to a data user what the quality of particular values in a dataset may be. Flags are typically stored in an array. They are aligned alongside values in a tabular format making them easy to cross-reference when managing data.

While flags are a simple concept they can sometimes be difficult to work with in practice. This difficulty is due to the general standardization problem in scientific organizations. Like many other aspects of data, flags are not standardized across organizations, with some orgs using string values to convey the quality of their data ("bad," "good," "suspect") while others use integers (0,1,2) [3]. This diversity of flag codes requires that Materia support flag standards which have differing numbers of flags, datatypes and names. In order to accommodate this plurality, Materia provides a function which allows users to affix a dictionary of their own flags to a DataSet object, visible in listing 1.2.

```
1  DataSet.flagcodes()
2      .are({
3      "None":"OK",
4      "Repeat Value":"Repeat Value",
5      "Missing Value": "Missing",
6      "Range": "Exceeds Range",
7      "SI": "Incosistent (Spatial)",
8      "LI": "Inconsistent (Logical)",
9      "Sp": "Spike"
10     })
```

Listing 1.2. How to define flag codes on a Materia DataSet.

After setting this dict on our dataset, subsequently extracted TimeSeries objects keep track of these key-value pairs so that users need only provide the key when executing a quality control test. An example of this can be seen in listing 1.5. By mapping the flag codes to strings in this way, Materia allows users to define their own keys for flags which enables them to access them in a more succinct way. Users can use "SI" instead of "Inconsistent (Spatial)."

3.3   Defining Tests

```
1  def rv_test(value):
2      n = 3
3      if not value.isnan():
4          if value == value.prior(n):
5              return True
6          return False
```

Listing 1.3. A QC test definition to check for multiple repeat values in a row using the Materia language.

The Materia language itself was designed around a functional paradigm, whereby users can define tests which operate on an abstract datapoint. Instead of developing functions which operate on specific arrays of values or contain internal iterations, users instead define Python functions with a single argument: "value." This argument represents a single datapoint in an arbitrary time series. This abstraction enables users to define tests in fundamental terms which mirror real test specifications. An example of a test definition using this syntax can be seen in Listing 1.3.

In this example, we can see how a series agnostic test definition syntax supports natural expression of a data quality control test. Instead of working with whole vectors or iterating over a vector passed in as an argument, we are instead expressing our repeat value test as "If a value is equal to the prior 3 values in this series, return that it failed the test (true)." We also see how the "value" argument provides significant functionality to support fluent test definitions.

```
 1  def spatial_inconsistency(value):
 2      comp_val = series_max_10.value().at(value)
 3      diff = value - comp_val
 4      avg = (value + comp_val) / 2
 5      threshold_p = 75
 6      exceeds_threshold = (abs(diff/avg) * 100.0 >
 7                                      threshold_p)
 8      if exceeds_threshold:
 9          return True
10      return False
```

Listing 1.4. A QC test definition in Materia which checks if the values within two related time series diverge beyond a specific threshold. series_max_10 refers to a time series pulled from the dataset prior to this function definition.

"Value" is itself an object which contains a scalar by default but can contain a vector of values. In order to support comparisons between vectors and scalars and maintain fluent test definition syntax it overloads the following binary conditional operators: "$==$," "$!=$," "$<=$," "$>=$," "$>=$," "$>$," "$<$". As can be seen with the "value.prior(n)" call, this object also keeps track of the context of the time series from which it was called. As many tests need to support comparisons between values preceding or following individual data points in a time series, its essential that Materia's test definition function provide an interface for retrieving contextually relevant data points.

```
 1  series.datapoint()
 2      .flag('Repeat Value')
 3      .when(rv_test)
```

Listing 1.5. How to execute a test on a time series object in Materia.

One further example of Materia's test definition syntax can be seen in Listing 1.4. In this test definition, we can see that our "value" abstraction also overloads mathematical operators to support binary operations between our value object, numeric constants, and other value objects. In order to support a diverse array of mathematical operations on various data types, these overloaded binary operators support standard operations between individual scalars and heterogeneous operations between vectors and scalars. Finally, we also see in Listing 1.4, that a method is provided by time series objects which allows us to pass in a particular value and get out a temporally aligned value from that series. This method enables users to define tests between related series of values within the same temporal range.

3.4   Calling Tests and the When() Function

In order to execute these tests within the context of a particular set of values, each TimeSeries object provides several methods which are chained together. These methods execute the test which is passed to "when()" as an argument and affix flags depending on the results of the provided tests. The syntax of these methods can be seen in listing 1.5.

This part of Materia implements a fluent syntax through chained method calls. This is done so that a test execution can mirror a natural language statement: "Flag a datapoint in [this] series with'repeat value' when the repeat value test fails." This syntax for calling tests in Materia further reinforces the mental model that tests are being called on singular data points in our TimeSeries.

In listing 1.5, we can see how the when() method operates from a user perspective; specifically, it is just a higher-order function which executes the "rv_test" argument passed to it. Internally, this when() function performs several tasks to execute passed tests and manage their results.

Immediately upon being called, the "when" function uses the Python library, inspect, for method reflection. Using getsource(), it stores the source code of the passed function so that this information can be later used to provide additional context to our flags. Second, it uses getargspec() to determine if an optional iterator argument was declared as part of our test definition. If there was, "when" will invoke the function with two parameters and not one. This secondary argument can be used for debugging or checking specific errors in the provided dataset.

After performing these reflective operations, "when" then declares a loop over the internal values array. At each iteration of this loop, a new "Value" object is created and passed in as an argument to the provided test function. This Value object is constructed with the contextual information of its surrounding values in the calling TimeSeries, its offset from the beginning of the array and its datetime index. By managing our loop inside of the "when" function and providing users with a robust object for testing values this simplifies test definitions significantly.

4   Evaluation and Discussion

Materia was formally evaluated within the scope of two categories which are commonly used to evaluate EDSLs: productivity and performance. The first, productivity, encompasses measures of how a language may improve a users ability to write effective programs in their domain. For Materia we first examined how closely a test definition in a language conforms to a natural language description of a test found in earth science handbooks and literature. Second, we compared specific implementations of QC tests in Pandas against functionally equivalent definitions in Materia. This enabled us to compare ease of mapping the domain problem space to the language of implementation. For performance, we compared the runtimes of test definitions in Materia against implementations in Pandas. For these tests we used 6 commonly found quality control tests as our benchmarks

4.1 Benchmark Tests

The six tests used as our benchmarks and foundations of syntactic comparison come from Campbell et al. missing-measurements, range, persistence, spatial inconsistency, internal inconsistency and change in slope [2]. Missing-measurements checks the difference between two date-times in chronological order, if they exceed a specified time interval that means a measurement is missing and a row must be inserted. Range tests seek to identify if a value exceeds some normal range in values. Persistence checks if a given value in a dataset is a repeat of one or more data points preceding it. With spatial inconsistency, two or more time series measuring the same data type in close proximity to one-another are compared. If one diverges it usually indicates a logging error. Internal consistency evaluates a break in a logical condition between two time series. For example: a minimum variable measurement at a certain time index cannot exceed a maximum reading from the same sensor for the same variable at the same time index. Finally, a change in slope describes a dramatic upward or downward change in our time series' slope over a short time period.

4.2 Productivity

For the first metric of productivity, natural-language similarity, we selected the following five descriptions of individual quality control tests:

- "No values less than a minimum value or greater than the maximum value the sensor can output are acceptable" – Range Test [6]
- "This test compares the present observation n to a number . . of previous observations." – Persistence [6]
- "A sharp increase or decrease [in slope] over a very short time interval (i.e., a spike or step function)" – Change in slope [2]
- "ensuring that the minimum air temperature is less than maximum air temperature" – Internal Inconsistency [2]
- ". . . data from one location are compared with data from nearby identical sensors" – Spatial Inconsistency [2]

These particular tests were chosen because they reflect the six cardinal tests enumerated by Cambell et al. [2], with the exception of "missing measurements."

"Missing measurements" was omitted from this examination because it was implemented as static method in the TimeSeries class. These descriptions were compared by the developers of Materia, through and informal side-by-side comparison. (These implementations are not included here for space reasons however they can be found at [10].)

From this examination, no clear consensus was found. Quantifying similarity on a four step range, from "very similar" to "similiar" to "dissimilar" to "very dissimilar" we determined that a Materia-based implementation of a persistence test, visible in Listing 1.3, was "very similar." We considered the very complex "Change in slope" test to be, by contrast, "very dissimilar". We found internal inconsistency and range tests "similar" and spatial inconsistency to be "dissimilar."

Overall, the results of similarity or dissimilarity seem to be most significantly related to how simple the mathematical or boolean operation used in a test may be. For the persistence test, syntactical similarity is bolstered by the fact that an natural language description explicitly mentions a comparison and that Materia provides a "prior" function which was designed to mimic natural language comparisons like this. On the other end of the spectrum, a Materia-based implementation of a slope test is over 30 lines of code and reflects the multiple calculations required to compare two slopes. The natural language description of this does not capture the complexity of these operations at all.

For the second metric gauging productivity, we compared Materia-based implementations of code against Pandas-based implementations. For Spatial Inconsistency, Spike and Range Tests, the details of implementation were approximately the same; although Pandas based implementations occasionally required the use of helper functions like np.logical_and to support vector-wise Boolean operations. With a logical inconsistency test we see more deviation with Materia's more simple definition where a user can simply declare "**return max_value < min_value**". In Pandas we require the use of a helper function, "np.where" to return an array of boolean values. Although not a significant addition, this does introduce some visual noise which can be hard to parse. Finally, we see significant divergence between implementations of a persistence test in Pandas and Materia. As can be seen in Listing 1.1, Pandas requires several different statements to identify and filter a time series down to identified repeat values. It requires further statements to express the result as a Boolean array. Without knowing what the code does, it's hard to understand even with a few minutes of exposure. By comparison, it takes mere seconds to understand what the persistence test in Materia (Listing 1.3) is doing.

4.3   Performance

Due to space limitations we cannot include a comprehensive breakdown of our performance measurements. It should be noted however, that across all tests, Materia did perform more poorly than the highly optimized Pandas. In general, Pandas, ran it's operations faster by an order of magnitude compared to Materia. We argue that this is not a fatal mark against this EDSL as the absolute runtimes of Materia never exceeded 2.5 s for more than 200,000 data points. When this represents 6 months of data and Quality Control is often done on a month by month basis or as part of prepossessing steps we deemed this an acceptably low runtime for practical use.

4.4   Additional Considerations

In addition to the above evaluation metrics, it should also be noted that Materia provides several features which support productivity but were not formally evaluated and are not found in more general purpose languages like Pandas. Specifically, Materia provides to users the automatic binding of flags to data, the automatic detection of column datatypes, alignment with a time series index, the graceful management of multivariate data stored in a tabular format, and the ability to handle nonstandard, highly variable header metadata.

In addition to these features, Materia also uniquely supports tracking Quality Control methods by storing the specific functions which were used to generate flags. This level of provenance is extremely important to creating comprehensive data products. With this feature, data producers are able to not only express that a datapoint may be suspect or bad but also *under what conditions* it was found to be suspect or bad.

5   Conclusions and Future Work

Over the course of this paper we introduced a topology of modern data quality control, existing solutions and the limitations of those solutions which motivate this work. We further introduced our embedded domain specific language: Materia, and detailed the design and implementation of it. This language was not shown to be superior in terms of performance compared to its most similar counterpart on the Python platform: Pandas. However, it was argued that it should be sufficiently performant for most typical quality control use-cases. In terms of productivity, Materia was developed to be more usable than Pandas for defining quality control specific tests and functions. Additionally, it was shown that Materia aggregates many features into one language which can positively impact the quality of life for developers building quality control applications and scripts.

There are many opportunities for future development of Materia. First and foremost, with the functional structure that was implemented for this prototype, its evident that many quality control tests are embarrassingly parallel. Accordingly, this language would significantly benefit from a deep embedding which could be exported to a GPU computing language. Many of these functions map 1 to 1 with a kernel that could be deployed to individual threads on a GPU architecture. In addition to this, with a deep embedding, Materia could support a deeper and richer syntax than is currently provided in this implementation.

Acknowledgements. We thank Michelle Strout for her invaluable input throughout the development of this project and editing provided for this paper. We would also like to thank Kate Isaacs for her input editing this paper and Chase Carthen for his input on the design of this language.

References

1. Brabyn, L., et al.: Accuracy assessment of land surface temperature retrievals from landsat 7 ETM + in the dry valleys of antarctica using iButton temperature loggers and weather station data. Environ. Monit. Assess. **186**(4), 2619–2628 (2013). https://doi.org/10.1007/s10661-013-3565-9. ISSN: 1573-2959
2. Campbell, J.L., et al.: Quantity is nothing without quality: automated QA/QC for streaming environmental sensor data. BioScience **63**(7), 574–585 (2013). https://doi.org/10.1525/bio.2013.63.7.10. https://academic.oup.com/bioscience/article-lookup/doi/10.1525/bio.2013.63.7.10
3. ESIP Envirosensing Cluster. Sensor Data Quality (2019). http://wiki.esipfed.org/index.php/Sensor_Data_Quality. Accessed 28 May 2020
4. ESRI. ArcGIS October 2017. https://resources.arcgis.com/en/communities/data-reviewer/
5. Gill, A.: Domain-specific languages and code synthesis using haskell. Commun. ACM **57**(6), 42–49 (2014)
6. Gouldman, C.C., Bailey, K., Thomas, J.O.: Manual for real-time oceanographic data quality control flags. In: IOOS (2017)
7. Georgia Coastal Ecosystems LTER. GCE Data Toolbox for MATLAB (2017). http://gce-lter.marsci.uga.edu/public/im/tools/data_toolbox.htm
8. Mckinney. W.: Pandas: a Foundational python library for data analysis and statistics. In: Python High Performance Science Computer (January 2011)
9. Campbell Scientific. LoggerNET, December 2017. https://www.campbellsci.com/loggernet
10. Scully-Allison, C.: Materia. Version 0.11, May 2020. https://doi.org/10.5281/zenodo.3870396https://github.com/cscully-allison/Materia
11. Scully-Allison, C., et al.: Near real-time autonomous quality control for streaming environmental sensor data. Procedia Comput. Sci. **126**, 1656–1665 (2018)
12. Scully-Allison, C.F.: Keystone: A Streaming Data Management Model for the Environmental Sciences. PhD thesis. (2019)
13. Sheldon, W.M.: Dynamic, rule-based quality control framework for real-time sensor data. In: Proceedings of the Environmental Information Management Conference, pp. 145–150 (2008). https://lternet.edu/wp-content/uploads/2010/12/eim-2008-proceedingssmall.pdf
14. Wilkinson, M.D., et al. The FAIR guiding principles for scientific data management and stewardship. In: Scientific Data (2016). https://doi.org/10.1038/sdata.2016.18

Models for Arabic Document Quality Assessment

Adnan Yahya[✉] [iD], Afnan Ahmad, Alaa Assaf, Rawan Khater,
and Ali Salhi [iD]

Electrical and Computer Engineering Department,
Birzeit University, Birzeit, Palestine
yahya@birzeit.edu

Abstract. Digital content has been increasing rapidly. This content can be generated, accessed and used by anyone and thus the need for quality assessment of web content before usage becomes an important issue. Devising methods to assess the quality of Arabic digital content is the focus of this paper. Our work was partially based on Wikipedia articles annotated into *featured* and *good* according to quality guidelines of Wikipedia. Our analysis was directed at finding features that can serve as best quality indicators. Using the defined features, we trained a high accuracy quality assessment model using machine-learning algorithms. Our work went beyond the Wikipedia documents to build a general model that can assess the quality of Arabic documents that lack Wikipedia metadata with acceptable accuracy. The model was trained and built using features from documents we collected from Arabic online news sites and blogs, and annotated in collaboration with university students.

Keywords: Document quality assessment · Arabic Wikipedia · Arabic information retrieval

1 Introduction

Due to the diversity of web content and the ease of posting on the web, one can expect diversity in web information quality and degree of trust. One cannot give the same trust to a social media post and an article in a well-known newspaper. The need for quality assessment of web content is paramount. But this is not a trivial task: manual quality annotation does not scale, so automatic quality assessment is needed.

In our research, we worked to build a model to assess Arabic document quality, first for the domain of Arabic Wikipedia articles characterized by abundant metadata but also large quality variations [10], then for general Arabic documents, that may lack Wikipedia style metadata.

Wikipedia has its own assessment system that classifies articles into quality classes according to specific criteria using manual judgement through a peer review process. The best articles are "featured" articles [1] and after that there are the "good" articles [2] that don't meet the criteria of featured articles but still of high enough quality. Articles that didn't undergo quality review and those that were not qualified to be high quality are called here "random" articles. The documents that are already assessed as

W. Abramowicz and G. Klein (Eds.): BIS 2020 Workshops, LNBIP 394, pp. 297–310, 2020.
https://doi.org/10.1007/978-3-030-61146-0_24

high quality (featured or good) by human experts constitute a very small fraction of the Arabic Wikipedia and will serve as the annotated data to train our Machine Learning (ML) based models for quality assessment of Wikipedia articles.

After building the Wikipedia model with high accuracy based on high quality Wikipedia articles, we collected documents from several popular online blogs and news sites. These documents do not have elaborate meta-data and are not classified based on quality. We manually annotated a number of such articles and trained an ML general model to assess document quality using features available for any article. We treated the quality level as a classification problem to place the article into one of the two quality classes: *high quality* and *random* [2].

For ML models we experimented with several algorithms. Based of performance results, Support Vector Machine –SVM- (Sequential Minimal Optimization –SMO-implementation) and Random Forest (RF) were the main classifiers adopted for our models. As usual in machine learning, four measures for the classification effectiveness are used: *Precision* (P), *Recall* (R), *F-measure* (or F-score) and *Accuracy* (A) which can be read from the confusion matrix of individual experiments.

The rest of the paper is organized as follows. In Sect. 2 we discuss related work on article quality assessment, general and Arabic. In Sect. 3 we outline our ML approach to article quality assessment and the feature selection. In Sect. 4 we give the results of our experiments for Wikipedia and General articles quality assessment and the effect of feature selection on the results. In Sect. 5 we give our conclusions and point to possible directions of future research.

2   Related Work

2.1   Article Quality Assessment

In [4], Blumenstock discusses the simplest metric that can be used to classify the articles of English Wikipedia to *featured* (Wikipedia highest quality articles) and *random* articles. The model achieved 96.31% accuracy when applying binary classification on the dataset for the threshold of word count equals to 2000 words, but as expected, this method has drawbacks and can be fooled easily. Lipka and Stein [9] give a more advanced step in identifying articles by analyzing a writing style feature which is character tri-grams. Their results improve on the word count (naïve) approach. Yahya and Salhi studied Arabic Wikipedia articles quality with emphasis on features from 3 groups; textual content features, non-textual content features and features related to contributors and editors but didn't use that to build models for quality based classification using machine learning [16].

Stivilia, Twidale, Smith and Gasser present seven information quality metrics: authority, completeness, complexity, informativeness, consistency, currency and volatility [13]. They define these metrics using 19 statistical measures from both the content and metadata of articles. The experiments show that the developed model can capture big differences between featured and random articles [13]. In [14], Warncke-Wrang, Cosley and Riedl did many experiments and investigations after Stivilia [12] to come up with an actionable model for assessing Wikipedia articles quality. They worked with completeness, informativeness, number of headings, article length and number of references features. The results of this model were close to the other models with a larger number of features. In [7], De La Calzada and Dekhtyar argue that not all articles in the Wikipedia are the same and define 2 article categories: *stabilized* and *controversial* articles and use different models to measure the quality of articles in each category. The stabilized model uses measures related to the structure and construction of articles while the controversial model depends on the history log of revisions for the article. Lim, Vuong, Lauw and Sun present two quality models: the basic model and peer review model to measure the quality of articles depending on the authorities of its contributors [8]. In [5] it is argued that the quality of content of medical Web documents is affected by domain features and the authors use specific vocabulary and codes and document type for improved results. In [15] the authors attempt to improve the quality of DBpedia by analyzing features and models that can be used to evaluate the quality of articles, providing foundation for the relative quality assessment of infobox attributes. In [6] Wikipedia article quality is assessed by analyzing article content rather than the feature set and utilizes NLP deep learning to achieve the reported results. In [11] a combination of the usual features and deep learning is used for better results in Wikipedia article quality assessment. All through, some of the features used are easy to calculate while the others are more sophisticated. We tried our best to find the best combination from different models and studied the importance and effect of these features in the quality of Arabic articles. We propose new features not mentioned in the above research which we found useful in the case of general documents.

3   Our Article Quality Assessment Approach

3.1   Classification Models Using Machine Learning (ML)

The main steps used to build a classification model using machine learning algorithms (implemented using software like WEKA) are as follows:

1. An annotated dataset (classified by humans) is used to train the model.
2. Feature extraction: the training dataset must be presented as features related to the classes of the model. When the classes are related to quality (high and low), then quality related features must be extracted from each instance.
3. Then a classification algorithm (classifier) is used on the extracted features for each instance in the dataset to build a model that can classify any external instance.
4. After that the model must be tested to get the evaluation measurements about the performance and decide if the model is acceptable or needs further fine-tuning.

Figure 1 shows the basic steps in building a classification model using ML. The same steps were used in building our two quality assessment models: for Wikipedia articles and for general articles. Next we describe the two machine learning models we built.

3.2   Wikipedia Based Quality Model

We experimented with features related to the textual content like writing style, spelling errors and metadata provided for Wikipedia articles such as links, multimedia content, edits, contributors and social media effects. The best combination of features was selected to help in building a high accuracy quality assessment model.

Naïve Approach: Word Count: We started with the simplest method, referred to as the *Naïve Approach* using word count as the sole measure for quality classification [4].

We applied this method to a balanced dataset with four classes of articles "Featured", "Good", "Random for featured" and "Random for good". As seen in Table 1, we found that the average length of featured articles is 9176 words, for good articles it is 4694, for HQ class (featured and good combined) it is 7038 and for the random class it is 653 words, indicating that word count seems to be a good indicator for Wikipedia article quality. We were interested in calculating the threshold value for word count that gives the minimum classification error rate for the two class cases: *High Quality* if the word count exceeds this threshold, *Random* otherwise. We considered word count threshold levels from 1000 to 3000 and the accuracy was calculated. Figure 2 shows the results. The threshold with minimum overall classification error rate is 2100 words.

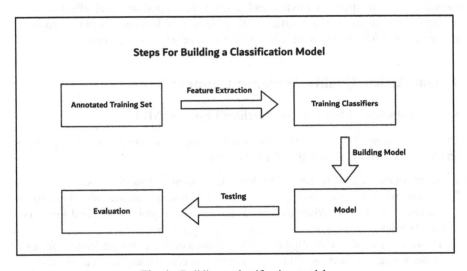

Fig. 1. Building a classification model.

Table 1. Average word count for quality class.

Class	Word count
Feature	9176
Good	4694
High quality (Featured+Good)	7038
Random for feature	749
Random for good	548
Random for high quality	653

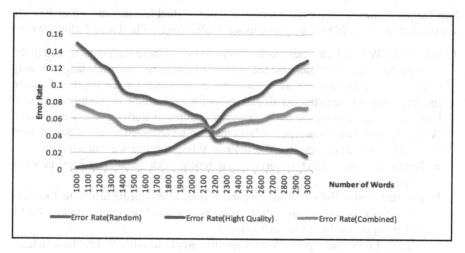

Fig. 2. Error rate for classification by word count.

The article word count was also used as a feature to train a classifier using the WEKA for the two classes High Quality vs. Random and Featured vs. Good. Many classifiers were applied in different modes of training and testing (10 folds cross validation, splitting data at 66%). As expected, the classifiers predicted class with high accuracy reaching 97.5% in the case of High Quality vs. Random. As expected, for the Featured vs Good case, the accuracy dropped to 83%.

Building our Dataset. We wrote a PHP code with Media-Wiki APIs [3] to extract 309 *featured* and 305 *good* articles from the Arabic Wikipedia. We considered two types of random articles: the first is arbitrary length (AL) random articles, the other has random articles close in length (CL) to the two high quality article subclasses (featured and good) to neutralize the effect of article length when needed. To maintain balance, we extracted 309 random articles of comparable length distribution to "featured" articles and called the set "random for featured" and another 305 of comparable length for the "good" case and called it "random for good". So, overall, we selected a total of 1228 random (of unknown quality) articles. For the AL case the selection was completely random, and for the CL case the selected articles were of comparable length and length distribution to the corresponding high quality subclass.

Analyzed Features: We studied the features in three main categories: textual features, non-textual features and contributors and editors. The following features were calculated using PHP code written with the Wikipedia APIs.

Textual Features: Textual content is the basic parameter to analyze in each article; we can use it to find specific parameters that may reflect the writing style, spelling errors and other quality features. The following textual parameters were extracted/studied: (1) Article Length in words, (2) Average Sentence Length, (3) Average Number of Words per Comma. (4) Number of Paragraphs. (5) Average Paragraph Length (in words).

Non-Textual Features: We considered several non-textual features and wrote the code to extract them from Wikipedia pages using Media-Wiki APIs. These features were:

1. Categories: Wikipedia articles can be tagged to one or more categories related to the topic of the article. We think that the number of categories an article tagged to may serve as a quality measure as it could be an indicator for the degree of article specialization, so we extracted all categories for each article in the training set.
2. Links: Links are among the most important features in studying any web content. Wikipedia itself has many types of links that can be analyzed to find if any of them has an effect on article quality. The links in Wikipedia that we extracted are:
   a. Language Links: links connecting an article with parallel articles in other languages.
   b. External Links: links that redirect users from a Wikipedia article to locations outside of Wikipedia. This feature indicates how much an article is connected to, and is supported by, other web content.
   c. Internal Links: links going from a specific article to other Wikipedia articles.
   d. Backlinks: links that come from Wikipedia articles to the Wikipedia article of interest. Back links may indicate trust in content (or authority of the article).
3. Multimedia Content: Multimedia content consists of tables, pictures, videos and sounds in articles. We conjectured that articles with multimedia content are expected to be of higher quality since multimedia elements can serve to support the article.
4. Number of High Quality Articles Related to an Article in Other Languages (OtherL_HQ): as we know, each article in Wikipedia is connected to other articles related in content but in different languages using language links. We think that if the parallel article is considered of high quality in the other language that will support the writers and help them improve the quality of the article. Therefore, the more high quality articles related to an article in other languages the better quality an article has.
5. Number of References: This parameter can be considered as one of the most important indicators of quality. More references may mean more work and effort the writer has done, and the more trust in the facts of the article.
6. Number of Shares in Social Media (Facebook, Twitter and Google+): Social Media is an essential part of today's life. Therefore, we think that the shares of an article in social media may be an indicator of quality. Code for counting the number of shares was written in PHP, using the APIs Facebook, Twitter and Google+ offer.
7. Age of Article: We looked at Wikipedia articles age as a quality indicator assuming that HQ articles must have an age that allows for modifications/improvements.

Edits and Contributors: Contributors to Wikipedia can be registered users, anonymous users or bots. Wikipedia stores all the users and their contributions to articles (known as revisions of articles) in the history log which can be a source of information about article contributors, edits and their numbers, sizes and dates. We studied the following features:

1. Number of Contributors: anonymous, registered and bot contributors, the latter being programs that perform specific actions to articles such as spelling correction.

2. Average Edit Size: articles average edit size was computed from the history logs.
3. Number of High Quality Articles Related to an Article by the Authors (NHQAA): We know that the writer's expertise can reflect strongly on the quality of his/her writing. The more high quality content an author publishes the more expertise he/she has. We took all the contributors to featured and good articles and for each author we counted the number of high quality articles he/she has contributed to, so we got a list that has all the authors and the number of high quality articles each author contributed to. Then we computed the sum of the number of high quality articles related to an article by its authors. This was implemented using PHP and Media-Wiki APIs.

3.3 General Article Quality Model

After building the Wikipedia based model, we moved to building a model to assess the quality of general text documents. For that we needed a dataset with articles from different sources to neutralize the effect of Wikipedia authoring guidelines, and limited features to those related to textual content alone so that the results are applicable to general web documents. Next we describe the dataset we collected and the features we studied and analyzed for use in the general article model.

General Dataset Collection: We collected articles in modern standard Arabic from popular news sites (https://www.alquds.co.uk/, https://www.aljazeera.net/, https://www.egyptwindow.net/, http://www.alriyadh.com/, https://elaph.com/ and prominent bloggers with large enough article sizes (http://ahmedjedou.blogspot.com/, https://www.essamalzamel.com/). The average word count was about 1050 words per document. Then we set to label the collected articles manually by volunteer university students. Each article had at least three assessments. We used a threshold for the average evaluation score to get a balanced dataset of 112 articles labeled as High Quality (HQ) and 113 articles labeled as Low Quality (Random).

Analyzed Features: The features we analyzed for the new dataset articles are:

Features Used in the Wikipedia Model: We used the same textual features used in the Wikipedia model available for general documents.

Other Features: The other features we used for the new model are:

1. Title Relation with Article Body: This feature describes the degree of overlap between the title and the body of an article. We calculated it by counting the number of occurrences of the title words in the article body.
2. Part of Speech (POS) Tagging: Each word in Arabic sentence can be classified to Verb, Noun, Pronoun, Adverb, Adjective and Numbers or others describing the word part of speech. To tag article words we used Stanford POS Tagger[12], which is a widely used open source tagger. We fed the sentences of each article and got the tag of each word. Then we computed the percentage of the occurrence of each tag in the article and considered them as attributes.

4 Results and Discussion

In this section, we present an analysis of the features studied. We also discuss the results of applying machine learning models using WEKA on features mentioned in the previous section for both Wikipedia and general articles models.

4.1 Wikipedia Model Features and Classification Results

First we present the results of studying the features mentioned in the previous chapter for the Arabic Wikipedia articles, and then we talk about the classification results for the quality assessment model for Arabic Wikipedia articles.

Results for Textual Features. Table 2 contains the results of analyzing the textual content features which are: word count, number of paragraphs, paragraph length, number of words per comma and sentence length. The average word count for the HQ class was 4914 words, for the random with CL to high quality articles the word count was 4889 words as expected but for the real random articles (any length) the average word count was 1261. This reflects the results of the naïve approach discussed earlier.

For the number of paragraphs and paragraph length, the result is distinguishable between the HQ class (with 57 paragraphs and 85 words) and the random of any length class (with 19 paragraphs and 68 words), but averages with random close in length are very close to high quality (with 57 paragraphs and 86 words per paragraph).

Table 2. Select features for wikipedia high quality, random of comparable-length (CL) and random any-length (AL).

Quality class → Feature↓	High quality (HQ)	Random with close length (CL)	Random with any length (AL)
Word count	4914	4889.00	1261.00
Paragraphs	57.46	56.99	18.62
Paragraph length	85.52	85.80	67.74
Words/Comma	18.36	22.88	23.15
Sentence length	26.85	26.26	25.84
Language links	60.34	66.70	19.47
Categories	13.13	10.62	6.74
External links	70.70	33.28	6.40
Internal links	382.00	271	93.00
Backlinks	459.00	394	118.00
Multimedia elements	25.24	12.37	3.63
References	126.20	50.11	7.76
OtherL_HQ	2.32	1.65	0.27

(*continued*)

Table 2. (*continued*)

Quality class → Feature↓	High quality (HQ)	Random with close length (CL)	Random with any length (AL)
Shares of article	47.20	71.67	8.32
Age of article	2587.00	2742.00	1856.00
Bot contributors	26.39	26.82	21.98
Registered contributors	40.95	42.32	29.01
Anonymous Contributors	46.03	58.72	26.95
Bot edits	82.44	82.85	61.04
Registered users edits	26.39	26.82	21.98
Anonymous users edits	40.95	42.32	29.01
Edit size (Bytes)	46.03	58.72	26.95
NHQAA	3154.00	2435.00	738.00

The high quality class has a lower number of words per comma (18) compared to the two random classes (23). Sentence length seems not distinguishable between the three classes with HQ having 27 words and the two random classes with 26 words.

For the case of HQ and CL random articles the results of text analysis are close, probably reflecting a common style of writing for Wikipedia enforced by editors or it may indicate that the effort used to write long articles is close to the effort used to write high quality articles. However, in the case of random articles of any length the results show differences in some features.

Results for Non-Textual Features: Table 2 also contains the results for analyzing the non-textual features which are: References, Number of High Quality Articles Related to an Article in Other Languages (OtherL_HQ), Shares, and Age of Article.

From the results we can see that the average number of language links for high quality articles is 60 and for random with close length (CL) is 66 while the random with any length (AL) has only 19 links.

Average number of categories for high quality is the highest with 13 categories while the two random classes score smaller results with 11 categories for random with close length and 7 for random with any length.

For external links, the high quality class has an average of 71 links while the random with close length has 33 links and the random with any length has only 6 links. For internal links, the high quality class has an average of 382 links while the random CL has 271 links and the random with any length has only 94 links. For backlinks, high quality class has an average of 459 links while the random with close length has 394 links and the random with any length has only 118 links. For multimedia elements, the HQ class has on average 25 elements while the random CL has 12 elements and the random with any length has only 4 elements.

For the number of references, we can see that the average number of references for HQ is 126.2 and for random with CL it is 50.11 while the random with AL it is only 7.76.

The average for the NHQAA for HQ is 2.32 and for random with close length 1.65 while the random with any length is only 0.27.

The average number of (social media) Shares for high quality articles is 47.24 and for random with close length 71.67 while the random with any length it is only 8.32.

The average age of articles for high quality is 2587 days and for random with CL 2742 days while the random with AL it is only 1856 days.

In summary, internal, external, backlinks, multimedia, references, OtherL_HQ and number of shares are different for high quality class and the two random classes. The features can be used as strong quality indicators.

Results for Edits and Contributors: The last part of Table 2 contains the results of calculating the averages for contributors, edits, edit size and Number of High Quality Articles Related to an Article by the Authors (NHQAA).

From Table 2, we can see that the average number of Bot contributors for high quality is 26 and for random with close length 27 while the random with any length is only 11. The average number of registered contributors for HQ is 41 and for random with close length is 42 while the random with any length is only 11.

The average number of anonymous contributors for high quality articles is 46 and for random with close length is 57 while the random with any length is only 12. The number of contributors for the random with close length is very close to the HQ class with a little difference in the number of anonymous contributors. Random with any length articles have a very different average, which may be directly related to the length of articles.

For bots edits, HQ class has an average of 82 edits while the random with close length has 83 edits and the random with any length has only 24 edits. For registered edits, HQ class has an average of 296 edits while the random with close length has 188 edits and the random with any length has only 38 edits. Regarding Edit Size, the HQ class has an average of 761 bytes edit while the random with close length has 781 bytes and the random with any length has only 297 bytes.

For Anonymous users edits, the HQ class has on average 72 edits while the random with close length has 94 edits and the random with any length has only 19 edits. The average for NHQAA for HQ is 3154 and for random with close length 2435 while the random with any length is only 738. This may mean that this feature, an indicator of writing quality of the authors, can play a big role as a quality indicator of articles. As seen from the comparison of edits averages between the random with close length and the HQ class, edits from bots are on average the same while edits from registered and anonymous users show differences. We can note that high quality articles have a higher number of edits from registered users compared with the two random classes.

To conclude our analysis, we can see clearly that the number of anonymous contributors, the edit size and the NHQAA are the main features with big differences between high quality and random articles and can thus be considered the stronger quality indicators compared to the other features.

Wikipedia Model WEKA Results: We trained the model with the extracted features, did many experiments with different feature combinations and different classifiers. Using 10-fold cross validation to evaluate the performance and Random Forest (RF) Classifier with default options with textual features only, the resulting model for the High Quality and Random with close length gave an accuracy of 68.5%, reflecting neutralization of article length. With the non-textual and the editors features the accuracy increased to 88.3%, a 20% improvement. The number of references played the main role in that (the gain was 13% while for the other features it was 5-9% only). When using SMO with normalized kernel with all features we reached an accuracy of 89.8%, which slightly better than the result for RF.

After many experiments with many combinations and classifiers, we reached the models with the best performance as shown in Table 3.

Table 3. Best models performance

Model	Close-length model		Any-length model	
Classifier ⊕	RF*	SMO**	R F*	SMO**
Precision	0.888	0.900	0.956	0.955
Recall	0.884	0.898	0.955	0.954
F-Measure	0.883	0.898	0.955	0.954
Accuracy	0.884	0.898	0.955	0.954

*Random Forest
** Normalized Poly Kernel 9 Exponent

In the first Wikipedia model (High Quality with Random) we notice that the values for precision, recall, F-measure and accuracy using SMO classifier (0.9, 0.898, 0.898 and 89.82%, respectively) are slightly higher than those for Random Forest. In the second Wikipedia model (High Quality with AL Random) the values for precision, recall, F-measure and accuracy measurements using Random Forest classifier (0.956, 0.955, 0. 955 and 0.955, respectively) are slightly higher than for the SMO. We ran our model on the new Wikipedia high quality articles listed as *featured* and *good* in the four months following the original dataset collection (we found about 50 such articles). We used these 50 articles to test the model and it classified 43 of the 50 instances as high quality, consistent with the accuracy using cross validation.

4.2   General Model Features and Classification Results

Next, we present feature analysis and results for the general, non-Wikipedia model. Table 4 contains the results for textual features and POS tagging. The average word count for the HQ class is 949 words Vs 1172 for the random, the sentence count average is 18.7 words for HQ Vs 34.2 for Random, while the sentence length for HQ is 144 words Vs 42 for Random.

Comma count is 57.6 for HQ and 69.5 for Random. Average number of words per comma in the HQ class is 52.1, Vs 30.1 in the Random class.

Number of shares in the HQ is 232 Vs 134 in Random. The relation between title and content is 40.1 for the HQ Vs 65.6 for random. Table 5 shows POS tagging results.

Table 4. Textual features for the general model

Class	HQ	Random
Word count	949.0	1172.0
Sentence count	18.7	34.2
Sentence length	144.0	42.0
Comma count	57.6	69.5
Words per comma	52.1	30.1
Shares	232.0	134.0
Title/Content	40.1	65.6

Table 5. POS tags relative frequencies

Class (POS)	HQ	Random
Nouns	0.6147	0.6071
Adjective	0.1286	0.1219
Adverb	0.0052	0.0056
Verb	0.1065	0.1137
Pronouns	0.0224	0.0244
Numbers and Others	0.0304	0.0361
Harf Jar	0.0921	0.0913

General Model WEKA Results: When we trained the general model using Random Forest classifier with default options and 10- fold cross validation we noticed that the accuracy for the model with textual features was only 74.5%. When the title/content attribute was inserted the accuracy improved by 5.5% to reach 80%. When we added different combinations of tags we found that the nouns, adjectives, pronouns, numbers had the most effect. The accuracy for the model with these features reached 84.5%, an improvement of 4.5%. SMO classifier performed worse. Table 6 shows a summary of WEKA results for the General Model. After that, we tested our model with 25 external articles from newspapers and blogs (13 High and 12 Low Quality). The model classified them with 80% accuracy. We finally tested the general model on Wikipedia articles using 100 articles (50 High Quality and 50 Random) which resulted in The model had 78% accuracy for this testing case.

Table 6. WEKA results for the general model

Classifier	Precision	Recall	F-Measure	Accuracy %
RF	0.845	0.845	0.845	84.5
SMO	0.727	0.725	0.724	72.5

5   Conclusions and Future Work

We developed models for Arabic article quality assessment in the presence and absence of extensive Wikipedia-style metadata. The basic limitation we had from the beginning of our work was the scarcity of annotated articles for training. The combination of Wikipedia articles and own annotated articles helped solve this problem. We succeeded in building a good model to classify Wikipedia and general articles based on textual properties and Wikipedia metadata when available. The idea of determining the quality of arbitrary Arabic articles can help a lot in many fields.

One can also investigate the application of our work to different types of web data: dialectal or mixed texts, shorter posts like tweets and other social media posts. One can look into many other features as potential quality indicators like the spelling errors, Arabic writing patterns, foreign language content, the use of dialect in writing, reference quality, mentions in more academic and general social media applications, document recovery, PageRank, site/author trust, Wikipedia infobox properties and access patterns. DL may also be a promising technology for quality assessment of Arabic articles.

The quality of shorter Arabic web content like tweets, comments and social media posts may be another interesting research issue to tackle.

References

1. Featured Articles, Arabic Wikipedia 2020 .3 .7 ويكيبيديا:مقالات مختارة. http://ar.wikipedia.org/wiki/ ويكيبيديا:مقالات_مختارة, Accessed 1 Nov 2014
2. Good Articles, Arabic Wikipedia 2020 .3 .7 ويكيبيديا:مقالات جيدة. http://ar.wikipedia.org/wiki/ ويكيبيديا:مقالات_جيدة, Accessed 1 Nov 2014
3. MediaWiki API: Main_page 12. 3. 2020. http://www.mediawiki.org/wiki/API:Main_page. Accessed: 2 May 2014
4. Blumenstock. J.: Size matters: word count as a measure of quality of wikipedia. In: Proceedings of the 17th International Conference on WWW, pp. 1095–1096. ACM (2008)
5. Cozza, V., Petrocchi, M., Spognardi, A.: A matter of words: NLP for quality evaluation of wikipedia medical articles. In: Bozzon, A., Cudre-Maroux, P., Pautasso, C. (eds.) ICWE 2016. LNCS, vol. 9671, pp. 448–456. Springer, Cham (2016). https://doi.org/10.1007/978-3-319-38791-8_31
6. Dang, Q., Ignat, C.: Quality assessment of wikipedia articles without feature engineering. In: Proceedings of the 16th ACM/IEEE-CS Joint Conference on Digital Libraries, pp. 27–30 June 2016, Newark, NJ (2016)
7. De La Calzada. G., Dekhtyar A.: On measuring the quality of Wikipedia articles. In: Proceedings of the 4th workshop on Information credibility, pp. 11–18. ACM (2010)
8. Lim, E., Vuong, B., Lauw, H., Sun A.: Measuring qualities of articles contributed by online communities. In: Proceedings of IEEE/WIC/ACM International Conference on Web Intelligence (WI 2006), Hong Kong, pp. 81–87 (2006)
9. Lipka, N., Benno, S.: Identifying featured articles in wikipedia: writing style matters. In: Proceedings of the 19th international conference on wwww, pp. 1147–1148. ACM (2010)
10. Safa, A.: Enhancing quality arabic web content through cross lingual information retrieval methods. Master Thesis, Birzeit University, Palestine (2019)
11. Shen, A., Qi, J., Baldwin, T.: A hybrid model for quality assessment of wikipedia articles. In: Proceedings of Australasian Language Technology Association Workshop, pp. 43–52 (2017)
12. Stanford Log-linear Part-Of-Speech Tagger. http://nlp.stanford.edu/software/tagger.shtml
13. Stvilia, B., Twidale, M., Smith, L., Gasser, L.: Assessing information quality of a community-based encyclopedia. In: Proceedings of the International Conference on Information Quality-ICIQ, Cambridge, MA, USA, 10–12 November, pp. 442-454 (2005)
14. Warncke-Wang, M., Cosley, D., Riedl, J.: Tell me more: an actionable quality model for wikipedia. In: Proceedings of the 9th International Symposium on Open Collaboration, ACM (2013)

15. Węcel, K., Lewoniewski, W.: Modelling the quality of attributes in wikipedia infoboxes. In: Abramowicz, W. (ed.) BIS 2015. LNBIP, vol. 228, pp. 308–320. Springer, Cham (2015). https://doi.org/10.1007/978-3-319-26762-3_27
16. Yahya, A., Salhi, A.: Quality assessment of Arabic web content: the case of Arabic wikipedia. In: 10th International Conference on Innovations in Information Technology (INNOVATIONS-2014), pp. 36–41. IEEE (2014)

Open Data Quality Dimensions and Metrics: State of the Art and Applied Use Cases

Soumaya Ben Hassine[1,3](✉) and Delphine Clément[2,3](✉)

[1] Covéa, Sèvres, France
soumayabh@gmail.com
[2] Microsoft, Paris, France
delphine.clement@microsoft.com
[3] ExQI, Paris, France

Abstract. While the economic benefit of open data is undeniable, its use as an asset in industrial processes is still a challenge. The lack of quality is indeed a typical argument for not leveraging open data. In fact, per the Data Office of the French government (ETALAB) in charge among others of the French open data initiative, only 9 out of the 34822 open datasets are tagged as *reference* datasets, that is supplied by certified publishers and which content can be reliable to be shared broadly, privately and publicly. Yet, no actual quality indicators are provided along with the metadata catalog of these 9 files.

What would then be the appropriate indicators for open data quality assessment, how would they differ from those used to assess DQ of traditional enterprise data? How can they be measured knowing the multiple reusability scenarios and how can they help users choose the datasets that best fit the purpose?

In this work-in-progress paper, we will answer these open data quality indicators questions and illustrate it with some case studies from the industry.

Keywords: Open data catalog · Linked open data · Reference data · Open data quality indicators · Open meta data quality indicators

1 Introduction

The Open data movement has evolved since the acceleration of its worldwide democratization in 2009, with Obamas's Open Government Initiative. In fact, its definition went from the need for "public institutions' data to be digitized, shared, accessible and freely usable by everyone" to include all sources of data (Governments, businesses, public, private, individual, etc.).

Since then, Europe, and specifically France, have reached a real maturity in open data transformation. To date, 15 to 20% of the open data activity in France is driven by the private sector [1] which is opening its data for reasons of innovation, research and development, transparency over its supply chain, brand or sustainability.

According to the latest publication of the economic impact of open data by the European Data Portal as an initiative of the European Commission [2], the private sectors

© Springer Nature Switzerland AG 2020
W. Abramowicz and G. Klein (Eds.): BIS 2020 Workshops, LNBIP 394, pp. 311–323, 2020.
https://doi.org/10.1007/978-3-030-61146-0_25

which demonstrates the highest potential of growth, 15% and more, thanks to open data, relate to the domain of agriculture, wholesale and retail trade, transportation, information and communication, finance and insurance and real estate. The assessment was made considering the level of digitization and data demand of the sector, the opportunity of open data supply to meet the demand and the economic impact of the potential. To this matter, three open data, open platform initiatives at the European level underline this trend well. In the area of agriculture, the "JoinData" cooperative [3], started as soon as 2013, allows farmers to centralize and share data coming from sensors amongst themselves and with companies from the agro industry. In the area of oil and gas, the "open industrial data" [4] project opens live operational data stream of a Norwegian oil and gas exploration company with the aim to create a digital representation of the industrial reality of one of its compressor in the North Sea. Last, in manufacturing, Microsoft and BMW launched in April 2019 an "open manufacturing platform" [5] with open industrial standards and open data model. The goal of this open initiative is to accelerate IoT developments and industry 4.0. solutions. The open data on this platform will be accessible to and re-used by an ecosystem of suppliers, partners, OEMs and other companies.

However, while the open datasets flock and the platforms that publish them proliferate, their usability is stifled by the lack of quality of the majority of these datasets. As in [6], even if gathering, opening, sharing and publishing such massive amounts of data is certainly a step in the right direction, data is only as useful as its quality.

In this work-in-progress paper, we will detail in Sect. 2 the state of the art of the open data quality indicators by interviewing the representatives of two of the major open data actors in France (ETALAB and OpenDataSoft). In Sect. 3, we will extend our study to a literature overview on the web data and specifically on the web linked data quality indicators. We will, then, detail in Sect. 4 some of the field feedbacks regarding the most important quality indicators and conclude in Sect. 5 by a proposal of a method for open data quality assessment.

2 Open Data in France: Actors and Portals

The Open Data movement has been favored, in France, by a set of official institutions (such as public institutions and local communities, ETALAB [7], the APIE [8]) as well as numerous associations (e.g. LiberTIC [9], FING [10]), activists (e.g. RegardsCitoyens [11]) and startups (e.g. Data Publica [12], OpenDataSoft [13]).

The maturity of the Open Data movement is such, today, that it has become less about evangelization since the topic is known, the benefits are understood, the best practices are disseminated and the how to get there is no longer new. Open data today is more opportunistic and leveraged as a catalyst to drive the understanding and the resolution of the hot topics of the moment, such as sustainability or obviously today COVID-19. In this respect, the coverage and level of quality of datasets in those areas are key indicators. For instance, on sensitive topics such as the one of the COVID-19 crisis, access to raw data rather than aggregate statistics is essential for trust and accuracy.

Even though in the space of government open data there is no obligation for calculating and publishing data quality metrics, there is a new open data trend in France in the commercial sector space where companies, for benefits of compliance, ethics or brand image, want to open their company data, which will require the quality of their open data to be certified. The obligation for data quality indicators of the open data may therefore come from this trend.

We chose to detail in this state of the art the contribution of ETALAB and OpenData-Soft, as they hold the two most popular Open Data platforms in France, respectively: *data.gouv.fr* and *data.opendatasoft.com*. We went through an interview of the leaders of these two institutions and asked about how they were dealing about open data quality in their respective platforms.

2.1 ETALAB: The Data Office of the French Government

ETALAB, the Data Office of the French government, is in charge of setting up and deploying the government's digital strategy that supports innovation, transparency and business effectiveness and opportunities. In fact, in 2005, the government issued a bill on open access to administrative documents and the re-use of public sector information that highlights 3 major provisions [14]:

- extending the scope of administrative documents that are freely available on the internet,
- extending the public sector information provided or disseminated may be freely re-used for purposes other than that of the public service mandate for which it was presented or received,
- and expanding the open data policy to include public and private entities, public service concession holders or entities whose activities are subsidized by the public authorities, and by providing streamlined access for Insee (National Institute of Statistics and Economic Studies) to some private databases for the purpose of mandatory statistical surveys.

Reference Datasets. One of the missions of ETALAB is the governance of the *Data Public Service*, a program that aims to publish reference datasets, that are public datasets with major social and economic impacts. These datasets should meet the critical need of French companies, administrations and government agencies of the availability of highly qualified sets of information. Yet, on the data quality front, ETALAB, does not aim to substitute itself neither to the publishers nor the re-users of the dataset. The publishers are required to ensure the quality of the metadata of the dataset they publish and the re-users are expected to define their own optimum data quality level suited for their use case. In fact, ETALAB acts as a facilitator between both. Indeed, there can be numerous different data usages from the same data set and the expectations in terms of level of quality could greatly vary; for some re-users, incomplete data can actually be better than no data at all. It is therefore difficult to define one single set of data quality indicators outside of any re-usability context. Having said that, each French government open dataset is published with some objective indicators around the popularity of the dataset through the number of downloads, the freshness (frequency of update), the geographical granularity of the data and the interpretability of the dataset through the completeness of the metadata. To complement those objective indicators, the re-user of the dataset can also provide feedback to the publishers around the quality of the data at usage. The feedback observed by ETALAB as the facilitator hub between the publishers and the re-users of the dataset aggregates around availability of the dataset, obsolescence of the dataset and data formatting issues. Unfortunately, ETALAB does not automatically mine this crowdsourced feedback around the quality of the data at re-usage to produce statistical recommendations to the dataset publishers or drive an open data quality improvement strategy.

Table 1 summarizes 3 quality indicators that are particularly tracked.

Table 1. Quality indicators used for reference datasets.

Quality indicator	Definition
Nb of downloads	Number of downloads of the file since its release
Nb of users feedbacks	Number of comments
Completeness of metadata description	Percentage of filled out metadata values

We note that it has to deal with the dataset's metadata quality rather than the data itself. Moreover, the two first indicators that help determine the file's reputation and its relevancy, thus its visibility and reusability, can only be assessed once the dataset is released. Therefore, the eventual bad quality of a file wouldn't be noticed beforehand.

Hereafter in Table 2 the mandatory metadata that have to be supplied along with the dataset according to the ETALAB standard.

Table 2. Mandatory metadata as per ETALAB.

Metadata	Definition
Title	The title of your dataset should be specific and the most precise possible. It needs to be searchable through search strings by re-users
Description	The description of your dataset allows potential re-users to get information on the content and the structure of the published data, the context related to its production, the contact references of the publishers, etc. The dataset description is typically the first thing that potential re-users read when they discover your dataset
Update frequency	Update frequency of the file

Source: [15]

Additionally, optional metadata could be added in order to help understand the file's content. Such metadata are: Keywords, Timestamp, License, Themes, Language, Modified, Data processed, Metadata processed, Publisher, Number of records, Size of records in the dataset (in bytes), Number of use cases that have been derived from the dataset, API call count, Download count, Attachments download count, File fields download count, and Popularity Score.

2.2 OpenDataSoft, the Leading Data Sharing Platform in France

OpenDataSoft (ODS) is the most popular Open Data aggregator in France. Beyond developing data sharing portals for French enterprises and government institutions, ODS holds one of the most famous open data portals in France (https://data.opendatasoft.com/pages/home/) with 20 456 datasets, 106 million downloads and 2.50 billion API calls.

So how does ODS assess the quality of the datasets it publishes? When interviewing the co-founder and CEO of the company, the answer was, just as ETALAB does, about the metadata quality. In fact, the most important quality criteria are the availability and accessibility of the dataset and the completeness of the metadata which can be regrouped into 3 classes [16]:

- **Standard metadata** *(mandatory)*: basic metadata displayed in the front office for the users. For example, *standard* metadata include the dataset title, the underlying license, etc.
- **Interoperability metadata** *(optional)*: standard used for metadata format (e.g. DCAT, INSPIRE), intended for automatic usage by other systems for interoperability purposes, or for regulatory compliance.
- Extra metadata

  - **Admin metadata** *(optional)*: intended for administrators usage only, therefore never displayed anywhere in the front office.
  - **Applicative metadata** *(optional)*: intended for specific applications, and not expected to be used by users directly. In most cases, the users will never have to access or edit them by themselves. They can be visible from the front office but also through the Opendatasoft Search API. For example

In fact, metadata management actually provides critical capabilities to leverage metadata which contributes to assessing the dataset's value for the benefit of the re-users [17].

3 Quality Dimensions for Linked and Open Data

While in the previous section we highlighted the re-usability of Open Data datasets as their major and fundamental characteristic, it is worth noting that these datasets are first and foremost web data and should, therefore, be subject to the same constraints and requirements as web and linked data in terms of quality expectations.

3.1 Quality Dimensions to Assess Linked Data

A comprehensive and exhaustive survey performed in [6] in the context of semantic web and linked data shows that data quality is supported by metadata related quality dimensions (licensing, interlinking, security, performance, trustworthiness, versatility) in addition to the traditional intrinsic data quality dimensions (accuracy, timeliness, consistency, completeness). These aforementioned metadata related metrics should help tackle the openness of the web data, the diversity of the information and the unbounded, dynamic, and sometimes as yet unknown, set of autonomous data sources and publishers behind these datasets. It is therefore important if not mandatory for that metadata to be of the highest quality, thus complete, fresh and accurate.

In this study 30 papers, 18 dimensions, 69 metrics and 12 quality assessment tools have been analyzed [6].

It is worth noting that some of these dimensions may partially overlap, such as Security and Trustworthiness, as the dimensions have been compiled from a bench of different papers.

3.2 An Application: The Quality of the Legal Entity Identifier (LEI)

The Global Legal Entity Identifier Foundation (GLEIF) was established in 2014 by the Financial Stability Board with the objective to create transparency in derivatives market post the last financial crisis. GLEIF is a supranational non-for-profit organization which facilitates the implementation and use of the Legal Entity Identifier (LEI) which is an open, standardized and high-quality legal entity reference data. GLEIF has endorsed the principles of the international open data charter [18]. Today, in the US and Europe, regulations require the use of LEIs to uniquely identify counterparties to transactions and thus LEI helps evaluate and manage risks, minimize market abuse and improve the accuracy of financial data.

Given this critical economic and financial usage and to the contrary of government open data initiatives for which context of re-use is multiple and unknown at the time of publishing, it is essential that LEI is considered as an industry standard, as an open and reliable data for unique company identification worldwide. This is why GLEIF runs a data quality management program and publishes monthly the "Global LEI Data Quality Report" [19]. The measurement of the quality of the legal entity identifier (LEI) is available to all re-users, current and potential. Below is the list of all Data Quality Dimensions used to calculate the monthly LEI total data quality score (Table 3).

Table 3. Data quality dimensions used by GLEIF data quality management program to measure the quality of Legal Entity Identifier (LEI) open data [20]

Quality dimensions	Definition
Accessibility	Data items that are easily obtainable and legal to access with strong protections and controls built into the process
Accuracy	The extent to which the data are free of identifiable errors; the degree of Conformity of a data element or a data set to an authoritative source that is deemed to be correct; and the degree to which the data correctly represents the truth about real-world objects
Completeness	The degree to which all required occurrences of data are populated

(continued)

Table 3. (*continued*)

Quality dimensions	Definition
Comprehensiveness	All required data items are included - ensures that the entire scope of the data is collected with intentional limitations documented
Consistency	The degree to which a unique piece of data holds the same value across multiple data sets
Currency	The extent to which data is up-to-date; a data value is up-to-date if it is current for a specific point in time, and it is outdated if it was current at a preceding time but incorrect at a later time
Integrity	The degree of conformity to defined data relationship rules (e.g. primary/foreign key referential integrity)
Provenance	History or pedigree of a property value
Representation	The characteristic of data quality that addresses the format, pattern, legibility, and usefulness of data for its intended use
Timeliness	The degree to which data is available when it is required
Uniqueness	The extent to which all distinct values of a data element appear only once
Validity	The measure of how a data value conforms to its domain value set (i.e. a set of allowable values or range of values)

We note that a bench of these quality dimensions focus on the data quality, few on the metadata quality (Accessibiliy and Provenance) unlike what was shown in the first subsection if this chapter. This is not surprising, as we have to deal here with an identifier quality assessment (the LEI) and a unique underlying dataset.

4 Case Studies: Applications from the Field

4.1 Quality Dimensions to Assess the Quality of the Company's External Data Catalog

This case study is about a project that has been held at a French insurance company in 2017 which goal was, inter alia, to promote and democratize the usage of external datasets, and specifically of Open Data throughout the company in order to enrich its knowledge about its customers as well as to maximize the risk control.

The project consists in developing a unique data catalog to store, centralize and share external datasets (brokers' files and Open Data datasets) that are used within the company's different departments and which feed several usages:statistics, risk prevention, data science, marketing, etc.

While brokers' datasets are of a good enough (because stable) quality with respect to the use cases that were planned for, open data represent an unexplored, new corpus of information of unknown provenance and quality.

Assessing the quality of such datasets is a challenge itself. In fact, data quality should, by definition, be "fit for use", whereas this evaluation exercise should be performed whatever the usage. Given the literature review above, we put in place the following ten dimensions that include the traditional quality dimensions to which we added contextual indicators such as the trustworthiness, the auditability, the visibility, the interlinking related to open data and multisource information (Table 4).

Table 4. Quality dimensions.

Quality dimension	Definition	Related metric
Auditability (new)	The dataset content is certified or secured by an agreement or a contract	The publisher is a reference dataset or an already known publisher for the company
Trustworthiness	Related to the provenance of the file, its publisher	A dataset is trustworthy when it comes from a certified public service* or when the publisher is highly popular (i.e. has a high number of followers or the publisher datasets have been downloaded massively)
Relevancy/Usability	Relevancy of the content	This indicator is estimated given the users feedbacks, so can only be determined post its release
Accessibility	Assesses the availability of the dataset	A dataset is accessible if - it is readable by a machine or a programming language (example: a delimited file) - it has an open license (for access and re-usability): ETALAB, ODbL - it has been acquired (bought or rented) - the download link is correct
Understandability	The content of the dataset is understandable for the end user either through a file description or a data model that details all the fields/columns and their meanings	If a data model (fields description) is supplied with the dataset
Visibility (new)	The dataset is easily identifiable	If the file content is described with a set of keywords, tags or a theme description

(continued)

Table 4. (*continued*)

Quality dimension	Definition	Related metric
Timeliness	The freshness of the file with respect to the use case needs	We distinguish 2 types of timeliness: - contextual/relative timeliness: timeliness with respect to the use case (for example: analyzing the postal addresses of 2011) - absolute timeliness: freshness of the file with respect to the current date
Coverage	The coverage of the file content with respect to its description (ex. file name on France with a departmental content)	Spatial/geographic coverage: country level vs department (or region) level
Uniqueness	Assesses the unique representativeness of a real world entity	Percentage of unique entities in the data file
Interlinking (new metric)	Data are easily "linkable" to other external datasets (i.e. measures the file capability of being enriched)	Number of potential key fields in the dataset (example: postal codes, SIRET code, etc.)

We note that, aside from the Uniqueness, all the quality indicators are related to the dataset's metadata. Moreover, two new dimensions, proper to the context of Open Data datasets quality assessment, are identified: Auditability and Visibility. Finally, the use case depicts a new metric to quantify the Interlinking dimension, more suitable to the company's initial goal to enrich internal datasets with external knowledge.

4.2 Linked Web Data Usage at Microsoft

The second case study is about mining the Bing Satori knowledge graph to compute corporate data structure recommendations. Company data from various open and proprietary web sources is ingested on CARD (Company Aggregated Relationship Data) master graph. The data ingestion step maps edges to graph ontology (e.g. <Tata motors: child of: Tata Group>, <Tata motors: industry: Automotive manufacturing>, etc.,).

Companies have properties like location, business operation, industry, services, etc., locations have next level properties like city, subdivision, country etc., similarly, the relationships have properties like start & end date of an event (merger, acquisition, partnership, joint venture, etc.,). If we consider joint ventures, it has its own set of next level properties like percentage of ownership, event date, etc. Additionally, information extracted through web structure mining have properties like number of hyperlinks (bond) between domains, date of extraction, etc. The data ingestion acts as a data quality gateway that ensures any record which enters the graph frames meet minimum metadata completeness requirements. Once the data is ingested to the company mastered graph CARD, it goes through conflation and gets optimized through complex network/graph techniques such as Triadic closure, path analysis, graph clustering and community detection. Those optimization techniques correspond to the interlinking linked data quality dimension as described in [6]. Once company mastered graph data is optimized, it becomes accessible to business processors and data stewards. In the data visualization layer, Company Structure recommendations are organized for intelligent and efficient review by the data steward from the highest probable recommendations which are based on a lower number of hops (shortest graph path between nodes), the highest number of internal transactions or highest business value tied to each recommendation to the least probable recommendations based on legal statuses (franchises, dealerships), low percentage of ownerships or higher number of hops (longer graph path between nodes). The goal is to provide data stewards with company context and as much relevant metadata as possible in the experience review so that the recommendation validation process is productive and does not require further additional manual research on the web.

This project concludes that working upfront on the quality of the web metadata at scale, in this case the links between corporate structure recommendations, helps provide trust into the data that is presented to data stewards for final manual review. The data stewards are human-in-the-quality-loop and use the metadata inherited from graph data ingestion and conflation to prioritize their backlog and to certify that the recommendations are good quality data (Table 5).

Table 5. Definition of the interlinking dimension, metrics associated and example of usage as part of the linked web data use case at Microsoft.

Dimension	Metrics	Description	Implementation as part of the Microsoft use case
Interlinking (refers to to the degree to which entities that represent the same concept are linked to each other, be it within or between two or more data sources.)	(1) detection of good quality inter-links	(i) detection of (a) interlinking degree, (b) clustering coefficient, (c) centrality, (d) open sameAs chains and (e) description richness through sameAs by using network measures, (ii) via crowdsourcing	Triadic closure, path analysis, graph clustering, community detection, centrality ("betweeness centrality"- number of pages reached through a link [shortest paths] and "closeness centrality" - how many pages are reachable in minimum number of hops from a page)

(*continued*)

Table 5. (*continued*)

Dimension	Metrics	Description	Implementation as part of the Microsoft use case
	(2) existence of links to external dataproviders	detection of the existence and usage of external URIs (e.g. using owl:sameAslinks)	
	(3) dereferenced back-links	detection of all local in-links or back-links: all triples from dataset that have the resource's URI as the object	Removal of dangling pages

5 Proposal and Perspectives

In order to start to make a proposal for an open data quality assessment method, we are overlaying the data quality dimensions referenced in the literature in [6] (Sect. 3.1) and used as part of the three applied cases described in this paper; the legal entity identifier (Sect. 3.2), the external data catalog (Sect. 4.1) and the company knowledge graph CARD (Sect. 4.2). When we come across a data quality dimension that was used by 3 overlaid elements or more, we are considering it as a relevant one to go to our proposed open data quality assessment approach (Table 6).

Table 6. Overlay of Open Data Quality Dimensions reference in the literature in [6] and used as part of the 3 applied use cases from this paper.

Quality dimensions	LoD [6]	LEI [20]	Use case 5.1	Use case 5.2
Accessibility/Availability	X	X		
Accuracy	X (security)	X	X (auditability)	
Completeness/Comprehensiveness	X	X	X (coverage)	X
Understandability/Interpretability	X	X	X	X
Consistency	X	X		
Currency		X		
Integrity		X		

(*continued*)

Table 6. (*continued*)

Quality dimensions	LoD [6]	LEI [20]	Use case 5.1	Use case 5.2
Provenance/Trustworthiness		X	X	X (interlinking)
Representation		X		
Uniqueness		X		
Timeliness	X	X	X	
Validity		X		
Licensing	X		X (accessibility)	
Interlinking	X		X	X
Conciseness	X			
Relevancy	X		X	X (interlinking)
Semantic accuracy	X			
Interoperability	X			
Versatility	X			

From this overlay analysis, six groups of data quality dimensions stand out and can then be considered as the most relevant for our proposal of Open Data Quality assessment framework: Accuracy (Auditability, Security), Completeness (Coverage), Comprehensiveness (Understandability, Interpretability), Provenance(Trustworthiness, Interlinking), Timeliness and Relevancy.

In terms of Open Data Quality indicators, they are more on the metadata than the data itself and they aim essentially at ensuring the visibility and the re-usability of the dataset. Indeed, whether in big data or open data at scale, the Data Offices today no longer have the time to go into the detail of the quality of the data in each dataset. Their role becomes one of making a comprehensive set of relevant datasets timely available for re-users, of organizing and managing those datasets through a data catalog and of being the warrant of a good quality of the metadata of the dataset. It becomes up to the re-users, for project CARD for example, the data stewards, to define and measure their expected level of data quality according to the business context.

We have described in this paper a brief state of the art of the quality dimensions that can help assess web and open data quality. We detailed two use cases from the industry that illustrate the underlying metrics, notably regarding the Interlinking dimension. Through ExQI (a French association dedicated to Information Quality and Information Governance) and its workgroup on open data quality, we will continue to further flesh out this outline of an open data quality assessment framework. We will test these six key data quality dimensions findings, prove their relevancy and fine tune the proposal based on a couple of re-use cases.

References

1. Entretien avec Jean-Marc Lazard ODS le 16/4/2020, http://www.exqi.asso.fr/. to be pub-lished soon
2. Huyer, E., Van Knippenberg, L.: The economic impact of open data - opportunities for value creation in Europe (2020). https://www.europeandataportal.eu/sites/default/files/the-economic-impact-of-open-data.pdf
3. JoinData Homepage. https://www.join-data.nl/about-joindata/?lang=en. Accessed 08 May 2020
4. What data is shared?, https://openindustrialdata.com/data/. Accessed 08 May 2020
5. Microsoft newsletter of 2019/4/2: Microsoft and the BMW Group launch the Open Manufacturing Platform. https://news.microsoft.com/2019/04/02/microsoft-and-the-bmw-group-launch-the-open-manufacturing-platform/. Accessed 08 May 2020
6. Zaveri, A., Rula, A., Maurino, A., Pietrobon, R., Lehmann, J., Auer, S.: Quality assessment for linked data: a survey. Semant. Web **7**(1), 63–93 (2016)
7. ETALAB HomePage. https://www.etalab.gouv.fr/. Accessed 31 May 2020
8. APIE HomePage. https://www.economie.gouv.fr/apie. Accessed 31 May 2020
9. LiberTIC HomePage. https://libertic.wordpress.com/libertic/. Accessed 31 May 2020
10. FING HomePage. https://fing.org/actions/open-data-impact.html. Accessed 31 May 2020
11. Regards Citoyens HomePage. https://www.regardscitoyens.org/#&panel1-1. Accessed 31 May 2020
12. Data Publica HomePage. https://www.data-publica.eu/. Accessed 31 May 2020
13. OpenDataSoft HomePage. https://www.opendatasoft.com/. Accessed 31 May 2020
14. Cap Collectif: Digital Republic Bill - Explanatory Memorandum. https://www.republique-numerique.fr/pages/digital-republic-bill-rationale. Accessed 08 May 2020
15. Publier un jeu de données. https://guides.etalab.gouv.fr/data.gouv.fr/publier-jeu-de-donnees/. Accessed 08 May 2020
16. ODS HelpHub HomePage. https://help.opendatasoft.com/apis/ods-search-v2/#metadata. Accessed 11 May 2020
17. De Simoni, G.: Adopting and Addressing metadata management as an enabler for effective digital transformation. In: Gartner Data & Analytics Summit 2017 (2017)
18. Open Data Charter Homepage. https://opendatacharter.net/. Accessed 08 May 2020
19. LEI Data - GLEIF Data Quality Management. https://www.gleif.org/en/lei-data/gleif-data-quality-management/about-the-data-quality-reports/download-data-quality-reports/download-global-lei-data-quality-report-april-2020#. Accessed 08 May 2020
20. GLEIF Global LEI Data Quality Report Dictionary, balDataQualityReportDictionary_2_v1.pdf. https://www.gleif.org/en/lei-data/gleif-data-quality-management/about-the-data-quality-reports/questions-and-answers. Accessed 08 May 2020

Synthesizing Quality Open Data Assets from Private Health Research Studies

Andrew Yale[1,4(✉)], Saloni Dash[2], Karan Bhanot[1], Isabelle Guyon[3],
John S. Erickson[1], and Kristin P. Bennett[1]

[1] Rensselaer Polytechnic Institute, Troy, NY, USA
a.yale9@gmail.com
[2] BITS Pilani, Goa Campus, Goa, India
[3] UPSud/INRIA University Paris-Saclay, Paris-Saclay, Paris, France
[4] OptumLabs Visiting Fellow, San Francisco, USA

Abstract. Generating synthetic data represents an attractive solution for creating open data, enabling health research and education while preserving patient privacy. We reproduce the research outcomes obtained on two previously published studies, which used private health data, using synthetic data generated with a method that we developed, called HealthGAN. We demonstrate the value of our methodology for generating and evaluating the quality and privacy of synthetic health data. The dataset are from OptumLabs® Data Warehouse (OLDW). The OLDW is accessed within a secure environment and doesn't allow exporting of patient level data of any type of data, real or synthetic, therefore the HealthGAN exports a privacy-preserving generator model instead. The studies examine questions related to comorbidites of Autism Spectrum Disorder (ASD) using medical records of children with ASD and matched patients without ASD. HealthGAN generates high quality synthetic data that produce similar results while preserving patient privacy. By creating synthetic versions of these datasets that maintain privacy and achieve a high level of resemblance and utility, we create valuable open health data assets for future research and education efforts.

1 Introduction

The inability to share private health data can stifle research and education activities. For example, studies based on unpublished electronic medical record (EMR) data cannot be reproduced, thus future researchers are not able to use them to develop and compare new research. This contributes to the reproduciblity crisis in biomedical research [3]. Making open data available for research can spur innovation and research. The public Medical Information Mart for Intensive Care datasets, MIMIC-II and MIMIC-III, are widely used with over 2000 citations reported in Google Scholar in March 2020 [7,10]. But since MIMIC-II and MIMIC-III focus on Intensive Care Unit patients in Boston hospitals, the resulting research may be biased and have limited generalization. Also since MIMIC requires users to undergo a training/approval process, it is not well suited for

© Springer Nature Switzerland AG 2020
W. Abramowicz and G. Klein (Eds.): BIS 2020 Workshops, LNBIP 394, pp. 324–335, 2020.
https://doi.org/10.1007/978-3-030-61146-0_26

classroom use. The cost and time required, along with re-identification risk concerns make de-identification only a partial solution to this problem.

Recent synthetic data generation methods provide an attractive alternative for making data available for research and education purposes without violating privacy. Deep learning approaches for synthetic data specifically show significant promise [1, 6, 8] In the future, synthetic data generation methods combined with automatic machine learning methods could enable synthetic versions of data to be released when research papers are published. Results could be reproduced and novel methods and analysis could be developed without compromising patient privacy. Fig. 1 illustrates one scenario for use of synthetic data. To accomplish this, synthetic data assets must have 1) privacy: how well does the synthetic generation data method preserve anonymity, 2) resemblance : whether the distribution of synthetic data is indistinguishable from the distribution of real data. 3) utility: can research studies be reproduced successfully with synthetic data and 4) efficiency: how practical is the training and generation pipeline.

In this paper, we report our experience of generating synthetic data using the process in Fig. 1 for two published research studies[14, 15] performed in the OptumLabs Data Warehouse (OLDW). The studies focus on the same cohort of children with ASD. The first focuses on the difference between gastrointestinal symptoms and oral antibiotic use in children with ASD and children without ASD. The second study investigates how different groups of children with ASD can be clustered based on their comorbid medical conditions (CMCs), and what that means about the different clusters. We utilize novel enhancements of the prior HealthGAN [4, 16–18] synthetic data approach, and then evaluate the privacy, resemblance, efficiency, and utility of the synthetic data.

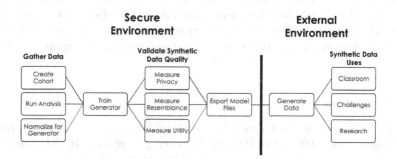

Fig. 1. Workflow used to generate synthetic data securely. The data is gathered in the same way as the studies did, processed, and used to train the generator model inside the secure environment. The synthetic data is then validated for privacy, resemblance, and utility. After being certified as private by the optumLabs staff, the model is exported. Finally, data is generated using the model and used for many types of applications.

This paper focuses on documenting and assessing the HealthGAN synthetic data generation method on real medical datasets in a secure environment and exporting the models outside of the environment. We used the OptumLabs Data

Warehouse as the source of the medical data. We report on the process of exporting the data from the OLDW[1]. We demonstrate how to verify privacy and resemblance using recently published metric. We make the tools for generating and evaluating synthetic readily available in a Python package. [2] The next section goes over synthetic data generation and evaluation methods. Following that the selected generation and evaluation methods are used to reproduce two studies in OLDW. The synthetic data for each of these datasets are evaluated for privacy, resemblance, and utility. Finally, we conclude and discuss future work.

2 Methods

2.1 Generation

We use HealthGAN [4,16–18], a generative adversarial network (GAN) method we developed to generate synthetic data for datasets containing categorical, continuous, binary, and time series data. This method is based on the Wasserstein Generative Adversarial Network [2,5] and uses a novel variant of the categorical encoding method from "The Synthetic data vault" (SDV) [9]. We added the ability to encode ordinal data to the original HealthGAN. By creating the SDV mapping using the inherent ordinal order and including values that might not exist in the original dataset, we improved resemblance of ordinal variables.

Part of the constraints for our generative model is the ability to export the model from the secure environment instead of synthetic data. To fulfill this requirement we needed to ensure that the model itself did not contain or require any real data to generate synthetic data. In "Privacy Preserving Synthetic Health Data" [17], after comparing multiple methods, we found that Health-GAN method best satisfies these constraints due to the fact that a neural network model does not store real data or require real data to run. For HealthGAN specifically, we export just the generator network and never export any actual data.

2.2 Evaluation

To test the synthetic data generated by HealthGAN we investigate privacy, resemblance, and utility. In "Privacy Preserving Synthetic Health Data"[17], we developed the concept of *nearest neighbor adversarial accuracy* and *privacy loss*. Nearest neighbor adversarial accuracy, shown in Eq. 1, compares the distance from one point in a target distribution T, to the nearest point in a source distribution S, defined as $d_{TS}(i) = \min_j \|\mathbf{x}_T^i - \mathbf{x}_S^j\|$, to the distance to the next nearest point in the target distribution, defined as $d_{TT}(i) = \min_{j, j \neq i} \|\mathbf{x}_T^i - \mathbf{x}_T^j\|$. By comparing this across all points, it gives us the adversarial accuracy. This metric can be interpreted much like balanced accuracy where the value is an

---

[1] As of 7/12, we successfully exported the model of the first dataset but approval on the second dataset is ongoing.

[2] github.com/TheRensselaerIDEA/synthetic_data.

average of the accuracy for each class. Therefore we are striving for a value of 0.5 where the synthetic and real data cannot be distinguished. If that is achieved then we can say that the synthetic data and real data have high resemblance.

$$\mathcal{AA}_{TS} = \frac{1}{2} \left(\frac{1}{n} \sum_{i=1}^{n} \mathbf{1} \left(d_{TS}(i) > d_{TT}(i) \right) + \frac{1}{n} \sum_{i=1}^{n} \mathbf{1} \left(d_{ST}(i) > d_{SS}(i) \right) \right) \quad (1)$$

Privacy loss is defined in Eq. 2 as the difference between the adversarial accuracy on the test set and the adversarial accuracy on the training set. As the ideal value for both of these is 0.5, the privacy loss should be 0.0 when privacy is completely conserved. In the case where the model is exposing data, the value of the training adversarial accuracy will be lower than 0.5, and therefore even if the test adversarial accuracy is 0.5, the loss will increase.

$$
\begin{aligned}
\textbf{TrResemblLoss } (Train\ Adversarial\ Acc.) &= E[\mathcal{AA}_{RtrA_1}] \\
\textbf{TrResemblLoss } (Test\ Adversarial\ Acc.) &= E[\mathcal{AA}_{RteA_2}] \\
\textbf{PrivacyLoss } &= Test\ \mathcal{AA} - Train\ \mathcal{AA}
\end{aligned}
\quad (2)
$$

Beyond nearest neighbor adversarial accuracy, we test the privacy of the synthetic data using a membership inference attack scenario. In this scenario an attacker attempts to determine whether a given record was used to train a model [13]. The attacker has black-box access to the model, meaning they have the ability to feed data into the model and observe the output of the model [11,12]. The original scenario doesn't exactly match what would happen with HealthGAN because the input to HealthGAN generator network is random noise, rather than real data. In HealthGAN setting the model the attacker has access to is just the generator and cannot train the model, only feed it random noise in order to generate data. Therefore, instead we show how using the synthetic data generated from the network and a variant of nearest neighbor accuracy can be used to assess vulnerability to this kind of attack.

In the attack scenario we are considering, an attacker has access to some real data R with incomplete records for each patient. We assume, without loss of generality, that the attacker has access to columns $[c_1 \ldots c_k]$, but not to columns $[c_{k+1} \ldots c_N]$. Simultaneously, the attacker has access to a synthetic (artificial) dataset A for which all columns $[c_1 \ldots c_N]$ are given, which allows him/her to create a predictor of columns $[c_{k+1} \ldots c_N]$ from columns $[c_1 \ldots c_k]$. Subsequently, this could allow him/her to predict the missing columns in real data, which could constitute a breach of privacy. This violation of privacy can be quantified in the membership attack scenario context by evaluating the fraction of real data records that can be identified after completing the missing data in R.

In the worst case scenario, the attacker has available a large fraction of the columns in R, making the attack simpler. We consider the limit case in which all the columns are available, and determine how easy it is to identify which real data records were used for training our data generative model. We construct R to be a random shuffle of the training and non-training data, and attempt to sort out if each point is from training or not, using a nearest neighbor classifier. We compute the distance from a sample in R to its nearest neighbor in A, then measure the AUC of prediction {training *vs.* non-training sample} using the measured nearest neighbor distance as a ranking measure. If the AUC is greater than 0.5 (chance level), then the model may be exposing private data by allowing the attacker to know which records are in training.

Finally, once the privacy and resemblance have been tested the utility of the synthetic data must be evaluated. The method for this varies based on what the real dataset was intended for, but in the case where we are replicating a published study we reproduce the analysis done on the original data on the synthetic data. For the two studies being reproduced in this paper we will use Cox regression and k-means clustering to analyze the performance of the synthetic data.

3   Research Studies Reproduced

We examine creating synthetic open-data for two research papers, including the synthetic generation method, and the approaches for validating the quality of the two generated datasets. Both of the original studies were performed in the OLDW. OptumLabs is an open, collaborative research and innovation center founded in 2013 as a partnership between Optum and Mayo Clinic with its core linked data assets in the OLDW. The database contains de-identified, longitudinal health information on enrollees and patients, representing a diverse mixture of ages, ethnicities and geographical regions across the United States. The claims data in OLDW includes medical and pharmacy claims, laboratory results and enrollment records for commercial and Medicare Advantage enrollees. The EMR-derived data includes a subset of EMR data that has been normalized and standardized into a single database. Access to the OptumLabs database is, justifiably, tightly controlled. Access is limited to partner researchers who are granted access to certified de-identified research data views. All work is completed in the secure environment in which OLDW is hosted. Any release of data or derived products like graphs and tables must be reviewed and approved by OptumLabs on a case by case basis. Data for the subjects in the studies are typically not released, which means the OptumLabs datasets used in published studies are typically not available.

3.1   Study 1: Gastrointestinal Symptoms and Oral Antibiotic Use in Children with Autism Spectrum Disorder

In "Gastrointestinal Symptoms and Oral Antibiotic Use in Children with Autism Spectrum Disorder: Retrospective Analysis of a Privately Insured U.S. Population" [15] the authors look at the relationship between taking oral antibiotics in early childhood and occurrences of later gastrointestinal (GI) diagnosis in children diagnosed with autism spectrum disorder (ASD). Previous studies have shown different rates of GI symptoms in children with ASD, but at least one study claims that the estimated odds of having a general GI complaint are

4.4 times greater for children with ASD than for children without ASD. One confounding factor in this comparison is the presence of oral antibiotics. Oral antibiotics in early childhood may cause long-term disruption in the gut microbiome's composition, leading to an increased number of GI symptoms in early childhood. Several studies have shown that on average children diagnosed with ASD consume more oral antibiotics than children without ASD. Therefore, this work looks at the hazard ratios of demographics, oral antibiotics, and ASD diagnoses. The authors found children with ASD had a greater rate of GI diagnoses than children without ASD. Examining hazard ratios, greater numbers of oral antibiotics significantly increased risks GI-related diagnoses for both groups.

EMR Data. The study looks at OptumLabs claims data from 1/1/2000 to 9/30/2015. The patients that are used in the dataset must have at least five years of un-interrupted data called continuous enrollment to be included. From there they are included in the ASD cohort if the patient has at least two different ASD related diagnoses in the time period.

In order to measure the effect of oral antibiotics, the study defines two periods for observation: early enrollment and late enrollment. Early enrollment is the first three years of the five year period, while late enrollment is the last two years. The number of GI related diagnoses are counted for each period and examined. This type of analysis separates separate short term GI symptoms related to oral antibiotics from longer term conditions.

The ASD cohort collected has 3,278 patients while the non-ASD cohort has 279,428 patients. Within that cohort 37% of the ASD cohort has GI related diagnoses and 20% of the non-ASD cohort has GI related diagnoses. The final demographics of the study population are tabulated in Table 2 of their paper [15]. Due to slight variations in the view we are using in OptumLabs, we are not able to get the exact same data. Specifically, our view is a zoomed out version of the census divisions, which means our data has four different regions instead of the nine in this study. However, the data is almost exactly the same despite this change. The nine regions in the original study can be cleanly aggregated into the four regions in our data, reflected in the similar cumulative counts for the four regions across both datasets. Our dataset contains a total of 283,462 patients, which is very similar to the overall total of 281,623 patients in the original study.

Synthetic Data. The first measure of quality for the synthetic data is our nearest neighbor adversarial accuracy metric [17]. The values for the *TrainResemblanceLoss* and the *TestResemblanceLoss* should be close to 0.5 to ensure that the resemblance is good, but most importantly the privacy loss, which is defined as *PrivacyLoss = TestResemblanceLoss - TrainResemblanceLoss*, should be as close to 0 as possible [17]. On our candidate model the *TrainResemblanceLoss* is 0.5236 and the *TestResemblanceLoss* is 0.5272. For the *PrivacyLoss*, we get 0.0036. Overall, these metrics indicate we are preserving privacy with a low privacy loss, but also maintaining resemblance with resemblance loss values close to 0.5.

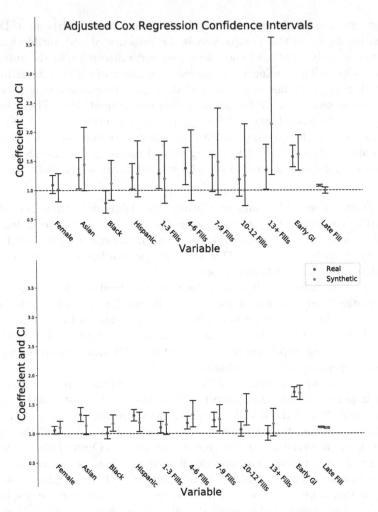

Fig. 2. Adjusted cox regression hazard ratio confidence intervals (CI) (95%) by different covariates observed on **ASD cohort** (top) and **population cohort** (bottom) for real and synthetic data.

Another measure of privacy is robustness against a membership inference attack [16]. This measured uses the area under the curve (AUC). Any AUC value above 0.50 in this test indicates a potential loss of privacy in a membership inference attack. To verify the baseline for this metric, we run the membership inference attack using the real data and obtain an AUC of 1.00 as expected. When computing this measure for the synthetic data, we get an optimal value of 0.50. This result further affirms previous privacy metrics, indicating privacy is also preserved in the membership inference attack scenario.

Finally, we assess utility by reproducing the study on the synthetic data. In the original study, the analysis used Cox regression to estimate adjusted and unadjusted hazard ratios (HRs). The original results from the Cox regression model show that GI diagnoses in late enrollment are more likely to happen if oral antibiotics are taken in early enrollment, regardless of an ASD diagnosis. In addition, the similarity in the HR of the two groups indicates that the effect is not exclusive to the ASD population. Another covariate was a diagnosis for GI symptoms in early enrollment. Understandably, this variable has a high HR as it likely indicates systemic issues with the GI tract.

We run the same Cox regression analysis on the synthetic data in order to calculate the hazard ratios of the same covariates. In Fig. 2 the ASD Cohort is on top and the Control Cohort is on the bottom, the real 95% confidence intervals are shown in black compared to the synthetic confidence intervals (CI) in red. If the CI overlap, then there is no statistically significant difference between the real and significant results. In the top of Fig. 2, the only variable that does not overlap in confidence intervals is whether the patient had any late enrollment prescription fills. In the bottom of Fig. 2, all the variables have overlapping confidence intervals, thus display no significant differences. In addition, the findings of the original paper state that the ASD and POP cohorts have similar hazard ratios for 7–9 fills. Our results verify this claim in both our real data with an ASD value of 1.25 CI (0.98, 1.61), and a POP value of 1.24 CI (1.11, 1.36), and in our synthetic data with an ASD value of 1.48 CI (0.92, 2.41), and a POP value of 1.24 CI (1.04, 1.49). These metrics indicate the synthetic data has high utility in terms of providing the same relationships demonstrated in prior work. Overall, the metrics computed demonstrate that the synthetic data created using our end-to-process result in data that retains privacy while maintaining high levels of resemblance and utility.

3.2   Study 2: Clustering of Co-occuring Conditions in ASD

In the paper "Clustering of co'occurring conditions in autism spectrum disorder during early childhood: A retrospective analysis of medical claims data" [14] the authors analyze patterns in diagnoses of comorbid medical conditions (CMCs) in patients with ASD. The study uses the same cohort of patients as the previous study. Based on the data, they are able to separate the patients with ASD diagnoses into three different clusters. The first cluster was 23.7% (n = 776) of the patients and encompassed a high rate of CMCs. The second cluster was 26.5% (n = 870) and contained patients with a higher rate of developmental delays. The third cluster, making up 49.8% (n = 1,632) of the data, contained low numbers of CMCs. Evaluating the data over time shows that the same patterns persist within these three clusters. The goal of this work was to help inform future treatment protocols for patients with ASD related to CMCs.

EMR Data. Membership in the cohort of patients ASD was determined with the same criteria as the previous paper, but with the addition of data regarding the CMCs. Seven different categories of CMCs were identified: auditory disorders, development delays, gastrointestinal symptoms, immune related conditions, psychiatric disorders, seizure disorders, and sleep disorders. These conditions were measured over time in six-month windows. For each patient, ten six-month windows were constructed for each of the categories, resulting in a vector of length 70 per individual with binary values indicating whether there was a diagnosis in the time window.

To analyze the CMC dataset the authors used principal component analysis (PCA) on the 70 CMC columns to transform the data to a continuous dataset, and then clustered the transformed data. This analysis was only done on the sub-cohort of patients with ASD diagnoses. Through k-means clustering they found the three different clusters of children with ASD defined previously.

Synthetic Data. Following the same process as the first paper, we train a new HealthGAN model on this dataset. We compute the privacy methods for the data and we evaluate the nearest neighbor adversarial accuracy metric. For our candidate model the performance on *TrainResmeblanceLoss* is 0.5416 and our *TestResemblanceLoss* is 0.5430. This value results in a *PrivacyLoss* of 0.0014. These metrics demonstrate that this synthetic data retains privacy as well as maintaining resemblance values close to 0.5. We also examine membership inference attacks using AUC. After verifying the baseline for the real data, the synthetic data had an AUC of 0.50, meaning no privacy loss in a membership inference attack scenario as well.

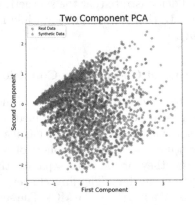

Fig. 3. PCA plot of real vs synthetic data for study 2

Next we examine resemblance of the synthetic data compared to the real data. For Study 2, we look at the population cohort, ASD cohort, and each of the three specific clusters. In addition to the demographic data, we look at the average number of CMC categories diagnosed per cohort. Visualizing the first 2 components of PCA for the real data and the synthetic data can further help understand any differences. In Fig. 3, PCA plots of the real and synthetic data (colored by cluster see below) show very similar distributions. We see that we have a good level of coverage over the real data. This is shown by the fact that we do not have many outlier values being generated in the synthetic data as well as the synthetic data not missing that many portions where the real data exists.

After measuring privacy and resemblance, we examine the utility of the synthetic data. Instead of Cox regression, the analysis of the original dataset is done with k-means clustering. To verify the utility of the synthetic data, this cohort is put through the same clustering method as the one used on the real data. The three emerging clusters are compared using characteristics of the CMCs for patients in each cluster.

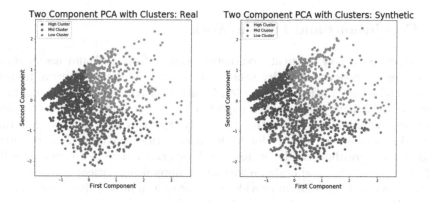

Fig. 4. PCA plot of real data and synthetic data with marked clusters

Using the PCA plot that was created to check the resemblance of the data, we can also check the utility by looking at how well the k-means clustering method works on the synthetic data. In Fig. 4, we have a PCA of the real data on the left and colored according to the clusters. On the right we have the PCA of the synthetic data colored by clusters found using the same k-means model found on real data. Visually we see that the clusters are very similar in both the real and synthetic plots which indicates a high level of resemblance and therefore utility.

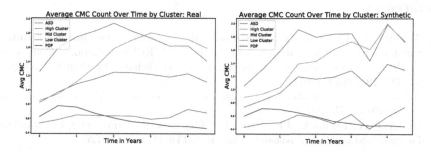

Fig. 5. Study 2: CMC over time by different clusters, real and synthetic

In addition to the pure size and shape of the clusters, we can also look at the average number of CMCs over time in the clusters. In Fig. 5, we can see a comparison of the average number of CMCs in each group over the time, in years, of the study. We see that there is a similar, if less smooth, trend in the groups. The high and mid clusters have a very similar relationship where they cross each other around the three year mark. The ASD population as a whole stays consistent in the middle of the graph, and the low and control populations have similar numbers of CMCs over time. Overall by replicating these various results from the original paper we can see that we achieve a high level of resemblance and utility while maintaining a high level of privacy[3].

4   Conclusions and Future Work

We demonstrate how to create synthetic datasets from real and assess their resemblance, privacy, and utility using a process that obtained approval from a commercial health EMR data provider. HealthGAN creates a GAN model in a secure environment, which is then exported to generate synthetic data outside the secure environment Recreating the analysis of two studies show that the method produces high quality for health informatics education. New research models and algorithms can be created on the synthetic data to create and evaluate new approaches and evaluate without compromising patient privacy.

Synthetic data generation provides an approach to make private data assets public. We recommend that empirical metrics for assessing privacy, resemblance and utility be utilized whenever synthetic data are generated, even if the underlying algorithms have theoretical guarantees of privacy. If synthetic data generated by automatic machine learning methods became a routine part of the publication process, scientific discovery and reproducibility would be accelerated and improved. Approaches could be developed on the synthetic data, and then evaluation on the real data in the secure environment. Since HealthGAN is designed to duplicate the underlying multivariate distributions while preserving individual privacy, it may not necessarily protect proprietary information represented in the data, such as business process and patterns.

References

1. Alzantot, M., Chakraborty, S., Srivastava, M.: Sensegen: A deep learning architecture for synthetic sensor data generation. In: 2017 IEEE International Conference on Pervasive Computing and Communications Workshops (PerCom Workshops), pp. 188–193. IEEE (2017)
2. Arjovsky, M., Chintala, S., Bottou, L.: Wasserstein GAN. arXiv preprint arXiv:1701.07875 (2017)
3. Begley, C.G., Ioannidis, J.P.: Reproducibility in science: improving the standard for basic and preclinical research. Circ. Res. 116(1), 116–126 (2015)

---

[3] Further analysis of the synthetic data for both studies is included in the supplemental material https://git.io/Jf3mK.

4. Dash, S., Dutta, R., Guyon, I., Pavao, A., Yale, A., Bennett, K.P.: Synthetic event time series health data generation. arXiv preprint arXiv:1911.06411 (2019)

5. Gulrajani, I., Ahmed, F., Arjovsky, M., Dumoulin, V., Courville, A.C.: Improved training of wasserstein GANs. In: Advances in Neural Information Processing Systems, pp. 5767–5777 (2017)

6. Gupta, A., Vedaldi, A., Zisserman, A.: Synthetic data for text localisation in natural images. In: Proceedings of the IEEE Conference on Computer Vision and Pattern Recognition, pp. 2315–2324 (2016)

7. Johnson, A.E., et al.: MIMIC-III, a freely accessible critical care database. Sci. Data **3**, 160035 (2016)

8. Krishnan, P., Jawahar, C.: Generating synthetic data for text recognition. arXiv preprint arXiv:1608.04224 (2016)

9. Patki, N., Wedge, R., Veeramachaneni, K.: The synthetic data vault. In: 2016 IEEE International Conference on Data Science and Advanced Analytics, pp. 399–410. IEEE (2016)

10. Saeed, M., et al.: Multiparameter intelligent monitoring in intensive care II (MIMIC-II): a public-access intensive care unit database. Crit. Care Med. **39**(5), 952 (2011)

11. Salem, A., Zhang, Y., Humbert, M., Berrang, P., Fritz, M., Backes, M.: Ml-leaks: Model and data independent membership inference attacks and defenses on machine learning models. arXiv preprint arXiv:1806.01246 (2018)

12. Shokri, R., Stronati, M., Song, C., Shmatikov, V.: Membership inference attacks against machine learning models. In: 2017 IEEE Symposium on Security and Privacy (SP), pp. 3–18. IEEE (2017)

13. Truex, S., Liu, L., Gursoy, M.E., Yu, L., Wei, W.: Demystifying membership inference attacks in machine learning as a service. IEEE Trans. Serv. Comput. (2019)

14. Vargason, T., Frye, R.E., McGuinness, D.L., Hahn, J.: Clustering of co-occurring conditions in autism spectrum disorder during early childhood: aretrospective analysis of medical claims data. Autism Res. **12**(8), 1272–1285 (2019)

15. Vargason, T., McGuinness, D.L., Hahn, J.: Gastrointestinal symptoms and oral antibiotic use in children with autism spectrum disorder: retrospective analysis of a privately insured us population. J. Autism Dev. Disord. 1–13 (2018)

16. Yale, A., Dash, S., Dutta, R., Guyon, I., Pavao, A., Bennett, K.P.: Assessing privacy and quality of synthetic health data. In: Proceedings of the Conference on Artificial Intelligence for Data Discovery and Reuse, pp. 1–4 (2019)

17. Yale, A., Dash, S., Dutta, R., Guyon, I., Pavao, A., Bennett, K.P.: Privacy preserving synthetic health data. In: Proceedings of the 27. European Symposium on Artificial Neural Networks ESANN, pp. 465–470 (2019)

18. Yale, A., Dash, S., Dutta, R., Guyon, I., Pavao, A., Bennett, K.P.: Generationand evaluation of privacy preserving synthetic health data. Neurocomputing (April 2020)

Author Index

Printed in the United States
By Bookmasters